CH00706487

MANUAL OF
APPLIED GEOLOGY
FOR
ENGINEERS

Institution of Civil Engineers, London, 1976

This volume was sponsored jointly by the Ministry of Defence and the Institution of Civil Engineers. It has been published separately by HMSO in full as "Military Engineering, Volume XV" and is published in this version with the agreement of the Controller, HMSO.

First published 1976
Reprinted   1978

© *Crown copyright, 1976. Reprinted by permission of the Controller of Her Majesty's Stationery Office.*

ISBN 0 7277 0038 3

AMENDMENTS

Amendment number	By whom amended	Date of insertion

CONTENTS

CHAPTER 12. GROUNDWATER AND ITS EXPLOITATION

ANNEX

FIGURES

MAPS

TABLES

MANUAL OF
APPLIED GEOLOGY FOR ENGINEERS

PREFACE

1. This manual is intended as a guide to geology for the practising engineer. Its aims are to enable him:

    a. To appreciate the relevance of geology to the engineering decisions he may be called upon to make.

    b. To make the right choice of sites for engineering works.

    c. To solve the more simple problems of geological investigation.

    d. To recognise those more complex cases where specialist geological advice is required.

It is hoped the manual will also be useful for instruction in first degree engineering courses and for courses in engineering geology.

2. The first seven chapters cover the basic geological knowledge and methods which an engineer should understand. The succeeding five chapters deal with the applications of geology in engineering, and include a chapter on Terrain Evaluation, a technique of growing importance in site selection. The application of hydrogeology to ground water flow and to water supply is also included.

3. The manual was sponsored jointly by the Ministry of Defence and the Institution of Civil Engineers. It was written by the following authors under the direction of and in consultation with a small steering committee composed of geologists and engineers:

Chapters 1 and 2	— Mr N F Hughes*, Department of Geology, Cambridge University.
Chapters 3 and 4	— Dr L R M Cocks*, Department of Palaeontology, British Museum (Natural History).
Chapter 5 Section 1 and Chapter 8	— Mr P J Beaven, Transport and Road Research Laboratory.
Chapter 5 Sections 2 and 3	— Dr R M S Perrin, Department of Applied Biology, Cambridge University.
Chapter 6	— Mr C R Cratchley, Institute of Geological Sciences.
Chapter 7	— Dr E P F Rose, Department of Geology, Bedford College, London University.
Chapter 9 Sections 1 and 2 and Chapter 10 Section 4	— Prof J L Knill, Engineering Geology Division, Imperial College of Science and Technology.

Chapter 9 Section 3
Chapter 10 Sections 2, 3,
   6 and 7, and   — Mr P F F Lancaster-Jones, Cementation
Chapter 11 Sections 2 and 3   Ground Engineering Company.

Chapter 10 Section 1   — Mr A M Muir Wood*, Sir William
                                 Halcrow and Partners, and
                                 Mr R G T Lane*, Sir Alexander Gibb
                                 and Partners.

Chapter 10 Section 5 ⎫
Chapter 11 Section 1 ⎬   — Mr I E Higginbottom, Wimpey Labora-
                                 tories Ltd.

Chapter 12   — Mr. G P Jones, Department of Geology,
                                 University College, London University.

*Member of Steering Committee

GLOSSARY

Included in the glossary are some terms which are not in the book but may be encountered in geological reports.

Acid rocks: Igneous rocks with relatively high silica and high potassium and sodium content.

Activity (soil): Ratio of plasticity index to clay fraction.

Adit: An access tunnel leading to a main tunnel or shaft.

Adsorbed water (pellicular water): Water only a few molecules thick that adheres to the surfaces of soil and rock particles.

Adsorption: The adherence of molecules or ions in solution to the surfaces of solids with which they are in contact.

Aeolian feature: Land feature caused by wind erosion.

Agglomerate (geol): Rock composed of a chaotic assemblage of coarse angular volcanic fragments.

Aggregate abrasion value: A measure of the resistance of aggregate to abrasion. The lower the AAV, the more resistant to wear is the aggregate.

Aggregate soundness test: A chemical test to measure the relative resistance of rocks to repeated wetting and drying and crystallisation of salts within pores.

Allophane: (*See* Section 5.2, paragraph 22).

Allowable bearing pressure: The additional pressure above that already existing which can be carried safely by a foundation material.

Alluvium: Deposit from river water.

Amphiboles: A group of igneous rock-forming minerals (*see* Section 4.3, paragraph 4).

Analogue model: A model subject to the same mathematical laws (usually expressed in the form of differential equations) as a prototype and hence permitting determination of behaviour of the latter by analogy.

Andesite: A fine-grained igneous rock (*see* Section 3.1, paragraph 2).

Angle of repose: The greatest angle to the horizontal made naturally by the sloping surface of a heap or embankment of loose material.

Anticline: An arch shaped fold.

Apatite: A calcium phosphate mineral, also containing fluorine or chlorine.

Apparent resistivity: (*See* Section 6.3, paragraph 1).

Aquiclude: A (relatively) impermeable barrier to groundwater flow.

Aquifer: Permeable layer of water-bearing ground.

Aquifer test: Test to determine the continuous rate of extraction of water obtainable from an aquifer.

Area of discharge: Area in which water flows out of an aquifer.

Area of recharge: Area in which water flows into an aquifer.

Arenaceous: Of sediments and sedimentary rocks with a large proportion of particles of sand grade.

Argillaceous: Of sediments and sedimentary rocks composed principally of particles smaller than sand grade.

Artesian: Relating to a confined aquifer (usually one for which the pressure surface is above ground level).

Ashlar: Masonry, accurately squared and dressed.

Asthenosphere: That part of the Earth's mantle beneath the lithosphere.

Augite: A dark-coloured silicate mineral found in igneous rocks (*see* Section 4.3, paragraph 6).

Axial plane (geol): The plane which contains the points of hinge of each bed involved in a rock fold.

Barchan dune: Sand-dune of crescentic shape formed by a single-direction prevailing wind in an area with restricted quantity of sand over a bare rock-surface.

Basalt: Fine-grained basic igneous rock of calcic plagioclase and augite, commonly extruded from volcanoes.

Base level: The lowest part of a river profile as it approaches the sea.

Basic rocks: Igneous rocks with relatively low silica and low potassium and sodium content.

Basin: In physiography, a basin-shaped area. In tectonic geology, the basin is filled with rock.

Batholith: A body of plutonic igneous rock, usually greater than 80 km^2 (sometimes many thousand km^2 in surface area).

Bauxite: Residual soil layer relatively rich in aluminium oxide, which is formed in tropical areas of high rainfall.

Bearing capacity (eng): Ability of a material to support a load normal to the surface.

Becker drill: A mobile power percussion drill for use in overburden containing large boulders and providing facilities through the drill rods for blasting or sampling.

Bedding: The layers seen in sedimentary rock.

Bedding plane: The 'plane' which separates adjacent strata in sedimentary rocks.

Bedrock: Solid rock underlying soil, sand and other unconsolidated material.

Bench: A platform in a quarry or mine face provided to assist operations on the face or for transport of quarried material.

Bentonite: An expansive clay formed from the decomposition of volcanic ash (*see* Section 3.1, paragraph 4b).

Berm (eng): A horizontal or sub-horizontal ledge on the face of a slope.

Biosphere (geol): Comprises all past and present living matter of the Earth.

Biotite: A dark coloured mica mineral which is common in acid igneous and metamorphic rocks.

Black cotton soil: A dark clay soil which shrinks and cracks in dry weather. Found in hot climates.

Blind zone: (*See* Hidden layer).

Borrow pit: An excavation to provide material usually for fill elsewhere.

Boulder clay: Local British term for till, considered inaccurate because neither boulders nor clay are essential constituents.

Breccia: A coarse-grained rock with angular fragments, may be igneous or sedimentary (*see* Section 4.3, paragraph 7).

Brickearth (geol): Originally a name for any material suitable for making bricks; now usually confined to river deposits of silt or finer material: believed to originate from the redeposition of loess in water.

British standard compaction test: A laboratory test for measuring the compaction of soils (*see* BS 1377: 1967).

Brown earth: (*See* Table 10 Group II).
Buried channel: Old channel concealed under later deposits.

Caisson: Structure for keeping water or soft ground from flowing into an excavation; sunk while digging down to good ground and later incorporated into the foundations.
Calcareous: Containing calcium carbonate.
Calcite: A mineral form of calcium carbonate.
Calcrete: Calcium carbonate concentrated by water movement and precipitated into a hard cement of a soil matrix.
Caliche: (*See* Calcrete. (Syn)).
California Bearing Ratio (CBR): The ratio of the resistance to penetration, by a plunger into the soil being tested, to a standard resistance.
Caliper log: Continuous record of borehole (more precisely of the distance between spring-loaded arms maintained in contact with the surface of a borehole).
Calyx or shot drilling: Boring deep holes by means of hard steel shot fed down a rotating hollow cylinder.
Cambering (geol): Convex curvature of an outcropping bed caused by settlement.
Catena: Natural sequences of related soils of which the distinction is governed by topographic and drainage factors.
Cation exchange capacity: Total negative electric charge not neutralised internally.
Cement (geol): Naturally deposited mineral matter of any composition forming the bond between fragments of a sedimentary rock.
Chalk: A soft, usually white, fine-textured limestone.
Chernozems: (*See* Table 10 Group IV).
Chert: A sedimentary rock composed chiefly of crypto-crystalline silica (*see* Section 3.2, paragraph 8).
Chlorite: Sheet-structure green mineral common as a secondary product of alteration in sedimentary and metamorphic rocks; also a clay mineral in some soils.
Chronostratigraphical: Applicable to the age, or comparative ages, of rocks.
Clastic: Used to describe sedimentary rock derived from mechanically broken components, rather than by chemical or organic deposition.
Clay: Sedimentary deposit composed of very fine flaky particles of clay minerals (*see* Chapter 5).
Clay fraction: The fraction by weight of a sediment of size less than ·002 mm effective spherical diameter; mineral composition is variable.
Clay minerals: A group of alumino-silicate minerals with characteristic sheet structures (*see* Section 5.2, paragraphs 18–23).
Cleavage (geol): The splitting of a mineral along its natural fracture planes or of a rock along closely spaced planes, other than the original bedding planes or joints.
Coastal morphology: The study of the shape and profile of coastal areas produced by natural forces.
Cobbles: Rounded stones of size greater than 75 mm diameter (*see* Table 7).
Collapsing clay: A clay of marine origin with open structure which may alter when subjected to a change of environment, particularly dehydration.

Colluvium: Sedimentary products of weathering washed ('hillwash'), or otherwise moved, down slopes.

Comminution: Size reduction by breaking, crushing or grinding.

Competent rock (geol): Rock which in the mass is stronger than adjacent rock.

Compressibility (eng): Variation of voids ratio with applied load.

Concordant (geol): Describes strata with bedding planes more or less parallel and probably conformable.

Concretion (geol): A compact, usually rounded, mass of rock within softer rock. The most common mineral of concretions is calcite.

Cone of depression: The roughly conical depression produced in a water-table or piezometric surface by pumping or artesian flow.

Confined aquifer: An aquifer of which the piezometric surface is above the top of the aquifer.

Conformable: Describes sedimentary rocks succeeding one another without signs of intervening tectonism or erosion.

Conglomerate: A coarse-grained sedimentary rock (*see* Section 4.3, paragraph 7).

Connate water: Water (often saline) trapped for long periods in rock pore-space, usually beneath the present or a pre-existing sea.

Continental crust: Less dense, more siliceous and thicker crust under continents; contrasted with oceanic crust.

Continuous flight auger: A rotary drill for use to limited depths in overburden which delivers a continuous but disturbed sample. Used where speed is more important than precision or the hole is more important than what comes out of it.

Coral: The calcareous skeletons formed in warm shallow seas by some species of sedentary coral animal (Anthozoa).

Core (I): Part of the Earth more than 2900 km beneath ocean surface, with a relative density of more than 10·0 as interpreted from the records of seismic shock waves.

(II): Cylinder of rock obtained by drilling.

Core barrel: A length of pipe next to the cutting bit of a core drill and which contains the core. A double-tube core barrel is a core barrel with two concentric pipes so arranged that in very soft rock the inner tube does not rotate and so damage the core.

Core logging: The systematic recording of material recovered from a core bearing machine so that the original position of any part can be identified.

Core recovery:
a. The percentage of a length of drilling which is represented by solid core samples recovered from the drill.
b. The retrieval and storing of rock cores from a core drilling machine.

Core-stones: Masses of relatively unweathered rock completely surrounded by weathered rock or soil.

Corrie (or cwm or cirque): Short steep-backed and flat-floored basins of glacial erosional origin flanking mountain ridges.

Corundum: A hard aluminium oxide mineral. Sapphire and ruby are gem varieties.

Counterfort drain: A deep cutting running down a slope, which usually contains a pipe, for easy removal of surface water and filled with stone which assists drainage and buttresses the slope.

Country rock (geol): The rock surrounding an igneous body.

Creep: Continuous deformation under load.

Cross-bedding: (*See* Current-bedding).

Crust (geol): Outermost part of the solid earth, of relative density 3·0 or less, with a maximum thickness of about 50 km.

Cryoturbation: Disturbance of soil or sediment caused by frost action.

Cryptocrystalline: Made up of crystals too small to be discernible optically (e.g. glass and chert).

Current-bedding (geol): Bedding which is formed at an angle to the horizontal by the action of swift local currents of water or air (syn: cross-bedding). Known also as false-bedding because its observed inclination cannot be used as an indication of the nature of Earth movements subsequent to deposition.

Cut-off: A barrier constructed below ground level to reduce water seepage.

Cut-off drain: A drain which collects and discharges surface water across the natural drainage slopes.

Cutter-liner system: A method of obtaining a continuous sample of soft sediments by attaching a plastic sheath to a core cutter.

Deflation (geol): Removal by wind of unconsolidated sediment accumulations.

Deformability, coefficient of: A figure expressing the relationship between stress and deformation of a rock mass.

Delayed yield: A discharge characteristic of a leaky aquifer.

Delta: Area of river deposits built out into the sea by a powerful river.

Dendritic drainage: River drainage pattern appearing on a map to resemble the branches of a tree.

Dennison sampler: A core barrel liner for use with large samples of weak materials to reduce losses and damage during and after recovery.

Density flow: A gravity induced flow of a relatively denser fluid under less dense fluid. Factors affecting density difference include temperature, salinity, and concentration of suspended sediment.

Depression curve: Record of profile of water-table as a result of pumping.

Diagenesis: (*See* Section 2.8, paragraph 2).

Diaphragm walling: (*See* Section 10.2, paragraph 6).

Diapiric: Describing an anticline or dome whose rocks have become ruptured by the upwards movement of less dense plastic material, e.g. salt.

Dilatancy (soil): Tendency of the volume to increase under increasing shear stress.

Diorite: A coarse-grained igneous rock (*see* Section 3.1, paragraph 2).

Dip: The angle that a bedding plane makes with the horizontal (*see* Figure 13).

Discontinuity (geol): Boundary between major layers of the Earth which have different seismic velocities.

(eng): Interruption to the homogeneity of a rock mass (e.g. joints, faults, etc).

Discordant: Describes relationship of a separate rock body to an organised sequence of rock layers with which the body is in contact.

Disturbed samples: Soil samples obtained in a manner which destroys the original orientation and some of the physical properties of the naturally disposed material.

Dolerite: Medium-grained basic igneous rock typically found in feeder dykes of volcanoes.

Doline: Solution hollow of mappable dimensions in weathered limestone.

Dolomite: A mineral form of magnesium/calcium carbonate, sometimes used incorrectly as a synonym for dolostone.

Dolostone: A sedimentary rock composed chiefly of magnesium/calcium carbonate (*see* Section 3.2, paragraph 8).

Dome (geol): A dome-shaped structure, usually associated with igneous or other diapiric cause.

Drawdown: Reduction in level of water-table as a result of pumping.

Drawdown curve: The trace of the top surface of the water-table in an aquifer or of the free water surface, when a new or changed means of extraction of water takes place.

Drift (geol): Term used in Britain and North America for superficial deposits produced during Pleistocene or Recent times.

Drowned valley: Lower part of a valley which has been inundated by a distinct rise in sea level.

Drumlin: Elongate hill of ice-moulded shape, composed of till and caused by ice-advance over previously deposited ground-moraine.

Dune: Isolated wind-blown sand accumulation.

Dutch cone penetrometer: The apparatus used in a particular form of field testing to establish the relative bearing and frictional strengths of soils at frequent intervals of depth.

Dyke: A relatively thin discordant sheet of igneous rock formed by intrusion into a fissure near the surface of the crust.

Earthquake intensity: A qualitative assessment of the degree to which shaking is perceptible to people, the amount of damage to man-made structures and the extent of visible deformation of the Earth. Measured on a 12 grade Modified Mercali Scale.

Earthquake magnitude: An absolute measure of earthquake size related to seismic energy released, determined by amplitude of elastic waves generated. Several different scales of magnitude are in use.

Effective porosity: Porosity (expressed as ratio of volume to total volume of ground) representing specific yield (q.v.).

Elasticity, Coefficient of (Young's Modulus): The ratio of tensile (or compressive) stress in a material to the corresponding tensile (or compressive) strain.

Electro-osmosis: Movement of a liquid under an applied electric field through a permeable medium.

Epeiric sea: A sea underlain by continental crust.

Epicentre: The point on the surface of the Earth directly above the focus of an earthquake.

Erratic (geol): A relatively large rock fragment, lithologically different from its surrounding rock, which has been transported from its place of origin (usually by glacial action).

Esker: Fluvio-glacial deposit formed in the stream-bed within an ice-sheet.

Eustatic: Describes change of sea-level which has been caused by a change in the standing level of the whole water body, rather than by a local land-level alteration.

Evaporite: A sedimentary deposit or soil composed primarily of carbonate, sulphate or chloride minerals produced from a saline solution by evaporation of the water.

Exfoliation (geol): The flaking of the outer layer of a rock in weathering.

Expanding electrode technique: (*See* Section 6.3, paragraph 2).

Exposure (geol): A visible rock outcrop.

Fabric (soil): Size shape and arrangement of the solid mineral and organic particles and the associated voids; similar in meaning to soil structure.

False bedding: (*See* Current bedding).

False colour film: Colour film with an additional emulsion sensitive to near infra-red wavelengths.

Fan deposit: Inclined deposit mass formed by intermittent torrential streams.

Fault: A planar crack in rock, with some displacement in the plane of the crack.

Feldspar: One of a group of abundant rock-forming aluminium silicate minerals, the chief minerals in most igneous rocks.

Ferralites: Prominently iron-rich residual soils.

Ferruginous (geol): Iron-bearing; frequently used in description of sedimentary rocks.

Field capacity: Moisture content of a free-draining soil when the rate of drainage has become negligibly small for practical purposes.

Fines (soil): Clay and silt fractions taken together.

Fissile (geol): Capable of being readily split along closely spaced planes.

Fissure flow: Flow of water through joints and larger voids.

Fissured clays: Clays which in their natural state show a system of fissures somewhat similar to a jointing system in a hard rock mass, on a reduced scale.

Flint: A variety of chert (cryptocrystalline silica), particularly formed in European chalk and also found redeposited.

Flocculation: Coalescence of fine colloidal particles into larger aggregates.

Flood plain deposit: Alluvial deposit of clay, silt, sand and gravel, filling a valley bottom.

Fluorescein dye: Substance such as Resorcinolphthalein dissolved in alkali to give a red or green fluorescence.

Fluorite: The mineral calcium fluoride.

Fluviatile: Of or pertaining to a river or rivers.

Foliation (geol): A general term for a small scale arrangement of planes in rock; most commonly applied to metamorphic rocks.

Fossil: The naturally buried remains of a once-living organism.

Fracture index: Ratio of seismic velocity for intact rock sample to seismic velocity for rock *in situ*.

Freestone: Informal term used for homogenous sedimentary rock with thick beds, which can therefore be freely shaped for building blocks.

Gabbro: A coarse-grained low-silica igneous rock found in plutons and consisting typically of plagioclase feldspar and pyroxene.

Gamma-gamma log: A continuous record of backscatter of gamma radiation from the ground adjacent to a borehole using a gamma-ray source of irradiation.

Gamma ray log: Continuous record of natural radio-activity of the ground adjacent to a borehole.

Garland drain: A perimeter drain at an intermediate level in a shaft or excavation.

Gault: A blue-to-grey clay lying between the Lower Greensand and the Chalk in the Cretaceous of Western Europe.

Geochemistry: (*See* Chapter 1).

Geode: Sub-spherical hole in a lava, often later filled with crystals.

Geohydrology: Science of the occurrence, distribution and movement of water below the surface of the Earth.

Geoid: Surface of gravitational equipotential at approximately mean sea level.

Geomorphology: (*See* Chapter 1).

Geophone: Microphone used for seismic prospecting.

Geophysical log: A continuous record of the variations in the physical features of the ground measured from a borehole.

Geophysical reflection profiling: Investigation process whereby time delays of reflection pulses of sound are used to map geological horizons and other features.

Geophysics: (*See* Chapter 1).

Geosyncline: A mobile downwarping of the Earth's crust, perhaps 100 km or more across, in which thick sediments (and often volcanic rocks) are deposited simultaneously with the downwarping.

Geotechnical: Pertaining to geotechnics, which is the application of scientific methods to problems in engineering geology.

Glacial deposit: Sediments deposited by ice sheets, glaciers and melt streams.

Glacial geology: (*See* Chapter 1).

Gleying: Formation of grey or green material in soil when stagnation of water results in exclusion of air and reduction of iron.

Gneiss: A banded metamorphic rock in which most of the constituent minerals may be seen with the naked eye (*see* Section 4.3, paragraph 8).

Goaf (in mining): The space from which a seam has been removed.

Gouge: Joint or fault filling of fine-grained material which may be produced for example by fault movement or glacial deposition.

Gouy layer: Zone in aqueous solution surrounding a micelle where concentrations of ions are different from those in the general pore solution (*see* Section 5.2, paragraph 28).

Graben: A trough-shaped parallel-sided valley usually bounded by faults.

Graded bed (geol): A single bed of sediment or sedimentary rock with a steady upward decrease of particle size.

Grading (geol): A 'wellgraded' sediment containing some particles of all sizes in the range concerned. Distinguish from 'well-sorted' which describes a sediment with grains of one size.

Granite: A coarse-grained igneous rock (*see* Section 3.1, paragraph 2).

Graphite: A finely crystalline form of carbon, sometimes occurring as a mineral in metamorphic rocks.

Gravel: Coarse sedimentary deposit including stones, sand and fines (*see* Table 7).

Greathead shield: A protective circular shield for hand tunnelling in soft ground; first used by Greathead in London in 1879.

Greensand (geol): A sediment or sedimentary rock containing the iron mineral glauconite which is green when un-weathered, but weathers brown or yellow.

Grit (geol): A coarse-grained sandstone with angular grains, but sometimes loosely used for indurated sedimentary rock.

Ground: All solid material below the Earth's surface.

Ground anchor: A tie or tendon anchored deep in the ground and stressed to provide a retaining force for a structure at the surface of an excavation.

Groundwater: Water occupying interstices, fissures and cavities in the ground.

Grout: Often neat cement slurry or a mix of cement and sand or other additives. Fluid for injecting into masonry joints, rocks or other ground.

Groyne: A wall built out from a river bank or sea shore to check or to increase scour.

Gulls: Tension gashes associated with cambering.

Gunite: A patent name for a cement and sand mixture which is sprayed from a gun by air pressire.

Gypsum: An evaporite mineral, a type of calcium sulphate (*see* Section 4.3, paragraph 7).

Hade: The angle of a fault plane from the vertical.

Haematite: A red iron oxide mineral ($Fe_2 O_3$).

Hardness of water:
a. Temporary hardness due to dissolved calcium bicarbonate, which is removed by precipitation on boiling.
b. Permanent hardness due to calcium sulphate or other dissolved salts, which remains after boiling.

Hardpan: Soil horizon which has become cemented by precipitation of iron oxides, calcite or silica; exposure gives rise to laterite crusts, calcrete and silcrete respectively.

Head (geol): Unstratified deposit produced by solifluction debris in periglacial conditions.

Hidden layer (syn: blind zone): A layer of ground with low seismic velocity which cannot be detected by seismic refraction.

Horizon (geol): Specified layer in a rock sequence.

Horizon (soil): Layer formed by natural processes within the soil; not necessarily horizontal when formed.

Hornblende: A dark-coloured silicate mineral found in igneous rocks (*see* Section 4.3, paragraph 6).

Hornfels: A fine-grained metamorphic rock resulting from contact metamorphism.

Horst: A high area bounded by faults.

Humus: Organic soil material sufficiently decomposed to have lost structural features.

Hydraulic conductivity (also known as permeability to the engineer): Rate of flow of water through a unit area at unit hydraulic gradient (*see* Figure 119).

Hydrogeology: (*See* Chapter 1).

Hydrograph: Time-based record of discharge of river, etc.

Hydrological cycle: (*See* Figure 110).

Hydrosphere: All the uncombined water above, on and within the crust of the Earth.

Hydrostatic pressure: The pressure exerted on a surface by a fluid at rest.

Hygroscopic: Tending to absorb water.

Igneous rocks: Those formed directly by crystallisation or super-cooling of a natural melt.

Illite: A clay mineral (*see* Section 5.2, paragraph 22).

Impermeable: Of a material through which water will not pass. (Often used in comparative sense).

Incompetent rock (geol): Rock which is weaker than adjacent rock and relatively more liable to deformation.

Induration: The hardening of a rock by the action of heat, pressure or cementation.

Infiltration: The downward movement of subsurface water under gravity from the land surface to the water-table, i.e. it is restricted to the zone of aeration.

Infra-red linescan: Detector recording middle (2–5 µm) or far (7–14 µm) infra-red wavelengths only.

Inlier (geol): Outcrop of older rocks surrounded by younger rocks.

Integral sampling: A technique of core drilling which provides knowledge of the original orientation of the samples recovered.

Intercalation: Lens of peat or other material occurring in sediment which is otherwise uniform.

Intercept time: Time lapse between seismic signals refracted by different interfaces.

Interflow: (*See* Section 12.2, paragraph 18).

Inter-granular flow: The idealized type of groundwater flow through porous media where, in the absence of fractures or fissures, the water moves in the interstices between the component grains of the aquifer material.

Intrinsic permeability: (*See* Section 12.3, paragraph 4).

Intrusive rocks: Rocks formed by the intrusion of magma into existing rocks.

Isoclinal fold: A fold with more or less parallel limbs.

Isostasy: State of level compensation between masses of rocks in adjacent sectors of the crust.

Isotropic: Having the same properties in all directions.

Jackhammer: A hand-held mechanical hammer drill for rock.

Joint: A planar crack in rock, usually part of a set of such cracks, along which there is no displacement in the plane of the crack.

Juvenile water: Groundwater believed to have come directly from magmatic sources.

Kame: Mound of fluvio-glacial gravel formed where a glacial stream flows out of an ice-sheet front.

Kaolinite: Clay mineral (*see* Section 5.2, paragraph 22).

Karst (Karstic): General term for a terrain of solution weathered rock, usually limestone.

Kentledge: Heavy material, usually concrete, stones or scrap metal, used as loading on a structure, e.g. to sink a cylinder caisson, or as a counterbalance for a crane

Keuper Marl: A fine-grained red siltstone forming the higher part of the Triassic in North West Europe, typically in Germany. The term 'marl' is here incorrectly used.

Lacustrine sediments: Sediments deposited or formed in a lake.

Lag gravels: The remaining accumulations of pebbles or larger stones after the finer dust and sand has been taken away by wind or water.

Lahar: A flow of water-saturated volcanic debris or the deposit of debris resulting from such a flow.

Land classification: (*See* Section 8.3).

Land element: (*See* Section 8.3).

Land facet: (*See* Section 8.3).

Land system: (*See* Section 8.3).

Landslide: Rock or soil displaced downhill by gravity.

Laterite: A horizon of tropical ferralitic soils with much re-precipitated iron-oxide that sets to a hard cement on exposure to air.

Lenticular: Shaped like a lens.

Lignite (Brown coal): A brownish-black coal, intermediate between peat and black coal.

Limestone: A sedimentary rock composed chiefly of calcium carbonate (*see* Section 3.2, paragraph 8).

Liquid limit: Moisture content at which clay becomes a slurry.

Lithological: Adjective from lithology, the descriptive study of rocks.

Lithosphere: That part of the Earth consisting of the Crust and the Upper Mantle to a depth of about 100 km, forming the crustal plates.

Lithostratigraphical: Applicable to description of rock type with no reference to age.

Loam: Soil with a wide range of particle sizes.

Loess: Silt-size dust-like deposit washed out of the atmosphere by rain and accumulating only in grass plain regions.

Longshore drift: Beach material transported by wave action and currents along the shore.

Mafic: Of ferromagnesian minerals present in an igneous rock; also descriptive of a rock with high proportion of these minerals.

Magma: Underground silicate melt from which igneous rocks crystallise.

Magnetite: Black iron oxide mineral with strong magnetic properties (Fe_3O_4).

Mantle: Part of the Earth at depths between 10 km under ocean surface (more under land) and 2900 km, with a relative density of between $3 \cdot 3$ and $5 \cdot 7$ as interpreted from records of seismic shock waves.

Marble: Metamorphosed limestone or dolostone (*see* Section 4.3, paragraph 8).

Marl (geol): Unconsolidated sediment of argillaceous and calcareous material.

Mass movement: Gravity controlled displacement of rock mass down a slope without complete disruption of structure.

Mbuga: East and Central African name for a land-surface depression containing fine-grained dark-coloured soils of alluvial and colluvial origin.

Meander: Strong repeated curves of a stream or river course, typically in the lower flood plain part of a valley.

Metamorphic: Of rocks altered by heat or pressure, or both, beyond consolidation diagenesis.

Metamorphic geology: (*See* Chapter 1).

Metasomatism: A progress of replacement of minerals in a rock by additional constituents to form fresh minerals.

Meteoric water: Water on or beneath the land surface which has passed there directly from the atmosphere.

Meteorite: Fragment of matter from the solar system which has struck the Earth's surface.

Mica: A group of rock-forming silicate minerals, all with good cleavage in one plane only (*see* Section 4.3, paragraph 4).

Micelle: An independent colloidal particle, usually of clay or humus (*see* Section 5.2, paragraph 24).

Migmatite: A composite rock which includes igneous and metamorphic constituents.

Mineralogy: (*See* Chapter 1).

Moisture content: Percentage by weight of water in a soil or rock sample.

Montmorillonite: Clay mineral (*see* Section 5.2, paragraph 21).

Mor: A surface horizon of acid and partly decomposed plant material characteristic of podzols.

Moraine: Glacial deposits of unsorted rock fragments relatable to the position of a former ice-sheet.

Mosaic: Assembly of adjacent air photographs without overlap for use as a map.

Mud (geol): Fine sediment of clastic particles.

Mudslide: (*See* Section 10.3, paragraph 6).

Mudspate: A mixture of soil and water moving in a manner resembling a fluid.

Mudstone: A non-fissile sedimentary rock of clay-sized clastic particles (*see* Section 4.3, paragraph 7).

Murram: Generally iron concretions formed in tropical soils, transitional to, or an early stage of, laterite formation.

Muscovite: A light-coloured mica mineral (*see* Section 4.3, paragraph 4).

N-value: The number of blows required to drive the sampler of the Standard Penetration Test its last 12 inches (300 mm).

Neritic deposits: Deposits formed below low tide level on the continental shelf.

Neutron log: A continuous record of backscatter of gamma radiation from the ground adjacent to a borehole as a result of controlled neutron radiation.

Normal fault: A fault in which the principal component of the relative displacement has been vertical. Also sometimes called a gravity fault.

Nucleide: A particular assemblage of neutrons (N) and protons (Z) in the form of an atomic nucleus defined by Atomic Number (Z) and Mass number (N + Z). Isotopes are nucleides with the same atomic number but different mass number.

Oil shale: Shale containing a sufficient proportion of hydrocarbons to be suitable for distillation to yield oil.

Olivine: A green iron aluminium silicate mineral found in basic and ultrabasic igneous rocks.

Oolite: Limestone composed chiefly of ooliths, which are small (0·25 to 2 mm diameter) spherical particles.

Open drive sampling: A method of obtaining relatively undisturbed samples of cohesive soils by driving an open cylinder fitted with a cutting edge and extension piece into the ground by sliding hammer or hydraulic ram (*see* Section 9.3, paragraph 31).

Oriented sample: A sample so marked that its original attitude in the rock or soil is known.

Orogeny: The process of mountain formation by major folding.

Orthotropic: Of a material which has different mechanical properties in one direction compared with the other two.

Outcrop: Rock seen at the surface or obscured only by superficial deposits (*see* also Exposure).

Outlier (geol): Outcrop of younger strata surrounded by older strata.

Overbreak: Rock excavated beyond a specified cross-section of a tunnel.

Overburden: Overlying consolidated deposits.

Over-consolidation (clay): Subjection to past effective over-burden pressure in excess of its present effective pressure.

Overcoring: A technique for isolating as far as possible a cylinder of rock carrying instrumentation from the surrounding rock; carried out with core drilling equipment.

Palaeobiology: (*See* Chapter 1).

Palaeomagnetic: Of a past magnetic field.

Palygorskite: Uncommon clay mineral (*see* Section 5.2, paragraph 22).

Parametric classification: Classification based on the direct measurement of appropriate factors.

Parent material (of a soil): The pre-existing sediment or rock from which a soil is formed by weathering.

Peat: A deposit largely formed of dead vegetation which may be in course of consolidation (*see* Section 5.2, paragraph 17c).

Ped: An aggregation or structural unit of soil.

Pedogenesis: The process of soil formation on the Earth's surface.

Pedology: (*See* Chapter 1).

Pedon: Smallest three-dimensional block fully representative of the characteristics of a soil type.

Pelagic deposits: Ocean sediments without land-derived material.

Pellicular water (adsorbed water): Water only a few molecules thick that adheres to the surfaces of soil and rock particles.

Peneplain: Plain (or almost a plain) which is the culmination of a cycle of erosion.

Perched water-table: A local water-table based on an impermeable rock-layer above a regional water-table.

Percolation: The movement of groundwater in the zone of saturation under hydrodynamic forces, generally with a dominant lateral component.

Performance test (for wells): Tests of drawdown for different rates of well pumping (also known as 'step drawdown test').

Peridotite: Ultrabasic igneous rock, formed essentially of the mineral olivine (of which peridot is the gem variety).

Periglacial conditions: Conditions in the area around a large ice-sheet.

Permafrost: Ground with the water content permanently frozen; observed in high latitudes and altitudes.

Permeability: The capacity of a deposit to transmit or yield water. (For more specific definitions *see* Intrinsic permeability and Hydraulic conductivity).

Petrology: (*See* Chapter 1).

Petroscope: An instrument for inspection and measurement of rock and joints within a borehole.

Photogrammetric machine: A device for plotting maps from air photographs using the principles of stereoscopy.

Phreatic: (Syn: piezometric).

Physiography: The shape of landforms, a synonym for geomorphology.

Piezometer: A device for measuring pore water pressures in soils or rocks.

Piezometric surface: The level of the water surface in an (imaginary) vertical well connecting with an aquifer.

Piping: An underground flow of water with a sufficient pressure gradient to cause scour along a preferred path.

Piston sampler: A tube with an internal piston used for obtaining relatively undisturbed samples from cohesive soils.

Plagioclase: A common rock-forming sodium calcium alumino-silicate mineral of the feldspar group (*see* Section 4.3, paragraph 4).

Planar structure (geol): Term used in geological reconnaissance to describe features, such as bedding, cleavage, schistosity, joints, faults and flow banding, until they can be identified by detailed investigation.

Planetismals: Small fragments of mineral matter in the solar system from which the Earth is believed to have been formed by accretion.

Plastic concrete: A concrete incorporating a constituent which can adjust to movement, such as bitumen or bentonite which provide flexibility at the expense of strength.

Plastic limit: Moisture content of clay or soil at which plastic properties are just discernible.

Plasticity index: The difference between the water contents of a clay at the plastic and liquid limits, i.e. the range of water content for which the clay is plastic.

Plate tectonics: A tectonic theory of lateral movement of lithosphere plates (of crust and upper mantle), to account for recorded movements of crustal sectors in geological time.

Playa: Flat area receiving intermittent drainage, often with salt concentrated by evaporation (in arid regions).

Plunging: Used to describe an anticline or syncline with an inclined axis.

Pluton: A larger mass of igneous rock inferred to have been formed by slow cooling in a pressure confined situation deep in the crust, e.g. a batholith.

Podzol: Soil type consisting of thin horizons with humus overlying leached horizons, formed chiefly under temperate forests.

Point load strength: A rock strength obtained by a portable point load tester to determine a comparative strength index for the rock.

Polished stone value: A measure of the resistance of rocks e.g. roadstone, to polishing. The higher the PSV number, the greater is the resistance to polishing.

Pore water: The water partially or completely occupying interstices in soil or rock.

Porosity: The ratio of the volume of voids in a rock or soil to its total volume.

Porphyry: A general term used rather loosely for igneous rocks which contain relatively large isolated crystals set in a finer-grained mass.

Pot-hole (geol): Circular hollow in river bed formed by grinding action of stones swirled by strong currents or by solution of limestone.

Pozzolan: Originally volcanic dust used at Pozzuoli, Italy as a hydraulic cement when mixed with lime. Made artificially by burning and grinding clay or shale.

Precipitation: Water falling as rain, snow, hail or in direct condensation as dew.

Pressure surface (syn: piezometric surface): The level of the water surface in an (imaginary) vertical well connecting with an aquifer.

Pyrite: Iron sulphide mineral, brassy yellow in colour.

Pyroclastic: Sediment composed of igneous rock fragments explosively ejected from a volcano.

Pyroxenes: A group of igneous rock-forming minerals (*see* Section 4.3, paragraph 4).

Qanats: Hand-excavated shafts and adits for water conveyance in the Middle East.

Quartz: The commonest mineral form of silica.

Quartzite: An indurated sandstone in which the pore spaces between the quartz sand grains have been filled with silica; in sedimentary quartzites the extra silica has arrived in percolating solutions; in metamorphosed quartzites the pores are filled by recrystallisation under pressure.

Quick clays: Clay sensitive to disturbance, whereby shear strength may be substantially reduced.

Radiometric age: Age in units determined from the study of the dissociation of radioactive isotopes of certain chemical elements.

Raised beach: Accumulation of former beach deposits, now distinctly above sea level.

Rayleigh wave: (*See* Section 6.2, paragraph 6).

Reaction co-efficient: (*See* Section 12.4, paragraph 4).

Recharge: Natural or artificial replenishment of an aquifer.

Reduction: Removal of oxygen from a chemical compound.

Rejuvenation: The downcutting action of a river in its own valley bottom; caused either by the relative uplift of the whole drainage area or by a fall in sea or lake level at the mouth of the river.

Relative density (general): The density of a substance compared with the density of water at 4°C. (Now the internationally accepted term, previously called Specific Gravity).

Relative density (soil mechanics): Percentage void ratio of a soil, defined as:

$$\frac{e_{max} - e}{e_{max} - e_{min}}$$

where

e = void ratio of soil sample

e_{max} = void ratio of soil loosely poured in laboratory

e_{min} = void ratio of soil when vibrated and compacted.

Remanent magnetism: Part of a past magnetic field detectable in certain igneous and sedimentary rocks.

Remote sensing: Measurements of the Earth's surface made from aircraft or satellite.

Residual shrinkage: Small amount of shrinkage in clay on drying past the shrinkage limit (*see* Figure 31).

Residual soils: Soils (in geological sense) remaining in their place of formation.

Resistivity log: A continuous record of electrical resistivity of the ground adjacent to a borehole.

Rest hardening: Reforming of shear strength in flocculated structures in clay with time lapse after mechanical disturbance.

Reversed fault: Fault, resulting from the action of compressional forces, in which the downthrow side lies underneath the fault plane.

Rheidity: Property of gradual flow in a material which will also fracture when rapidly stressed.

Rheotropy: Reduction of shear strength due to mechanical disruption of flocculated structures in clay.

Rhyolite: Fine-grained acid igneous rock typically with orthoclase quartz and biotite.

Rip rap: Stones of irregular shape within specified limits of size, tipped or placed to protect an embankment from scour.

Rock: (*See* Chapter 1).

Rock bolt: A stressed bolt anchored deep in a rock mass which applies a compression through a plate at the rock face.

Rock dowel: A means of pinning rock across one or more planes of weakness.

Rock mechanics: Study of the mechanical properties of rocks, especially properties of significance to the civil engineer.

Rock quality designation (RQD): The total length of cores in solid pieces exceeding 100 mm in length, expressed as a percentage of the drilling length from which the core run was obtained.

Rock roller bit: A drilling bit with two to four freely rotating toothed rollers. Used for drilling in hard rock.

Rockhead: The top surface of a rock which has not undergone recent disturbances. Above may be loose boulders or residual soils or alluvium.

Sabhka: A term originating in Arab countries to describe a coastal salty flat; also to describe the salt-bearing soils which form these flats.

Saline soils: Soils with an appreciable content of soluble salts (generally 1%).

Saline wedge: Intrusion of saline (ground) water beneath fresh water.

Saltation: Form of sediment transport whereby particles are plucked from the surface and redeposited downstream or downwind.

Sand: Sedimentary deposit with coarsest material between 2 mm and 0·02 mm diameter (*see* Table 7).

Sand fraction: The fraction by weight of a sediment of size range 2·0 to 0·02 mm effective spherical diameter.

Sand run: Loss of sand from the side or top of an excavation, generally from a confined bed of sand and often initiated by water flow.

Sandstone: A sedimentary rock of sand-sized particles (*see* Section 4.3, paragraph 7).

Scarp slope: Relatively steep erosional slope formed parallel to the strike of uplifted strata.

Schist: A metamorphic rock with parallelism of many of the minerals (*see* Section 4.3, paragraph 8).

Schistosity: A parallelism occurring in metamorphic rocks, usually due to the parallel arrangement of platey and ellipsoidal minerals.

Scree: Large scale accumulation of talus near to the angle of repose.

Sector piling: A method of providing a continuous cut-off by overlapping circular concrete bored poles.

Sedimentary rocks: Consolidated sediments, usually conspicuously bedded or layered.

Sedimentology: (*See* Chapter 1).

Sediments: Deposits formed of particles accumulating on a land surface or in water or by crystallisation at temperatures below 100°C.

Seismic shock wave: A shock wave originating from an earthquake focus or from an artificial explosion.

Seismic refraction technique: (*See* Section 6.3, paragraph 9).

Seismicity: Pertaining to earth vibrations or disturbances produced by earthquakes.

Seismology: The study of natural or artificial earthquake waves, the methods of recording them and the interpretation of the records.

Sensitivity: Ratio of disturbed to undisturbed shear strength of a soil.

Shale: A fissile sedimentary rock of clay-sized particles (*see* Section 4.3, paragraph 7).

Shear strength: Shear stress of a material at failure.

Shear wave: (*See* Section 6.2, paragraph 6).

Sheet jointing: Joints which probably develop from the unloading of a rock mass as the cover is eroded. The joints are roughly parallel to the ground surface especially in plutonic igneous masses.

Shell and auger: A basic drilling rig for use to limited depths in overburden. The auger loosens up the materials, the shell recovers them for inspection. In soft materials the shell can be adapted to do the work of the auger.

Shotcrete: Sprayed concrete (in USA includes Gunite).

Shrinkage limit: Moisture content of drying clay below which air enters the pores.

Silcrete: Silica concentrated by water movement and precipitated into a hard cement of a soil matrix.

Siliceous: Referring to rocks principally composed of silicate or silica minerals.

Sill: A mainly concordant sheet of igneous rock intruded between strata near the surface of the crust.

Silt: Sedimentary deposit with rounded particles not more than 0·02 mm in diameter (*see* Table 7).

Silt fraction: The fraction by weight of a sediment of size range 0·02 to 0·002 mm effective spherical diameter.

Siltstone: A sedimentary rock of silt-sized particles (*see* Section 4.3).

Slake durability test: Measurement of rate of degradation of rock in water to obtain an index of comparative weathering performance.

Slaking: The breakdown of rock structure by water.

SLAR: Side-look airborne radar.

Slate: A fine-grained metamorphic rock with well-developed planes of cleavage (*see* Section 3.3, paragraph 10).

Slickensides: Polished and grooved surfaces resulting from shearing along fault planes.

Sliding (geol): The gravity sliding of a rock mass after the rock has consolidated (contrast slumping).

Slip circle:

  a. A portion of a circle on a vertical section along which it is assumed (to facilitate calculation) that sliding will occur.

  b. An approximately circular line of a section where slipping has occurred.

Slumping (geol): The sliding down of an unconsolidated sediment mass on an underwater slope.

Smectite: A group of clay minerals with swelling properties (*see* Section 5.2, paragraph 21).

Soil: Term used differently in geology and engineering (*see* Chapter 1).

Soil log: A systematic record of soil types sampled at various depths at a single location.

Soil profile: The whole pedological sequence from surface down to unweathered sediment or rock.

Soil texture: Field term descriptive of soil grain size distribution.

Solid: Term used by geologists for rock underlying drift. Note, however, that the rock need not necessarily be consolidated.

Solifluction: Creep of water-saturated soil on sloping ground, often in regions subjected to alternate periods of freezing and thawing.

Solum: Collective term for the layers or horizons within a pedological soil, other than the parent material.

Sonde: An instrument lowered into a borehole to obtain continuous records (logs).

Sonic log: A continuous record of velocity of sound through the ground adjacent to a borehole.

Sonic velocity test: A method of measuring the velocity of propagation of sound waves through a material.

Specific retention: Ratio of volume of suspended water to volume of associated voids.

Specific yield: Ratio of voids not occupied by suspended water to the total volume of the associated ground.

Spontaneous (or self-potential log): Continuous record of difference in potential between a probe and a fixed earthing point.

Stabilisation (soil): Process of mechanical and/or chemical treatment of a soil to increase its strength or its other properties of practical importance.

Stalactite: A conical structure usually of calcium carbonate deposited from dripping water and hanging from the roof of a cave.

Stalagmite: A conical structure usually of calcium carbonate, developed upwards from the floor of a cave by the action of dripping water.

Standard Penetration Test (SPT): The most commonly used *in situ* test to measure in relative terms the resistance of soil to deformation by shearing.

Step drawdown test: (*See* Performance test).

Stock: A body of plutonic igneous rock, usually less than 80 km^2 in surface area, discordant from the surrounding rock (contrast batholith).

Storativity: (*See* Section 12.3, paragraph 5).

Stratification: The arrangement of sedimentary rocks into parallel or sub-parallel strata separated by bedding planes.

Stratigraphy: (*See* Chapter 1).

Strike: The azimuth of a horizontal line drawn on a bedding plane. Strike is at right angles to the direction of true dip (*see* Figure 13).

Structural geology: (*See* Chapter 1).

Sub-artesian: Of conditions similar to artesian, but in which the pressure level remains below the land surface.

Submerged forest: Remains of former vegetation normally in the form of peat and tree-trunks, now found below high tide marks of the sea.

Subsidence: The sinking or caving in of the ground, or the settling of a structure to a lower level, essentially as a result of removal of support, e.g. by coal mining.

Subsoil: The use of this term should be discontinued; *see* CP 2001 paragraph C165 (quoted in Chapter 1).

Suspended (vadose) water: Water held in the ground by molecular and capillary attraction above the level of the water-table.

Swallow-hole: Vertical hole caused by solution in limestone country and down which streams flow underground.

Swelling pressure: Pressure exerted by confined swelling clays when moisture content is increased.

Syncline: A trough-shaped fold.

Tacheometric alidade: A sight rule for a plane table which can be used to measure distance.

Talus: Loose and incoherent deposits, usually at the foot of a slope or cliff.

Tectonic geology: (*See* Chapter 1).

Tectonics: Term applied to large scale crustal features and movements both visible and inferred, and to their dynamic interpretation.

Terrace: Relatively level area, relict of earlier river alluvium on the side of a valley above the present flood-plain.

Terrace deposit: A dissected remnant on a valley side of an old flood plain deposit.

Terrain evaluation: Assessment of an area of ground for engineering or for other purposes.

Theis non-equilibrium equation: (*See* Section 12.8, paragraph 3).

Thixotropy: The property of liquefying under shearing motion and gelling on standing.

Throw: The vertical component of displacement which has occurred along a fault plane (*see* Figure 14).

Thrust fault: A fault with a large hade, in which one set of rocks has been pushed over another set; an extreme type of reversed fault.

Till: Universal term for any sedimentary deposit resulting directly from the melting of an ice-sheet.

Topsoil: Upper humic part of soil profile; *see* CP 2001 paragraph C170 (quoted in Chapter 1).

Tor: A prominent exposed granite hill-top (term originating in SW England).

Translocation: Movement of material in solution or colloidal suspension through soil.

Transmissivity: Rate of transmission of water through unit width of an aquifer under unit hydraulic gradient (*see* Figure 119).

Transpiration: Water-loss from leaves and other plant organs to the atmosphere.

Transported soils: Soils (in engineering sense) which have been transported from their place of formation.

Triaxial test: Axi-symmetrical loading test of cylindrical soil or rock sample under all round cell pressure.

Tsunami: Japanese term for sea wave, usually confined to long period (60 second) waves, produced by a submarine earthquake or similar disturbance.

Tufa: Precipitated limestone deposit found around springs issuing from a limestone formation.

Tuff: A medium or fine-grained volcanic ash.

Turbidite: A sedimentary deposit from a turbidity current (*see* Section 3.2, paragraph 7).

Unconformity: A surface of erosion which separates rocks of two substantially different ages.

Undisturbed samples: Samples of engineering soil from a borehole or trial pit which have been disturbed so little that they can be reliably used for laboratory measurements of their strength.

Unfissured clays: Clays which in their natural state do not contain a system of fissures.

Uplift (geol): The elevation of parts of the Earth's surface which initiates a cycle of erosion.

(eng): The pressure exerted by water in the joints of a structure.

Vacuum well point: One well, of a number, used for groundwater lowering, connected to the suction side of a pumping plant.

Vane test: A measurement of the *in situ* shear strength of a soil by a rotating vane (*see* Section 9.3, paragraph 44).

Varves: Thinly laminated fine-grain glacial lake sediments, reflecting seasonal changes in deposition.

Vermiculite: Clay mineral (*see* Section 5.2, paragraph 22).

Vesicular lavas: Lavas containing gas bubbles which have been trapped during solidification, e.g. pumice. The bubbles may later be filled by minerals.

Voids ratio: Ratio of volume of voids to volume of solids (of porosity).

Volcanic ash: Deposit of sand grade and finer fragmental material derived from a volcano in eruption.

Vulcanology: (*See* Chapter 1).

Wagondrill: A chassis-mounted mechanical hammer drill for rock.

Washboring: Sinking a casing or drive pipe to bedrock by a jet of water within the casing, sometimes helped by aid of driving. Used for casing off loose ground before diamond drilling starts in the solid rock, but not for reliable soil sampling.

Water-table: Level up to which all rock pore-space is filled with underground water.

Weathering: The process by which rocks are modified *in situ* by surface agencies, such as wind, rain, temperature changes, plants and bacteria (*see* Section 3.4).

Wireline equipment: Core-boring equipment which provides for the recovery of core tubes and cores by wire suspension through the drill rods.

Wrench (or tear) fault: A fault in which the fault plane is more or less vertical, but where the principal movement of the rocks has been horizontal along the fault plane.

Zonal soils: Mature pedological soils, formed by the long continued influence of one climate, often with distinct horizons.

Zone (geol): Group of rocks distinguished by consistent fossil or mineral content.

Zone of aeration: That part of the ground in which the voids are not continuously saturated.

Zone of saturation: That part of the ground in which the voids are continuously saturated.

CHAPTER 1

THE SCOPE OF GEOLOGY

1. The study of earth materials goes back to the first use of metals, but the interpretative study of the minerals and an understanding of the true origin of rocks dates effectively from the beginning of the nineteenth century. The study of geology has subsequently become enlarged by the growth of numerous sub-disciplines, some like geophysics being now as well known as geology itself. The term earth science (or geoscience) has been coined to embrace once again the whole field.

2. The various sub-disciplines which are defined below are concerned with:

a. Basic description of earth minerals.

b. The application of other sciences to earth materials.

c. Methods of obtaining information for historical synthesis of earth evolution.

d. Practical applications to human demands and problems.

3. Definitions:

(1) **Mineralogy** is essentially descriptive of the naturally occurring compounds and chemical elements that in various combinations constitute the rocks of the earth; the investigation and classification of these minerals is mainly chemical and crystallographic.

(2) **Petrology** is first descriptive (petrography) of rocks that are naturally occurring associations of minerals. Secondly from the nature and distribution of rocks, the sequence of their mode of formation is inferred (petrogenesis).

(3) **Geochemistry** is the application of chemical methods and theory to explaining the pattern of distribution of minerals and rocks, and it makes a major contribution to the prediction of occurrence.

(4) **Geomorphology** is descriptive of the present exposed surfaces of the rocks of the crust of the earth, and seeks to interpret these surfaces in terms of natural processes (chiefly erosion) which lead or have led to their formation. The term 'Physical Geology' has also been used in this sense.

(5) **Pedology** (Soil Science) is concerned with those parts of the present earth surface which have become weathered or otherwise modified

in situ by solar energy and by the effects of organisms to form a soil which is of primary importance to man in agriculture.

(6) **Sedimentology** comprises the description and interpretation of the products of erosion (deposits) and of large assemblages of organisms which result in accumulations of their mineral skeletons. The study is extended to include processes of sedimentation, the interpretation of past deposits of this kind and of the natural changes they have subsequently undergone (diagenesis).

(7) **Stratigraphy** (Historical Geology) includes first the description of rocks locally in their observed sequence of formation. Secondly a composite reference scale of layered rocks is selected to represent a continuous period of time. Thirdly the synthetic activity is the time-correlation of any other set of rocks with the reference scale of rocks by comparison of evolutionary (or occasionally unique) events inferred from observations on the two sets of rocks.

(8) **Palaeobiology** (Palaeontology) is concerned with recording details of all traces of past plant and animal life. Usually an attempt is made to interpret the mode of life of the organisms (mostly extinct); this is however only a step towards establishing the evolutionary history, for its use in stratigraphy, of the group of organisms concerned. Palaeobiology is currently the principal evolutionary basis of stratigraphic correlation.

(9) **Tectonic geology** is the study of the gross arrangement of major rock bodies in the crust of the earth, and the elucidation of the origin and development of the vertical and horizontal movements that have led to this arrangement.

(10) **Structural geology** is the study of the effects of forces causing flow folding and fracture structures to develop in previously consolidated rocks. Interpretation from the observed fractures and folds and their patterns, of the forces and of their directions and constraints, is the main basis of tectonic reconstructions.

(11) **Metamorphic geology** is the study of the effect of natural heat and/or pressure on the mineral content of previously formed rocks in terms of the chemical and physical stability of individual minerals under these conditions, but short of complete melting.

(12) **Vulcanology** is the study of high-temperature melted rock phenomena resulting from internal earth processes, either confined beneath the observable crustal surface, or sometimes unconfined at the surface in the form of volcanoes.

(13) **Geophysics** is the study of all the gross physical properties of the earth and its parts, and is particularly associated with the detection of the nature and shape of unseen subsurface rock bodies by measurement of such properties and property contrasts. Small scale applied geophysics is now a major aid in geological reconnaissance.

(14) **Seismology** is a branch of geophysics concerned with recording of earthquakes and the interpretation of these records. Increasingly important is the art of prediction of earthquakes with the eventual possibility of ameliorating their surface effects.

(15) **Hydrogeology** is the study of the natural (and artificial) distribution of water in rocks, and its relationship to those rocks. In as much as the atmosphere is a continuation of the hydrosphere, and is in physical and chemical balance with it, there is a close connection with meteorology.

(16) **Glacial geology** is the study of the direct effects of the formation and flow under gravity of large ice masses on the earth's surface. Glaciology is concerned with the physics of ice masses.

(17) **Impact geology** is the study of the effects of collisions of extra-terrestrial masses with the earth's surface, and has derived great impetus from study of similar features of the moon.

Use of terms 'rock' and 'soil'

4. Unfortunately engineers and geologists have developed distinct uses of these terms which cannot be reconciled and which will cause confusion in the present context unless each use is prefixed.

TABLE 1. RECOMMENDED USES OF THE TERMS
'SOIL' AND 'ROCK'

Current geological usage	Recommended geological usage	Example of a typical section	Recommended engineering usage	Current engineering usage (Refs to CP 2001: 1957 appx)
(a)	(b)	(c)	(d)	(e)
Soil (pedology definition)	Soil profile	Temperate weathered profile	Topsoil (humic)	Topsoil (C170) (humic)
'Drift' deposits	Unconsolidated deposits	Glacial till (Pleistocene)	Engineering soil	Soil (C161) (with boulders)
(Solid) Rock	Semiconsolidated rock with concretions	Oxford Clay with concretions (Jurassic)		
	Consolidated rock	Coal Measure shales (Carboniferous)	Bedrock	Rock (C155) (Bedrock)

NOTE: It is recommended that 'soil' and 'rock' should not be used *in writing* without qualification.

3

As indicated in Table 1, geologists have accepted the normal pedological definition of 'soil' meaning the whole weathered profile down to unweathered rock; they have also classified as 'rocks' all ancient deposits, whether consolidated or not (because consolidation is an accident of local crustal circumstances). Engineers on the other hand have used 'topsoil' for the upper humic part of the pedological soil profile, and 'soil' for everything underneath that is movable by normal earth moving equipment; 'rock' or 'bedrock' was sufficiently consolidated to require blasting or comparable effort to break it up. The usage of these terms is as described in the relevant paragraphs of the appendix to CP 2001: 1957:

> *C155. Rock.* In the engineering sense, hard and rigid deposits forming part of the earth's crust, such as sandstones, limestones, metamorphic formations and igneous masses, as opposed to deposits classed as soil. Geologists define rock as any naturally occurring deposits be they hard or soft, but excluding topsoil.

> *C161. Soil.* In the engineering sense, any naturally occurring loose or soft deposits forming part of the earth's crust, particularly where they occur close enough to the surface of the ground to be encountered in engineering works, but excluding topsoil. The term covers such deposits as gravel, sands, silts, clays and peats. It should not be confused with the agricultural or pedological soil which embraces only the topsoil and subsoil as here defined.

> Pedological soil may come within the meaning of the word soil in the engineering sense when it is excessively deep as in some tropical and continental regions, but in the British Isles the two conceptions of the word can be kept distinct.

> *C165. Subsoil.* The weathered portion of the earth's crust that lies between the topsoil as here defined and the unweathered material below. The term is sometimes used to refer to the soil in the engineering sense (e.g. subsoil drainage) but its use in this sense is deprecated.

> *C170. Topsoil.* The superficial skin of the deposits forming the earth's crust that has, by processes of weathering and the action of organic and other agencies, been transformed into material capable of supporting plant growth. The term thus embraces the upper or humus-bearing horizons of the soil of pedology.

The differences between the two well established procedures are clear. In the long term it will be essential to avoid these difficulties of confusion by developing modifications of both words which add more precision to any use, and to persuade all users to subscribe to this practice, particularly in any written statement or record. This persuasion will take time and will only be accepted if it is simple; the uses shown in the centre of Table 1 are therefore suggested and will be used where appropriate in this book.

5. *Engineer, geologist and prediction.* Geotechnical subjects with particular relation to engineering are discussed in the later chapters of this book; in the early chapters however, an attempt is made to isolate those elements of the whole of

geology which are necessary to a general understanding of 90 per cent of the problems in this field likely to be encountered by an engineer. This applies particularly to the direct finding or tracing of either rocks or underground water.

The geologist is trained to predict conditions by means of a very wide range of observations, by understanding of processes, and by integration of results. It cannot be too strongly stressed that field observations which have no obvious direct relevance to engineering, such as the collection of fossils, may be crucial to a geologist in making an effective assessment of a problem; the geologist will normally need all the assistance he can get in recording such observations.

REFERENCE LIST—CHAPTER 1

CP 2001: 1957   Site Investigations. Code of Practice 2001, British Standards Institution, London (under revision).

CHAPTER 2

GENERAL GEOLOGICAL THEORY

SECTION 2.1. INTRODUCTION

Every geological problem facing an engineer contains elements derived from the whole history of the earth. That history is best considered as a continuous evolutionary succession of events that are all interconnected. Both the use of the expression 'events', however, and any isolation of some of them for study, are purely conceptions for the better convenience of the limited senses of the human observer. Despite strongly held beliefs, for instance, in the periodicity of some of these events there is no convincing evidence for such an arrangement. The order of treatment therefore in this chapter could not be entirely logical unless it were arranged in historical sequence beginning with the formation of the earth. Because for the present purpose it is only feasible to begin with the known or observable features of the earth, the subdivisions employed tend to follow the man-made handling pattern for geoscience referred to in Chapter 1, and do not call for any particular reading sequence.

It follows that, however apparently complete an interpretation is provided in any field, the distance in terms of geological time or space from direct human observation should be included in any estimate of confidence limits.

SECTION 2.2. THE EARTH

1. *Observation.* The deepest borehole yet made from the earth's surface is to about 8 km. Natural disturbance of the crust in geological time has brought to view some rocks believed to have been formed at about 20 km depth. Some volcanic material may come from about 50 km depth. Otherwise the composition and structure are not directly known, but sufficient circumstantial evidence is available to limit speculation. Interpretation of the transmission rates of seismic shock waves (from earthquakes) is the principal source of data. Although an engineer is most unlikely to be directly involved with rocks beyond the depth of 8 km, there is no logical separation of visible rocks from what is below, down even to the centre of the earth. Historical explanation of the nature and relationships of rocks depend on the theory of the whole earth, and thus prediction about distribution of surface and near-surface rocks does require some knowledge of the whole, although obviously progressively less about the more central parts of the earth.

2. *Temperature.* The surface temperature of the earth remains within a small range in absolute terms and is mainly determined by solar radiation, the effects

of which do not penetrate far into the poorly conducting surface rocks. That the inner parts of the earth are at a very much higher temperature is indicated by observed average temperature gradient in mines and boreholes of 3°C per 130 metres, by measurable heat flow through the crust and by volcanoes. It is suspected that any simple concept of a central source of heat should be modified by the irregular distribution in some of the outer parts of the earth of heat from radioactive isotopes such as those of uranium and potassium.

3. *Formation of the Earth.* The current theory of the formation of the earth requires 'cold' accretion of planetismals with a subsequent heating caused perhaps partly by gravity differentiation of the core from the mantle (originally there would have been no distinction) and partly from high radiogenic heat production from short-lived nucleides in the early stages. Such an origin calls for expansion in the volume and surface of the earth during its subsequent history; whereas this is not proved, the opposite (contraction) is not considered to be likely. A very long period in balance with no change also seems unlikely, although as mentioned above, the surface temperature conditions must have remained in a surprisingly small section of the absolute scale.

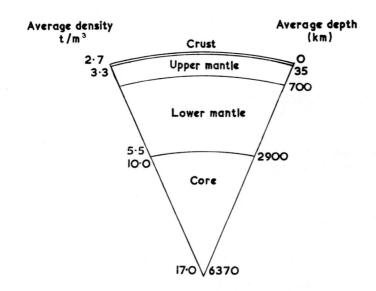

Fig 1.   Diagram of layered structure of the earth as interpreted from seismic data (approximately to scale)

4. *Mantle and core.* The interpretation of the earth (Figures 1 and 2) as consisting of a core surrounded by a very thick mantle, in turn covered by an extremely thin crust, depends on the recorded observations of an international

	Earth layers		Depth to base of layer (km)		Seismic velocity (P) km/s	Density inferred t/m³	Composition inferred
			Ocean	Land			
CRUST	Upper crust		4	–	1·7	1	water
			5	10	3·5	2·0	sediments
	[CONRAD DISCONTINUITY]	LITHOSPHERE		20	5·6	2·7	acid igneous
					6·3	3·0	
	Lower crust						basic igneous
	[MOHOROVIČIĆ DISCONTINUITY]		10 —	40	7·0	3·1	
					8·1	3·3	
MANTLE	Upper mantle						ultra-basic
	[20° DISCONTINUITY]	ASTHENOSPHERE		400		3·5	
	Lower mantle						high density minerals
	[GUTENBERG DISCONTINUITY]		2900		14	5·7	
						9·4	
CORE	Outer 'fluid'						nickel/ iron
	Inner 'solid'						

Fig 2. Table of data (amplifying Figure 1). All information except that from the upper crust, is derived from seismic, volcanic and meteorite studies

network of seismic stations. The different rates of transmission of shock waves are taken to indicate passage through regions of material of calculable density with distinct interfaces of relatively sharp velocity (and thus by inference density) changes. The calculated density range for the core can be satisfied by a nickel/ iron composition; many meteorites which are believed to have been derived from the break up of a planet similar to the Earth, are of this composition. The greater proportion, however, of meteorites observed to fall are siliceous (chondrites) and these could correspond conveniently with the supposed density and composition range of the mantle.

5. *Crust.* The thickness of the crust is about 1 per cent of the mantle thickness. The predominant material is silicate; it is the oxidised outer layer adjacent to the oxygen bearing atmosphere. The explored crust is 47 per cent oxygen and 28 per cent silicon by weight. To judge from the composition of extruded volcanic material, the composition of the lower crust and upper mantle from which it originates is higher in iron and magnesium and lower in silicon than the surface rocks.

6. *Minerals.* Only the chemical elements, oxygen, silicon, aluminium, iron, calcium, sodium, potassium and magnesium, in decreasing order of importance, form more than 1 per cent by weight of the crust. These elements are combined in numerous common minerals, each with a structure and composition determined by the temperature and pressure of the environment (level in the crust) of their formation. The occurrence of these minerals in rocks can therefore be used directly to indicate the level of origin of the rocks concerned.

7. *Rocks.* Rocks formed by freezing (crystallization) of a high temperature silicate melt are termed igneous; they are characterised by complete intergrowth of the crystals which thus individually have irregular margins. In contrast rocks formed in lower temperature and pressure conditions from the products of erosion are termed sedimentary; they are composed of discrete particles which are each rounded or subangular and which may be close-packed but not inter-grown.

SECTION 2.3. **EVOLUTION OF THE CRUSTAL SURFACE**

1. *The atmosphere.* The earliest fossils, which are from rocks more than 3000 Ma (million years, formerly m.y.) old, were simple aquatic plants; since that time evolution of living organisms in the sea has been continuous, which indicates surface temperatures remaining between about 0° and 40°C for this long time. The early atmosphere of the earth probably contained methane, hydrogen, water vapour and carbon dioxide etc, but no oxygen. Oxygen is thought (Figure 3) to have built up very slowly from the photosynthetic action (intake carbon dioxide) of marine plants (algae) and to have reached an adequate level to support marine animals by about 650 Ma B.P. (million years before present). Dependent on the oxygen level was the proportionate quantity of ozone in the outer atmosphere, acting as a protective screen against the otherwise unlimited ultra-violet

(U.V.) solar radiation. Apparently by about 400 Ma B.P. the process of increasing oxygen in the atmosphere and ozone in the outer atmosphere was sufficiently advanced to permit animal and plant life on land. It is likely that evolution in this field of atmospheric composition continues, in addition to the very recent changes resulting from human activities.

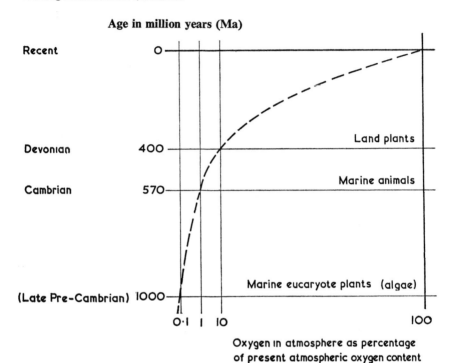

Fig 3. Supposed relationship of major events of biological evolution to the progressive increase in the percentage of oxygen in the atmosphere

2. *The atmosphere and the hydrosphere.* The rivers and oceans are one system; their content of dissolved gases is physically in balance with the atmosphere, and correspondingly the water vapour content of the atmosphere is in balance with the hydrosphere (*see* also Section 2.7, paragraph 1).

3. *Weathering of rocks.* The exposed parts of the crust are continually attacked chemically by the combined atmosphere and hydrosphere contents, controlled only by the physical characteristics of these agents under a gradient of solar radiation determined geographically by the mode of rotation of the earth. The chemical activity is therefore greatest equatorially and least effective at the poles.

Stratigraphic scale divisions (periods)	Radiometric dates ($\pm 5\%$) Ma	Some rocks dating from these periods	
		In Britain	Elsewhere
Holocene		Fen silts	Loess; USSR, China
	·01		
Pleistocene		Boulder clay	
	1·5		
Neogene Pliocene Miocene			Alpine uplift & deposits
	30		
Palaeogene Oligocene Eocene Paleocene		Bembridge Limestone London Clay	Brown coal; Germany Nummulite limestone; Egypt
	65		
Cretaceous		Chalk	Deccan lavas; India
	135		
Jurassic		Portland Stone Oxford Clay	Lithographic limestone; Bavaria
	190		
Triassic		Cheshire salt	Muschelchalk; Germany
	225		
Permian		New Red Sandstone (North sea gas field)	Karroo coals; South Africa
	280		
Carboniferous		Coal Measures Carboniferous Limestone	Coal measures; Europe USA, USSR
	345		
Devonian		Old Red Sandstone	
	395		
Silurian		Wenlock Limestone	Limestone at Niagara Falls
	430		
Ordovician		Skiddaw Slates	Orthoceras limestone; Sweden
	500		
Cambrian		North Wales slates	Leningrad clay; USSR
	570		
(Pre-Cambrian)		Schist and gneiss of Northern Highlands	Schist and gneiss of Canadian shield, Grand Canyon, Africa, Peninsular India, Central Australia etc
	(4600)		

Fig 4. Main divisions of the global stratigraphic scale and some examples of rocks formed in the periods listed

11

4. *Topography*. If the surface of the earth was homogeneous (composed of a single rock type), the rivers would tend to flow out radially from the hills. Each river would have a standard profile, steep near the source and gradually changing to a negligible gradient as it reached the sea. The observable complications in practice are due to:

a. Heterogeneity of variety of surface rocks with different physical properties.

b. Changes of relative level of land and sea which alter the base-level of the standard river profile.

5. *Deposition of sediment*. The products of rock weathering are transported by water, ice, and to some extent by wind, and accumulate as sediment. This sediment usually becomes consolidated subsequently into rocks.

6. *Control by biosphere*. The nature and composition of the biosphere representing all life at any one time, controls through the chemistry of the atmosphere and hydrosphere the whole of those rocks of the crust produced under surface temperature and pressure conditions. It can be demonstrated that the nature of these rocks has changed during earth history by evolution in the whole interlocked system of surface processes.

7. *Stratigraphy*. Since a crust was formed on the earth, parts of it have been uplifted and eroded, providing deposits which accumulated in layers, youngest uppermost. The plants and animals of the biosphere were continuously evolving and therefore different fossils of them came to be included in the successive deposits and are taken to characterise them. The resulting sequence in time has been described (Figure 4) in terms of named sub-divisions (periods) and is known as the stratigraphical scale.

8. *Radiometric ages*. Radiometric ages are calculated from the known rate of decay of certain isotopes of potassium, rubidium, uranium and thorium, and the amount of decay product remaining. They are given in years before present (B.P. is taken from 1950). There are technical difficulties in the chemistry which result in an uncertainty of from ± 1–10%, and in most cases there are additional geological uncertainties. Dates given therefore for the rock boundaries and other events are approximate and cannot yet be improved.

Section 2.4. VERTICAL MOVEMENTS IN THE CRUST

1. *Changes of land level*. With respect to mean sea level it is clear that parts of the land have moved either up or down both in historic times as indicated by old geographic maps and in earlier times as proved by the finding of marine fossils on hill-tops and even near the tops of the highest mountains. This effect is seen to be local even in Great Britain where Scottish and northern coasts are characterised by raised beaches while southern England has drowned valleys such as the estuaries of Devon and submerged forests at many points (e.g. Pett level, near Hastings).

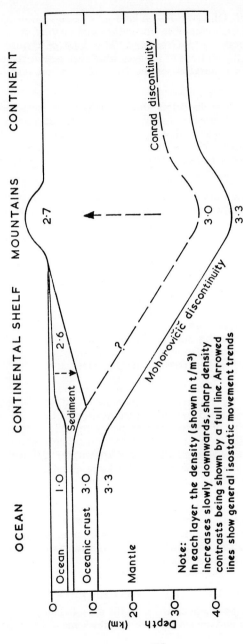

Fig 5. Diagrammatic section of the crust and upper mantle across a continental margin into an ocean area

13

2. *Changes of sea level.* Observations on land may be complicated by true rise and fall of the sea level itself caused by formation and melting of large ice caps such as those of Antarctica and Greenland. This kind of change known as eustatic is typical of the fluctuations of ice ages which are rare in geological time, although one such change characterises the latest one million years and is still continuing.

3. *Deltas.* Boreholes made in deltas show hundreds or even thousands of metres of sediment all laid down virtually at sea level to judge from fossil and other characteristics. This suggests that the crust in the area of the delta sinks at about the same rate as the sediment accrues. This further suggests that such a sector of crust floats like a raft on something below and sinks in response to loading by sediment.

4. *Mountains.* Mountain areas suffer rapid erosion and judging from details of river drainage and temporary lakes, they rise as would a floating raft with some of the load removed. The melting of a considerable ice cap which fifty thousand years ago covered Scandinavia has also resulted in a rising of the crust there which continues today when almost all the ice has gone.

5. *Isostasy.* These observed effects in deltas and in mountains suggest that sectors of the crust are independently in delicate vertical balance. It has also been observed that high mountain parts of the crust appear to be made of the lightest materials (Density 2·7) and low ocean floor parts of the crust of high density material (3·0), gravity measurements however are relatively uniform and show that equal areas of the Earth's surface are underlain by approximately equal masses of crust. The state of balance is known as isostasy, with a compensation level at the base of the lithosphere (*see* Section 2.6, paragraph 2).

6. *Interpretation.* As shown in Figure 5, a mountain range on this model is taken to have a root of light material which causes it to float higher with off-loading. The root when formed by lateral compression at the same time as the mountains, was of course of finite size and will ultimately become exhausted when as a consequence the mountains will cease to rise. The isostatic mechanism which causes a delta to cease accumulating is not known; the sediment is probably considerably less dense than 2·7 as it contains large quantities of water in the upper layers. The total effect of erosion, transport and deposition on the crust is to maintain balance in the crust. The origin of the whole flotation is uncertain, but it is probably connected with a change of state of minerals at the top of the mantle at a critical temperature and pressure; it remains a problem. The crust is unexpectedly sensitive to loading, to an extent that even large engineering works could produce a detectable effect; in certain circumstances elastic compression of rocks may also be an important factor.

SECTION 2.5. HIGH TEMPERATURE PROCESSES

1. *The Upper Mantle.* In whole-earth terms the crust (which may be observed) is very thin (5 to 50 km) and covers the 2900 km thick mantle which is the main

feature of the Earth and has not yet been observed directly. The low thermal conductivity of the crust separates the heat felt at the surface of the crust, which is mainly from solar radiation, from the heat of the mantle; the rising temperature gradient (*see* Section 2.2, paragraph 2) indicates a probable temperature of the order of 700°K at the top of the mantle. The specific density of the upper mantle material is inferred to be 3·3 from its transmission rate of seismic shock waves; this is greater than that of the volcanic lavas (3·0) which appear to have originated at this depth.

2. *Volcanoes.* These vents for escape of molten rock or magma from below the solid crust vary from large linear fissures as in Iceland to isolated tubular conduits associated with the conical volcanoes. The form and arrangement of the igneous products depends on the viscosity of the magma locally involved; magma with a high silica content flows less easily forming steep cones and is liable to freeze locally thus blocking the vent and leading to explosive activity and pyroclastic products. Magma with a low silica content, which has come from the upper mantle or low in the crust, flows easily from fissures, forms only very low cones and seldom explodes; the product basalt of this type of activity forms most of the crust in ocean areas and in large land areas such the the Deccan and Karroo plateaux. The oldest rocks (3800 Ma) known in the crust are probably of volcanic origin and a large part of the total crust including the ocean floors has been so formed through geological time. In volcanic areas there are usually a large number of sheet intrusions of magma into fissures in the upper crust, the intrusions being known as dykes and sills, when subsequently exposed by erosion.

3. *Plutons.* Erosion of the crust has also exposed many large igneous masses of solidified magma with steeply inclined or vertical sides and of unknown depth; although these plutons are emplaced among other rocks of all kinds, there is no evidence of lateral displacement of these other rocks to accommodate the pluton. The composition of the pluton is usually of highly siliceous rock such as granite, but the lower parts of a large one may be of the much less siliceous rock (gabbro) with gradations between. Two major possible modes of formation have been suggested:

a. Hot (gabbroic) magma from the upper mantle has 'eaten its way' up into the crust, ingesting the more siliceous crustal rock of lower melting point and thereby becoming itself more siliceous, towards a limit of composition (granite).

b. Crustal rocks have been melted in extreme metamorphism and re-crystallised as granite.

In the former case the margins would be sharp (*see* Figure 6), and in the latter they would be imprecise; both kinds of margin are quite common in separate plutons and it is believed that both processes operated in separate cases, probably at different depths. In the continental areas the total proportion of crust of plutonic origin is large in area, and may be much greater in volume at depth. The

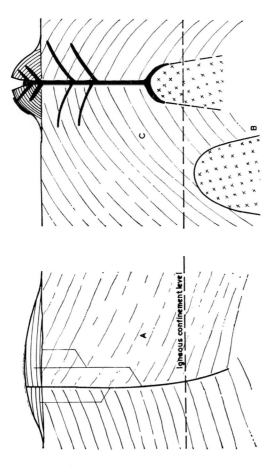

NOTES:

A. Large basaltic domed volcano (e.g. Iceland) over deep fissure through the crust

B. Granite pluton formed slowly without displacement of surrounding rocks, below igneous confinement level (due to gravity effect of overburden); these plutons may subsequently (after solidifying) be exposed at surface by uplift and erosion (e.g. Dartmoor granite)

C. Steep volcanic cone (e.g. Vesuvius) of relatively viscous andesite or rhyolite with interbedded ashes, formed when appropriate magma chamber lies above igneous confinement level

Fig 6.   Diagrams to illustrate origin of igneous masses

local causes of initiation of volcanoes or plutons are not always clear but most can now be explained in the general plate-tectonics model (*see* Section 2.6, paragraph 2).

4. *Igneous rocks.* Igneous rocks derive therefore originally from the upper mantle of which the chemical composition (sometimes known as mafic, which is rich in magnesium and iron) is believed to be close to the mineral olivine. Rather rarely, olivine-rock (peridotite) occurs at the Earth's surface and is taken to represent a small slice of mantle caught up in the crust in a disturbed area; such rock is described as low in silica. All other igneous rocks are to some extent the products of melting mantle rock or of 'contamination' of this mantle material by the more siliceous (oxidised) crust through which it has to pass; they are progressively higher in silica to a limit (e.g. granite) at which the silica content has risen above the crust average. Peridotite-gabbro-diorite-granite may be thought of as a series in this respect, but the names have simply been allotted to arbitrary divisions of chemical constitution. When greater precision of statement is required more subdivisions may be named. For the engineer it is only important to know in which major division a previously unknown name or sub-division falls. In such a complex chemical environment as the crust there are other rare possibilities, but they need the attention and interpretation of a geologist.

5. *Terminology.* For many years in geological literature gabbro has been termed 'basic' and granite 'acid'. The main difference between these two as stated above is in silica content and is thus not related to the more normal chemical use of 'acid and basic'; such terms should be avoided. Rocks such as gabbro and granite are coarse grained (crystals of 1 centimetre length and more); this suggests slow cooling and freezing to a solid, and is part of their interpretation as being formed in plutons far beneath the surface of the crust. Magmas of such composition (and all others) may by local accident be extruded from volcanoes; the resultant quick cooling produces basalt and rhyolite which are simply fine grained equivalents, both mineralogically and chemically, separately named as rocks in the field long before their identity could be demonstrated microscopically. Dolerite is a name for a rock of intermediate grain size between gabbro and basalt which would be named more simply 'micro-gabbro'.

SECTION 2.6. **LATERAL MOVEMENTS IN THE CRUST**

1. *Continental drift.* Alfred Wegener suggested in 1912, on the basis of similarities of palaeoclimate inferred from fossils and of fit of continent shape, that most of the present continents had been formed as one land mass together until 200 Ma ago and had then moved apart; unfortunately although many geologists supported the idea, geophysicists were unable for some years to find an acceptable mechanism for such displacement of continental masses.

2. *Plate tectonics.* Since 1965 there has been developed the concept of crustal plates approximately one hundred kilometres thick and thus embracing part of the upper mantle; these plates have sufficient theoretical rigidity to allow the

17

postulated lateral movement. The term lithosphere is applied to the plates, and the term asthenosphere to the mantle area beneath them which is contrasted in not being part of this rigidity. The plates cover the whole surface of the earth and largely carry oceanic crust; some of these plates bear areas of continental crust as well. Although the continents move with their plates, they are thought to be incidental in this movement. The plates are believed to be continuously evolving in that new material is being added at one edge at rates of up to 200 mm per year, and at the other edge plate material is carried down into the mantle. It is not known whether the formation and destruction are in balance or whether the total earth surface is very slowly increasing.

3. *Plate formation.* The mid-Atlantic oceanic ridge which runs the whole length of the ocean on its mid-line has been shown to be locus of formation of new crust from the mantle by solidification of partly melted upper mantle material along a line of mantle upwelling, due possibly to some form of convection within the mantle. The new crust is added symmetrically on the two sides of the ridge in the form of an increment to each plate (North America and Eurasia) as the plates move apart. It has been possible to demonstrate this because the material involved retains remanent magnetism from the magnetic field at the time of cooling; as there have been complete reversals of polarity of the field every few million years through geological history, successive strips which have been added are alternately normal and reversed in their remanent magnetism. A ship-towed magnetometer was used to demonstrate this, but it is also apparent in Iceland which lies astride the ridge and is simply a high-point of a ridge which is elsewhere submerged.

4. *Plate destruction.* The plate formation can only continue if it is accommodated by removal of some material at the leading plate edge where it faces another plate. The edge of a plate of oceanic crust will turn down in what is called a Benioff zone underneath the edge of a continental plate, and the material so hidden is eventually reabsorbed in the mantle. This is presumed to occur for example under Japan. At such a continent to ocean margin there is normally prolonged sedimentary deposition from the edge of the continent into the adjacent sea forming a large filled trough known as a geosyncline. As indicated in the diagram (Figure 7) this tends to be very strongly deformed into a fold mountain belt, in the further course of the movement described.

5. *Seismic evidence.* Apart from the palaeomagnetic strips mentioned in paragraph 3, which have been identified over much of the ocean area of the world, the main evidence is from earthquake distribution which when plotted for the world very clearly indicates the belts of activity along plate margins. There is also a distribution of much deeper-focus earthquakes (down to 700 km) obliquely under the plate destruction zones. In general the considerable and almost continuous seismic activity of the crust and upper mantle fits the crustal plate theory.

6. *Rheidity.* At first sight circulation flow movements in the upper mantle, and deep earthquakes caused by fracture of solid material in the same region,

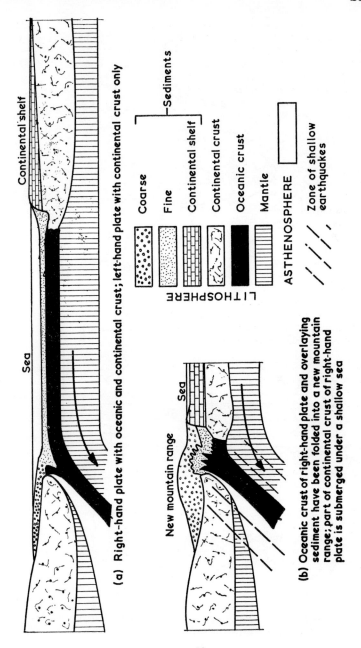

(a) Right-hand plate with oceanic and continental crust; left-hand plate with continental crust only

Continental shelf

Sea

Sediments
- Coarse
- Fine

Continental shelf
Continental crust
Oceanic crust
Mantle

LITHOSPHERE

ASTHENOSPHERE

Zone of shallow earthquakes

New mountain range

Sea

(b) Oceanic crust of right-hand plate and overlaying sediment have been folded into a new mountain range; part of continental crust of right-hand plate is submerged under a shallow sea

Fig 7. Two stages of development of a plate destruction zone, where the right-hand plate passes under the left

19

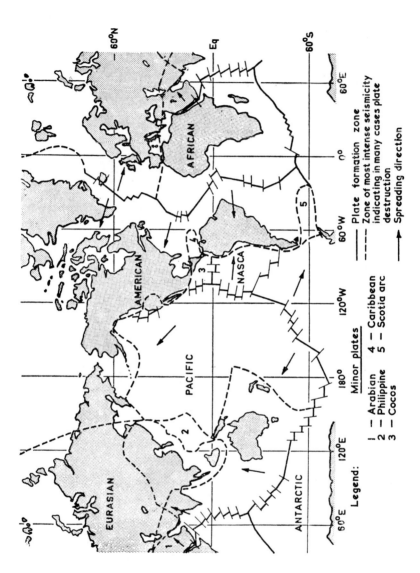

Fig 8. Lithosphere plate pattern

appear incompatible. The explanation appears to be the rheid state of the upper mantle material in which fracture occurs when a force is applied quickly, and flow when it is applied slowly. The materials with comparable viscosity properties at surface temperatures and pressures are ice and pitch.

7. *Pattern of plates.* As will be seen from the map (Figure 8) the principal plates are Antarctic, African, Indian, Pacific, East Pacific, American, European/ Asian, South-east Asian, but in several places there are additional smaller ones. The whole theory has been developed so recently that exploration in many remote parts of the world is insufficient to explain all the details of plate margin relationships; but major difficulties which might influence the whole theory are not expected on this account.

Section 2.7. WATER

1. *Earth source of water.* All the water of the surface hydrosphere and the atmosphere is assumed to have originated as a differentiate which has migrated upwards from the mantle. Igneous rocks and natural melts (magma) contain some water and it is assumed that new water (juvenile water) continues to reach the crustal hydrosphere, although probably only in small quantities. Water from hot spring activity is not necessarily juvenile, and is more likely to be surface water recycled by heating.

2. *Hydrosphere.* All water in oceans, seas, lakes, rivers and in pore-space underground is one continuous body of fluid with gravity-controlled distribution (Figure 9). Water occurs underground to a depth determined by the limits of pore-space which is reduced progressively downward by compaction by the column of rock above, and is negligible below an average depth of about 8 km. In a sense therefore the whole hydrosphere rests on the rock without pore-space at this level.

3. *Circulation.* Water is in continuous circulation by evaporation from exposed water-surfaces and transpiration through growing land plants, clouds and rain, percolation through air-filled soil and rock pore-space vertically under gravity to the main water-table, and down-slope migration in the water-table to sea-level. This meteoric water forms the principal circulation of the hydrosphere. Some of this water known as connate water may remain 'locked-up' underground for long periods of geological time and may be to some degree saline.

4. *Water-table.* The water-table is the level in rock underground below which all pore-space is filled with water, and above which much pore-space is air-filled except for a capillary zone and some percolating water passing down freely under gravity. The water-table is domed under hill regions following in a subdued way the generalised contour of the hills; the domes fall off to sea-level in the lowland areas. Water-table levels fluctuate slightly with seasonal changes in rainfall and percolation.

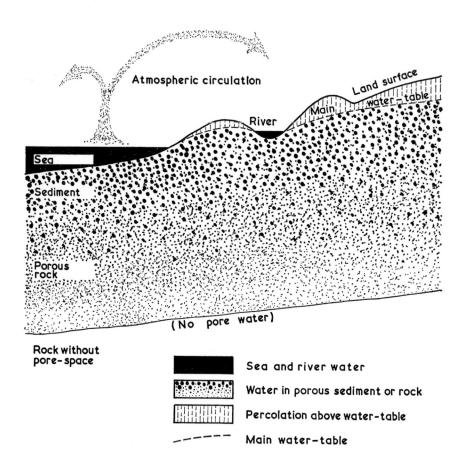

Fig 9.   Diagram to indicate general theoretical distribution of underground water. (In practice porosity also varies locally with type of rock represented)

22

The main water-table is based on mean sea-level and all rocks below that level with any pore-space will generally be saturated; igneous and other rocks with fissure space only in joints will have that space saturated; rare rocks that are even without fissure space will simply be surrounded by water in other rocks. A perched water-table is a subsidiary one in hill terrain above a main water-table, and is caused by a bed of relatively impermeable rock. Percolating water will build up a 'dome' over this impermeable bed, and will escape over the margin of the bed; this escape will be in the form of springs if the margin of the bed outcrops. All the naturally exposed water in rivers, lowlands drains, lakes etc, forms a surface expression of the water-table; in densely inhabited countries however most rivers have been canalised and this point applies in general but less precisely than in an undisturbed environment. There is normally a limited capillary fringe of partial saturation immediately above any water-table.

A water-table 'dome' depends for its height on the permeability of the rocks in which it occurs; theoretically in a rock which had infinite permeability to water, there would be no dome and the main water-table would be 'flat' at sea-level. Also theoretically if the rainfall and the percolation in an arid region were cut off entirely the dome would subside to sea-level. The natural water-table can only be below sea-level where rift-valley faulting has lowered the local land surface below sea-level, and where there is also intense aridity to evaporate the lake which would otherwise fill up to sea-level (e.g. the Dead Sea). True sand desert oases may be at sea-level or at least appreciably below the surrounding terrain because they are due to removal of the sand by wind until the water-table is intersected by the sand surface. A most interesting early appreciation of water-table features is indicated by the qanats of Iran (*see* Chapter 12).

5. *Caves.* Caves are formed inland in areas of limestone rock by running water in the percolation zone above the water-table. Flooded caves or caves intersected by the water-table indicate a change in water-table level since they were first formed. Sea-caves are related to present or past sea-level.

6. *Water divining.* The possibility of human possession of perception outside the acknowledged senses is beyond the scope of this book. However, endowed or would-be diviners will be much more successful if they appreciate the main points given above about water distribution. It may also occur to them that geological reconnaissance and physical laws provide greater probability of success, regardless of how that success may be presented.

7. *Ice.* As well as relatively thin floating ice, part of the hydrosphere at the present time (notably in Greenland and Antarctica) is frozen into land-based ice-caps. These ice-caps form only in mountain areas when the annual winter snow-fall exceeds the annual summer melting. The resulting ice flows out radially down the valley as glaciers and often onto plains or into the sea as ice-sheets. The balance of the process turns on the word 'exceeds', however small the amount of excess. Consequently fluctuation is historically frequent; glaciers in the Alps and elsewhere have been melting back since about 1830, but a reversal to this trend

could occur in the time of a human generation; in Roman times and well into the Middle Ages, there was less ice than now. Icebergs on the oceans are by-products of the land ice-sheets.

There are two major geological consequences:

a. A mountain area loaded by ice sinks isostatically as the ice mass grows and recovers (a little more slowly) as it wanes; the bases of both the Greenland and Antarctic ice masses lie in places below sea-level from this cause.

b. The total volume of water in the oceans decreases as the ice grows and increases as it melts; the consequent changes in sea-level are termed 'eustatic' and the current rise in level is an important factor in planning sea-defence work in this century.

8. *Permafrost.* In high latitudes such as in Arctic Canada and in northern Siberia, all underground water may be frozen to varying depths, e.g. up to 300 metres in about 75°N. In each summer season anything from the top few centimetres to several metres at the lower latitude edge, thaws out forming undrained extremely wet mobile soil. If such material is removed, more thawing occurs to the same depth from the new surface. The construction difficulties are great and obvious (*see* Chapter 11).

SECTION 2.8. PHYSICAL AND CHEMICAL CHANGES AFFECTING ROCKS

1. *Joints.* Almost all rocks at the Earth's surface show joints which are cracks along which no slip movement has taken place; they are usually widened and made more obvious by surface weathering but they occur with similar frequency underground. The detail of most quarrying operations depends on joints, and an understanding of their causes and probable patterns is important in many other engineering works. Although the resulting joints may appear similar in the field, the causes are distinct and are often associated with certain types of rock:

a. Volcanic rocks become jointed by contraction on cooling after solidification.

b. Plutonic rocks expand from offloading (of overburden pressure) during uplift and during erosion.

c. Sedimentary rocks are usually domed during uplift, and offloaded by erosion from the previous consolidation pressure.

d. Metamorphic rocks particularly, but many folded sedimentary rocks also, may become jointed during compressional and shearing movements.

2. *Diagenesis.* Sedimentary rocks and volcanics which have been formed at the Earth's surface, suffer progressive alteration as they become buried under increasing thickness of subsequent rocks formed above them. Pore-space is eventually eliminated, at a depth of 8–10 km, by physical close-packing of the

rock particles under compression. Solubility in water of some constituents increases with the pressure and temperature gradient, and some minerals may become re-distributed in the diminishing pore-space even at times before deep burial.

3. *Folds*. Layered rocks may be seen to have been folded in response to lateral pressures, in all forms from simple up-and-down flexures to complex overfolds. The precise style of fold is determined partly by the nature of the rocks themselves, partly by the crustal stress pattern in the area and partly by the depth of burial of the rocks concerned at the time of folding. Folds produced in rocks that were very deeply buried at the time have a more regular plastic appearance, because the rock material has slowly flowed into folds under these extreme conditions. Folds originating in less deeply buried rocks will be less regular in detail, and in many cases the rocks will have failed by brittle fracture along faults and/or joints.

4. *Faults*. In a disturbed area of crustal rocks, faults (fractures) may result in both vertical and horizontal displacements in either sense. They are usually local, extending for not more than a few kilometres. Although faults occur in most rocks their apparent frequency on maps is high in mining areas where they have been encountered and studied in workings, and usually low elsewhere as they may not have been detected. There are in addition some fundamental faults of up to several hundred kilometres in length which delimit crustal plates or large sectors of plates; such faults are likely to be still in active movement.

5. *Metamorphism*. The extremes of diagenesis may in practice be difficult to separate from the early stages of regional metamorphism (Figure 10), which is theoretically distinguished as being due to the compressional and thermal effects of lateral compression on a large scale or of very deep burial. The results are usually expressed in the development of new minerals (hence the name of the process), representing the very high temperatures and pressures reached in the formation of such rocks as schists and gneisses. On a much smaller scale simple effects of high temperature only in the rocks immediately adjacent to igneous bodies are known as thermal metamorphism.

6. *Diapirs*. In some large scale disturbances of the crust, or even during diagenesis, some less-dense rocks may come to be buried under denser layers. In certain of these cases the lighter rock may eventually 'flow' up and break out of any available fault or fissure, reaching the surface as an irregularly disturbed mass. The most usual substance involved is rock salt, although in all but arid climates, such material would soon be re-distributed from the surface in solution. Certain (less dense) acid igneous rocks such as granite often also occur in diapirs and reach the surface for the same reason, although because the density contrast is less the effect is slower.

7. *Mineral concentrations*. Most metallic minerals are only available to economic expenditure of energy when they have been locally concentrated by

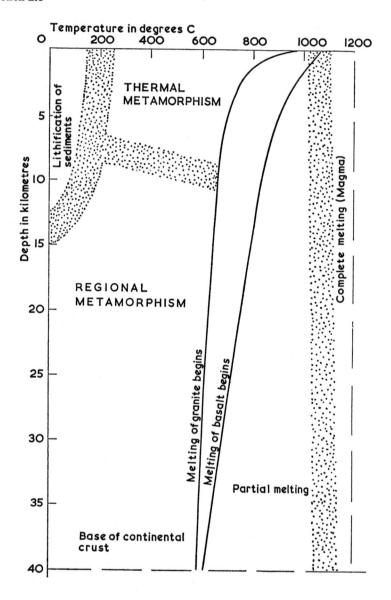

Fig 10.   General temperature and pressure (3 kbars per 10 km depth) fields of
crustal rock metamorphism

some natural process well above their average level in the crustal composition (*see* Section 2.2, paragraph 6). This may be a high-temperature igneous concentration or a low-temperature sedimentary one or both, but the mechanism must be understood in each case before a reasoned search can be made. Copper is associated with certain volcanic lavas; uranium occurs in certain rare igneous rocks and secondarily in certain organic shales; manganese is concentrated in certain deep ocean sedimentary deposits associated with submarine lava flows. Some of these conditions permitting these geochemical processes are known to have occurred so rarely in geological history that there is legitimate cause for alarm at the rate at which human civilisation is dispersing such valuable concentrations.

8. *Hydrocarbons.* Coal is known to have been formed from the accumulation of land-plant debris in swamps. Petroleum and natural gas are thought to have been formed from remains of very large numbers of microscopic marine (or aquatic) plants and animals buried with mineral matter in sedimentary rocks. Some natural gas, as in the southern North Sea, probably originated from diagenetic changes in deeply buried coal seams. Probably all petroleum and natural gas results from diagenetic heat and pressure effects on the source material. The resulting hydrocarbon, being less dense than the water which fills all the sub-sea-level pore-space, migrates upwards through the water and is usually lost in scattered surface seepages back to the hydrosphere and atmosphere. Hydrocarbon reservoirs (oil and gas fields) only occur when geological irregularities underground place an impermeable seal locally in the way of this universal upward migration. Hence the search for oil and gas is a search for subsurface geological traps.

9. *Earth magnetism.* Although much has now been discovered about the Earth's axial dipole field, linked with its rotational axis, the most unlikely fact was that accurate records of past magnetism were preserved like fossils in the rocks. When a lava-flow cools or when certain sediments accumulate, magnetic mineral grains are oriented in the direction of the Earth's field at the time and then permanently 'frozen' in. Land masses have clearly moved with crustal plates throughout the history of the Earth; palaeomagnetic observations provide the latitudes (but not the longitudes as there is no reference) for rocks of past ages. A further surprising observation of palaeomagnetism is that the Earth's dipole polarity has reversed many times, and that the patterns in time of these reversals aid in interpreting geological history.

10. *Earthquake prediction.* Various countries like Japan and Southern California are placed astride large fundamental faults, with resulting numerous earthquakes. If all such faults could move regularly and steadily in response to the forces causing them, there would probably be none of the serious earthquakes which occur when energy is stored at a fault unable to move and then suddenly released. Both prediction of the stored energy situation, and even measures to effect its more gradual release, appear now to be feasible; these could transform the position of countries unlucky enough to be sited on or near the crustal plate margins of the world (*see* Section 2.6, paragraph 3).

SECTION 2.9. INFORMATION FROM MOON EXPLORATION

1. The exploration of the moon has, so far, as expected, had more affect on the geological theory of the earth than on practical matters. The moon appears to lack a biosphere, atmosphere, surface hydrosphere, and oxidised crustal rocks. It perhaps represents what the earth would be like deprived of these features, and consisting therefore of the mantle covered by a crust resembling the earth's present oceanic crust (less the sediments), punctured by volcanoes. There are chemical differences in the distribution of elements in the moon rocks which have yet to be explained, but they may simply reflect the moon's different evolution.

2. *Impact phenomena.* Some large impact craters on the earth have been known for many years, but both study and discoveries have now been greatly extended, following work on the moon where such craters remain unweathered indefinitely.

3. *Age of the moon.* Radiometric dates appear to show approximately the same age for the moon as for the earth. This has rendered unlikely the speculation that the moon had been formed from the earth at a later stage.

CHAPTER 3

THE FORMATION OF ROCKS

SECTION 3.1. HOW IGNEOUS ROCKS ARE FORMED

Introduction

1. Igneous rocks may be defined as those rocks which have solidified from liquid melts, or magmas as they are termed in geology. These rocks are classified in two ways, firstly by their chemical composition and secondly by their grain size. The grain size is largely dependent on the speed at which the rock has cooled; if the cooling has been very slow then the individual crystals will be large and the rock will be coarse-grained; if the cooling has been fast then the crystals will be small, even microscopic, and the rock will be fine-grained. Thus molten lava quickly extruded from a volcano will become a fine-grained igneous rock when it has cooled and solidified.

2. The chemical composition of igneous rocks varies considerably and the usual method of expressing this is by the percentage of silica (SiO_2) contained. A rock with a high silica content is termed 'acid' and one with a lower silica content is known as 'basic'. These terms apply whether the grain-size is fine or coarse. It is not necessary for a rock to be sent to a laboratory for chemical analysis to determine its chemical composition; the various types may be determined, approximately enough, by field characteristics for almost all engineering purposes. The most common types of igneous rock are as follows (their properties and field identification are dealt with in Chapter 4):

	Coarse-grained	*Fine-grained*
70% approx SiO_2 (Acid)	Granite	Rhyolite
60% approx SiO_2 (Intermediate)	Diorite	Andesite
50% approx SiO_2 (Basic)	Gabbro	Basalt

Igneous rocks are formed either from a complete melting of previously-formed rock (of any type; igneous, metamorphic or sedimentary or any combination of these) or more often from magmas from deep in the Earth which had not previously been near the surface, and which had been molten (or solid-state above their melting point) for very long periods. If solid rock is forced down to a sufficient depth by any means then it will melt. If subsequent forces push it to the surface again then it will appear as an igneous rock.

3. Acid magma is much more viscous than basic magma, and this accounts for the fact that the two most commonly found types of igneous rock are granite (coarse-grained) and basalt (fine-grained), since basic basalts are extruded more

easily than acid rhyolites. Thus wide areas, such as the Deccan of India or much of Northern Ireland, may be covered by sheets of basalt, which have spread out quickly over the land or ocean floor before cooling. On the other hand acid magmas tend to cool slowly at depth. When erosion subsequently occurs the granites thus formed are exposed to view, such as the chain of granites at Dartmoor, Bodmin Moor, St Austell, Land's End and the Scilly Isles, all in South-West England (which may be connected at depth), or the immense Sierra Nevada granite in the Western USA.

4. When igneous rocks appear on the surface as outcrops, the magma has intruded the previously-existing rock in a number of different ways, giving rise to differently shaped igneous bodies, the chief types of which are:

a. *Dykes and sills* (Figures 11 and 12). Both of these sheet-like structures vary from a few centimetres to more than 100 metres in thickness, a most usual thickness being about 1 to 5 metres for dykes and rather thicker for sills. They are both injections of magma from underground into the overlying rock, a dyke cutting more or less vertically through the bedding of the older rock and a sill intruding roughly parallel to the bedding. Often the heat from the liquid magma alters the surrounding rock for a short distance on either side of the contact (rock is usually a rather poor conductor of heat) and this phenomenon is known as 'contact metamorphism'. Dykes and sills are always fine-grained at their edges, but in thicker intrusions the rock may be coarse-grained towards the centre of the intrusion, where the magma has cooled slowly.

b. *Volcanic flows and ashes.* These come from an active volcano and may occur over land or over sea floor, or even flow from land to sea. As explained above, flows of basic composition are much more common than acid flows. When later preserved in a sequence of rocks, flows may be distinguished from sills in two ways; (1) rocks above a flow show no contact metamorphism, and (2) flows do not transgress vertically from bed to bed as sills

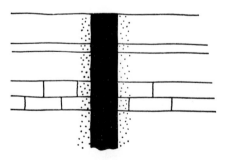

Fig 11.   Cross-section through rocks intruded by a dyke. (The stipple indicates the area in which contact metamorphism is sometimes found)

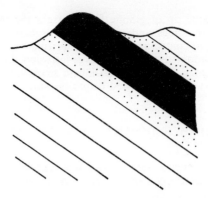

Fig 12.   Cross-section through dipping rocks intruded by a sill. (Stipple as in Figure 11)

sometimes do (*see* Figure 71). Whilst some flows are of liquid magma, others are density flows of hot gases, which can move large quantities of ash and cinder at high speeds, and which on solidification have an aerated texture when geologically young (sometimes termed ignimbrites). After some geological time and burial, these small gas holes usually become filled with secondary recrystallisation products. Ashes are also ejected from volcanoes, sometimes to great heights, and come to rest some distance from the volcano, both on land and out to sea. Beds of ash can eventually solidify into rock, varying from many metres to less than a centimetre in thickness. Such rocks may have similar engineering properties to mudstones, siltstones or sandstones of equivalent grain size, although in some sequences of solid sedimentary rocks there are very fine-grained thin ashbands, known as 'bentonites' (for uses *see* Section 10.2, paragraph 6c and Section 12.7, paragraph 6).

c. *Volcanic necks.* These are near-circular structures in plan, varying from about 100 metres to over 1 kilometre in diameter, up which magma has passed, usually into a once-active volcano. They may be filled with a variety of igneous rocks, often with a conglomeratic or brecciated selection of many different rocks. Sometimes the neck contains coarse or fine-grained igneous rock crystallised from magma, which has come up at a late stage in the volcano's history and solidified before reaching the surface.

d. *Batholiths.* These are much larger structures than necks, often elongate in plan and ranging from one or two kilometres up to many hundreds of kilometres in length. These usually represent the once-molten roots of mountain chains which, after melting, have moved upwards, first to crystallise and solidify and finally to become exposed on the surface when the covering rock above has been eroded away. Batholiths are invariably made up of coarse-grained igneous rocks. The largest ones are usually acid in

31

composition, made up of granite formed from the melting of continental crust. Around the sides and at the roofs of batholiths (when these are exposed) a zone of contact metamorphism is usually seen, with a thickness of metres or even kilometres around a large intrusion. In addition large blocks of pre-existing rock sometimes fall into a magma whilst the latter is still molten, and may be much altered before the surrounding magma solidifies.

Section 3.2. HOW SEDIMENTARY ROCKS ARE FORMED

Introduction

1. The distinction between igneous and sedimentary rocks is that whilst the former came from beneath the Earth's surfaces and crystallised from hot magmas, sedimentary rocks were formed at the Earth's surface at normal atmospheric temperature.

2. There are two main types of sedimentary rock, which will be considered separately:

a. Those whose constituent particles have been transported to the place of deposition, known as 'clastic' rocks.

b. Those which have been formed from nearby, either by aggregation of organic matter or by chemical deposition.

Clastic rocks

3. These rocks are classified according to their average grain size as follows: coarse-grained rocks are Conglomerate and Breccia (both consolidated) and Gravel (unconsolidated); medium-grained rocks are termed Sandstone and finer-grained rocks are known as Siltstone, Shale or Mudstone. The properties and identification of these rocks are given in Chapter 4.

4. Most clastic rocks are formed under water, but some are formed on land, in particular in desert conditions and also deposited by ice sheets in colder latitudes. Desert sandstones are usually well-sorted, that is most of the constituent particles are of approximately the same size, due to sifting by the wind. On the other hand glacial deposits have been dumped haphazardly by melting glaciers or ice sheets, and are usually un-sorted, with rock fragments of all sizes and shapes jumbled to form tillites or breccias, which may be interspersed with lenses of finer-grained rocks.

5. The clastic sedimentary rocks formed under water are usually of fairly uniform grain size since they have been sorted to a greater or lesser extent by the water currents which have transported them. The material is carried in the first instance by rivers. Nearly all the transport of sediment occurs only at periods immediately following heavy storms, since only quickly-moving water is capable of carrying anything but the finest sediment. It is only at times of really violent

storms, such as occur perhaps two or three times in a century, that the largest blocks are moved, e.g. Lynmouth disaster of 1953. Eventually the sediments will come to rest in a lake, delta or sea and will compact into rock when the weight of the overlying sediment has caused consolidation, thousands or even millions of years later. The degree of consolidation of a rock bears no relation to its geological age, for example Cambrian rocks laid down over 500 million years ago near Leningrad, USSR, are much softer and weaker than rocks of the same grain size laid down in the Alps less than 20 million years ago.

6. Sedimentary rocks formed under fresh water are often indistinguishable from those formed under sea water, apart from the fossil animals and plants which they may contain. Some types of animal and plant thrive in fresh water or on land, others in sea water; only a small proportion of living things can tolerate both fresh and salt water. Fossils are also useful in determining the age of the rock, since different animals and plants existed at different evolutionary stages in the Earth's history. However the age of the rock is not usually of direct importance to the engineer, since its physical properties do not correlate with its age.

7. Although much rock-forming sediment is deposited directly by rivers, ocean currents and waves are also important in shifting loose sediment on the sea floor, and even eroding submarine and coastal rock outcrops. Banks of sediment near the edges of ocean basins or continental shelves are often unstable, and the addition of more sediment, storm conditions, or perhaps an earthquake shock, may cause such banks to slide into the depressions. Such subaqueous slides often form turbidity currents of sedimentary particles in water suspension which can reach speeds up to 50 km/h and thus spread the sediment evenly over a large area, perhaps some tens of kilometres wide. The rocks eventually formed from these deposits, 'turbidites', usually consist of a mixture of grain sizes, each turbidite bed having coarser particles at its base and finer particles at its top, termed a 'graded bed'. The tops and bottoms of beds in a turbidite sequence are usually more parallel than other types of sedimentary rock, with each bed extending as far as the original turbidite flow.

Rocks formed in situ

8. These fall under four headings, each with a different mode of origin:

a. *Limestones, dolostones and cherts.* The first two rocks are mainly composed of the carbonate minerals calcite ($CaCO_3$) and dolomite $CaMg(CO_3)_2$ respectively. Most are organic in origin, being largely made up of the remains of fossil animals and plants; some are inorganic, made up of chemicals precipitated from sea water in shallow high temperature conditions, some are a mixture of the two. However they are often recrystallised during subsequent geological time, so that both organic and inorganic limestones may be considered to be alike by the engineer. Many limestones contain small spherical objects, varying much in size but often about 1 mm in diameter, called ooliths. These are carbonate aggregations formed by

33

crystallisation around a minute central particle, and kept rounded by continual wave or current action during growth. Some limestones, such as the Jurassic limestones forming much of the Cotswold Hills in England, are largely made up of ooliths. Chalk is another type of limestone, formed by immense numbers of algal skeletons so small as to be invisible to the human eye (about 10 microns diameter).

Chert, including its variety flint, consists of amorphous silica (SiO_2) often occurring as bands or nodules within limestone sequences. The silica is deposited originally under sea water, and concentrated into the bands or nodules during the rock-forming period subsequent to deposition.

b. *Evaporites.* When sea water evaporates, usually by the sun's heat acting in an enclosed, or semi-enclosed basin, then the various salts remain behind, which can eventually accumulate as rock. This process occurs on the shores of the Persian Gulf today, when sea water occasionally covers wide coastal areas and subsequently dries out. Thick beds of salt (NaCl) and other evaporite minerals exist under many parts of the Earth. They are not seen at the surface except in arid areas, since the salts will have been dissolved and carried away by circulating groundwater.

c. *Coal.* Some large rivers, such as the Nile, form deltas at their mouths where sediment spreads out over a considerable area. At some periods in the Earth's history very large deltas have been covered by vegetation, which on death has become buried. Thick deposits of dead vegetation become coal seams when the subsiding deltaic deposits subsequently become preserved as rock. These seams are usually interbedded with the sand, silt and mud brought down by the river. Coal-forming forests flourished at particular times, for example nearly all the coal in Britain and the eastern USA is of Carboniferous age (about 300 million years ago) and is formed from large extinct trees and fern-like plants. However coal of other geological periods from the Devonian onwards is to be found in many parts of the world, for example in the Tertiary of Spain.

d. *Oil and natural gas.* Since they occur in rock, oil and gas are mentioned here for completeness as geological phenomena. Both are formed from the decay of microscopic marine animals and plants without hard parts, leaving organic hydrocarbons which are deposited in most types of sedimentary rocks. To become of economic importance they have to be concentrated by subsequent geological processes into traps capable of being tapped by drilling. Gas often originates by emigration from coal deposits.

SECTION 3.3. **HOW ROCKS MAY BE CHANGED UNDERGROUND**

Faulting, folding, jointing and earthquakes

1. When rocks lie at an angle to the horizontal they are known as 'dipping rocks' (*see* Figure 13). The dip angle is expressed in degrees from the horizontal, followed by the compass orientation of the maximum dip, for example 'the limestone dips 45 degrees at 120 degrees' or sometimes simply 'Dip 45/120'.

'Strike' is by definition the direction at right angles to the maximum dip when seen in plan view, so that on a geological map the line of outcrop at a level surface of a thin rock bed follows the strike.

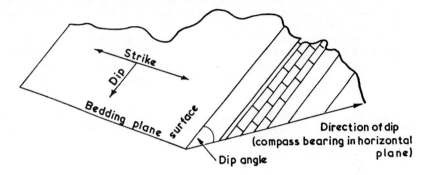

Fig 13.   Block diagram to demonstrate dip and strike

2. When rocks fracture and move this is termed 'faulting' (*see* Figure 14). The plane of fracture is termed a 'fault'. The amount by which the rocks appear to have been displaced is termed 'throw', and is expressed in linear measurements, usually metres. There are many different types of fault (*see* also Chapter 7), but the most common are:

a. *Normal (or gravity) fault* (as in Figure 14), where relative movement between two blocks of rock is vertical along the fault plane.

b. *Tear fault*, where relative movement between two blocks of rock is horizontal along the fault plane.

c. *Thrust fault*, where one block of rock has been driven over another.

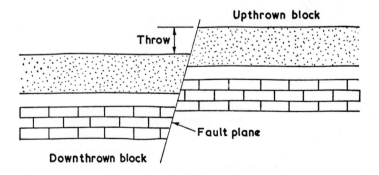

Fig 14.   Cross-section through faulted rocks

35

3. In the first two types, the fault will appear as an approximately straight line on a small-scale geological map; in thrust faults the line on the map will be most irregular, since the fault plane will be approximately horizontal, and its outcrop pattern will be very dependent on the local topography. Some tear faults may be extremely large, for example the San Andreas fault in California USA is over 1000 km long, and has a lateral throw of more than 200 km separating two continental plates (*see* Chapter 2). Movement along this fault caused an earthquake, followed by a disastrous fire in San Francisco in 1906, when the relative lateral movement was 6 metres at one time along a large part of the fault. However many faults may be seen in quarry faces which are completely stable today, and whose throw may be as little as 10 mm.

4. Often when rock strata have been subjected to pressure, usually horizontal; instead of breaking they will have buckled. This is termed 'folding' (Figure 15). The scale of folding varies widely, on the one hand it is possible to find a piece of gneiss perhaps 100 mm across with twenty or thirty small folds across its face; on the other hand some folds have amplitudes of several hundred kilometres. A syncline and anticline are illustrated in Figure 15. Sometimes folding is associated with faulting, at other times there is no such relationship (*see* also Chapter 2). When pressure has occurred from more than one direction, then domes and basins are formed, the first with the rocks dipping outwards on all sides, the second with the rocks dipping inwards towards the centre. When the core of a dome is formed of a material such as salt which is plastic at the depth and pressure concerned, this may be squeezed upwards into the overlying rock to form a diapiric structure.

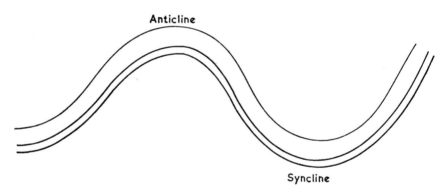

Anticline

Syncline

Fig 15.   Cross-section through folded rocks

5. When folding takes place, very often cracks develop in the strata, even though no movement may take place along them. Such cracks are termed 'joints', and joint patterns often make up a meshwork of cracks, usually at right angles to each other. The cracks are often enlarged, particularly in limestones, by the

subsequent passage of groundwater. Joints are usually at right angles to the bedding planes, and can also form in thick beds without external pressure, simply due to the internal stresses which exist in the very first stages of rock formation, at the same time as the pore water is driven out. Jointing can also be formed when igneous rocks cool, a common form being the hexagonal-sided columnar jointing to be seen in many basalts. This is the origin of the impressive structures at the Giant's Causeway, Ireland, and Fingal's Cave, Scotland.

6. After rocks have been folded, perhaps faulted, uplifted and partially eroded, then very often they are subsequently submerged and again have fresh sediment deposited upon them. When this second sedimentary sequence has been changed into rock, both sequences together may be brought up to the surface again and exposed as a rock outcrop. The junction between the first sequence and the second sequence is known as an 'unconformity' (Figure 16). An unconformity is recognisable in even a small outcrop by the difference in the angle of the dips above and below the plane of unconformity. The rocks beneath an unconformity do not necessarily have to be sedimentary rocks; where sedimentary rocks rest upon igneous or metamorphic rocks, that is also known as an unconformity. The difference in age between rocks below and above an unconformity varies considerably, anything between perhaps 50 000 years and thousands of millions of years.

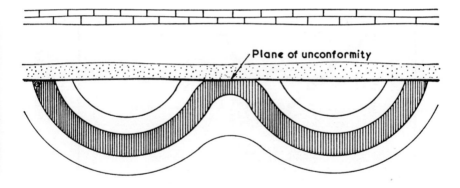

Fig 16.   Cross-section through an unconformity

7. Earthquakes are sudden tremors in the Earth, and usually result from the movement of faults. The chief occurrences of earthquakes may be mapped out as a series of belts (*see* Figure 17). These earthquake belts mark the margins of continental plates (*see* Chapter 2). Thus in stable areas, not near the edges of plates, such as Britain, the danger from earthquakes is small; but in areas of high seismic activity, such as Japan or Greece, then extra care is needed in the design and construction of any permanent, or even semi-permanent structure.

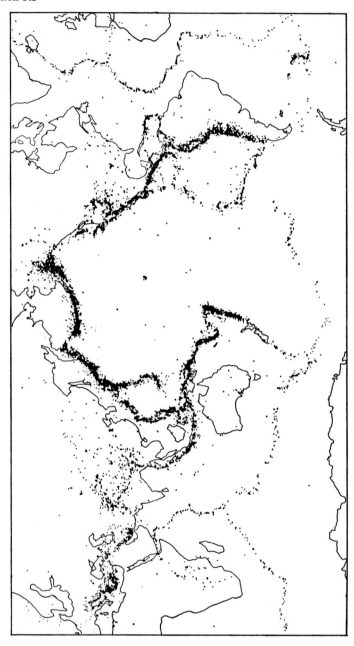

Fig 17. Distribution of earthquakes between 1962 and 1967 (after BARAZANGI and DORMAN 1969)

8. Earthquakes are measured either by local intensity or by magnitude. The scale used for intensity is the modified Mercali scale, which is a largely subjective scale based on the actual effect of the earthquake at the place of measurement. The scale runs from I (detected only by seismographs), through VI (slight damage), to XII (catastrophic). Thus the intensity of a single earthquake will differ in value from place to place, depending on its varied local effects. The magnitude scale, on the other hand, reflects the total energy released by an earthquake, and will only have a single value for each shock. The magnitude is calculated by complex equations relating the ground motions recorded by a seismograph to the distance of the instrument from the epicentre. Several magnitude scales have been proposed, but that most commonly used is by Gutenberg and Richter, which is logarithmic. Small detectable disturbances, with energy releases of about $6 \cdot 3 \times 10^5$ergs, have a magnitude of 0, whilst the largest, with an energy release of about 2×10^{25}ergs, have a magnitude of $8 \cdot 5$. There are more than a million earthquakes each year, ranging from an annual average of one earthquake with magnitude 8 or more, to a majority with magnitude less than 3.

Metamorphism and hydrothermal activity

9. When rocks of all types are subjected to very high temperature and pressure, they will melt and remobilise to form igneous magmas. However since the melting points of each of the many constituents within a rock varies widely (and also the melting point increases with increase of pressure), some constituents will recrystallise and reform before others. Rocks which have been noticeably altered, but which still reflect some traces of their original bedding and structure are termed 'metamorphic' rocks. A special case of metamorphism, contact metamorphism, has been mentioned above, but this is only a local phenomenon, occurring close to intruded igneous rocks. Most metamorphic rocks are found in much larger areas, which have been pushed down to substantial depths in the Earth's crust, perhaps 20 km deep or more, and then raised again and finally eroded in some subsequent period of Earth history. The whole of the Highlands of Scotland is one such area, where regional metamorphism has occurred, interspersed by a few granites where the rock has been completely melted and remobilised upwards into the metamorphic horizons. Much larger areas of metamorphic rocks occur for example in Canada and Central Africa.

10. The four principal types of metamorphic rock are as follows:

a. *Slate.* Shales may have the orientation of their constituent particles altered by a relatively small increase in pressure and temperature to form slate. When this has occurred the rocks will become hard and brittle, and split along directions often at angles unrelated to their original bedding planes. This property is termed rock cleavage. Slate from North Wales was used extensively as roofing material during the nineteenth century, and is still common.

b. *Schist.* Finer-grained rocks which have become completely recrystallised are termed 'schist'. The minerals recrystallise under high temperatures and pressures parallel to one another; especially conspicuous in most schists

39

are shiny flakes of mica minerals, but many other minerals also occur. The schistosity of a rock is similar to the cleavage of slates, but on a coarser and less uniform scale. Sometimes the schistosity parallels the bedding planes, at other times it lies oblique to the bedding planes, depending on the direction of the pressure at the time of recrystallisation.

c. *Gneiss.* When rock is coarsely recrystallised it is termed 'gneiss' (pronounced 'nice'). The crystal size may be as coarse as igneous rocks such as granite, but gneiss is distinguished from granite by banding, which may reflect the relic structure of bedding planes and by the fact that the crystals are preferentially orientated in one direction, as opposed to the crystals in granite, which grow slowly from small nuclei in the liquid magmas and have random orientation.

d. *Marble.* When limestones are metamorphosed, they recrystallise to form marble. The streaks often seen in true marble are caused either by the local separation of original impurities in the limestone, excluded on recrystallisation of the calcite, or else by the cracking or brecciation of the limestone under stress. This geological usage of 'marble' is more restricted than the commercial use, which incorporates any kind of limestone used for ornamental work under the term, whether metamorphosed or not. It is even sometimes erroneously used to describe any sort of polished stone, perhaps granite or even slate.

11. Rocks may be slightly affected by hydrothermal activity which is the action of water circulating deep enough to become first hot and then often superheated under pressure, penetrating the pore spaces of rocks and acting as a lubricant in joints. Occasionally water reaches the surface in the form of hot springs, which are usually of local occurrence and not necessarily associated with active volcanic areas. Geysers, which are hot fountains of waters ejecting at sporadic intervals, are usually confined to volcanically active areas.

Section 3.4. HOW ROCKS MAY BE CHANGED NEAR THE SURFACE

1. Rock decay with little or no transport of the products is termed 'weathering'; when the rock is simultaneously removed this is termed 'erosion'. The effects of weathering on a rock may be considered on the large scale and on the local scale. The large scale is discussed in Chapter 5 in the description of landforms, which also covers erosion. The local scale is also discussed in Chapter 5 under the formation of soils.

2. All rocks near the surface are affected to a greater or lesser degree by weathering. The principal types of weathering are:

a. *Surface weathering.* The zone of surface weathering varies greatly, from less than 1 mm in some rocks to perhaps as much as 200 metres in another. Climatic conditions affect the depth of weathering; in tropical climates the zone of weathering tends to be deeper than in temperate climates, but there is also a difference in weathering between extremes of climate. Weathering is achieved firstly by physical means, in which rocks may be shattered by temperature changes, by gravity effects, and by the pressure of growing organisms in cracks; and secondly by chemical means, in which the individual minerals that make up a rock are dissolved or decomposed by the water, oxygen and carbon dioxide of the atmosphere, and also by the chemicals produced by live or decaying organisms. The usual effect is to make a weathered zone structurally much weaker from the engineering point of view, and thus in all site work it is most important firstly to differentiate between weathered and unweathered rock, and secondly to try to find out the depth of weathering, which may vary between different parts of the site.

b. *Weathering below the surface.* Joint systems occur in most types of rock, both igneous and sedimentary, and joint planes make natural zones of attack along which and down which water and chemicals in solution can penetrate. Limestones are particularly prone to subsurface weathering of this sort, since calcium carbonate is water soluble, and large caves and channels can form underground. Well-known examples include the Cheddar Caves, England, and the Carlsbad Caverns, USA. Such caves in limestones are an engineering hazard, since the building of any heavy surface structure can cause the sudden collapse of cave roofs.

3. Diagenesis is the name given to the processes which alter the character of a sedimentary rock after it has been deposited, either by the reactions between the various constituent minerals and particles with each other, or by the reactions of the various constituents with pore or circulating fluids. Such reactions are termed diagenetic at the lower temperature range, and metamorphic at the higher temperature range, but there is no rigid division between the two. Diagenesis occurs at two principal times, firstly when the original deposit is still in contact with sea or lake water soon after the time of formation, and secondly, after this period, when such direct contact with the original water has been removed. Diagenesis usually goes on until the constituents and pore fluids are all in chemical equilibrium.

4. Water also percolates through all types of rock, firstly through pores in the rock itself and secondly along fault zones and other surfaces, such as bedding planes and unconformities, which can lead to planes of weakness in apparently massive rock. Professional geological advice should always be sought in regard to large civil engineering works.

REFERENCE LIST—CHAPTER 3

BARAZANGI M and DORMAN J, 1969 — World Seismicity Map of ESSA Coast and Geodetic Survey epicenter data for 1961–1967. *Bulletin of the Seismological Society of America, Volume 59, No. 1.*

CHAPTER 4

PROPERTIES AND IDENTIFICATION OF ROCK TYPES

Section 4.1. INTRODUCTION

1. This chapter is intended as a first guide for an engineer who finds himself on a new site without immediately available professional geological support. It is stressed that on any major project, professional geological advice should be sought as a matter of course. Many occasions have arisen in the past where difficulties or disasters, both during and after construction, could have been avoided if a geologist had been consulted during the earliest planning stages.

2. All the visible rock at the site should be examined briefly before any individual rock is studied in detail, since frequently several rock types are present. If no rock is visible at the surface because of vegetation or soil cover, then look around the site for a radius of perhaps half a kilometre for exposure, for example, in stream beds. If solid rock still cannot be found, then it must be discovered by digging or drilling.

3. Chapter 7 deals with geological information sources. The availability of geological maps and other data should be ascertained before reconnaissance starts, and an assessment of the literature made in the office. But however good the pre-existing geological data may be, a detailed site survey will still probably be necessary for two reasons, firstly because many rocks vary greatly over small distances, and this variation will not be shown on the scale of any existing map, and secondly because the topsoil and subsoil may also vary greatly, and their thickness and content are unlikely to have been previously known in detail.

Section 4.2. ASSESSMENT OF OUTCROP

1. The larger the rock outcrop the better for the geological understanding of the site. In addition to observing the type or types of rock, it is also important that the degree of coherence and the strength of the rock as a whole should be assessed, and possible zones of fracture should be sought. The rock outcrop should be examined closely for traces of bedding (*see* Section 3.2). If the rock is massive, with no trace of bedding (although jointing may be present) then an igneous, rather than a sedimentary, rock may be suspected. Well-bedded rock is nearly always of sedimentary or metamorphic origin. If the outcrop consists of more than one type of rock, then a sample of each type should be taken and an attempt should be made to assess the relative proportions and dispositions of the various rock types present. Each type should be detailed in the subsequent

written report, for example a natural outcrop or quarry might be described in the report as 'Alternating shale, siltstone and sandstone units, with shale beds varying from 100 mm to 1·0 metre, grading up into siltstones of about the same thickness; sandstone beds roughly constant at about 500 mm; total shale about 40 per cent, siltstone 40 per cent, sandstone 20 per cent. Stratigraphical thickness of rock exposed totals 32 metres. Uniform dip of 30 degrees at bearing 095 degrees.' It is important that samples of each rock type (in separate numbered bags, each with a detailed locality label) should be kept, both as a reference collection for later consultation, and for possible submission to a professional geologist.

2. Since amounts of dip and directions of strikes usually vary slightly from place to place within a rock outcrop, it is often sufficient for most purposes to record them only to the nearest 5 degrees. For slope stability purposes, however, greater accuracy may be needed. No special instruments are needed to measure dip and strike. It is often useful to lay a pencil or any straight edge on the rock surface and to move it about until the maximum dip angle from the horizontal is obtained, and then to measure the angle of the pencil. A compass will give the bearing of the dip, and the angle can be measured using a protractor and plumb line if a clinometer is not available. In the record of dip, only the dip angle and bearing is usually given, the strike is not normally mentioned since it is by definition at right angles to the dip. The dip seen on the walls of a gorge or steep-sided excavation is only the apparent dip, unless that surface is by chance vertical and at right angles to the strike. All apparent dips are shallower than true dips (*see* Chapter 7). Another possible source of error is hill-creep, in which rocks near the surface on the side of a hill have been moved either by gravity sliding or by the shearing action of ice, resulting in a different dip from the main mass of rock deeper inside the hill.

TABLE 2. **WEATHERING AND ROCK GRADE CLASSIFICATION FOR SOME AUSTRALIAN GRANITE**

Grade	Type	Description
I.	Fresh	No visible deterioration.
II.	Slightly weathered	Yellow-brown limonite staining and some decomposition of feldspars.
III.	Moderately weathered	Considerably altered, but pieces of NX (50 mm diam) core cannot be broken by hand.
IV.	Highly weathered	Pieces of NX (50 mm diam) core can be broken by hand.
V.	Completely weathered	Sample disintegrates when placed in water.

3. Rock near the surface has usually been altered by the climate; this is termed weathering. In general weathered rock is structurally much weaker than fresh rock, and thus for example deep weathering is an advantage when rippability is being considered, and a disadvantage, sometimes a dangerous disadvantage, when buildings or dams are to be built upon it. As an example Table 2 shows grades of weathering for an Australian granite. Other tables can be prepared for other rock types.

4. Rock usually changes colour upon weathering. One of the commonest changes is oxidisation from the green or grey colour of ferrous iron to the red or brown colour of ferric iron; however reddening does not automatically denote recent weathering, many unweathered sedimentary rocks being made up of constituents which were red. In some rocks the weathering goes very deep, occasionally even tens of metres, and if this is suspected then fresh rock may have to be drilled for on an important engineering site. Weathering is generally deeper in tropical climates, an extreme example is the granites in the western part of the New Territories of Hong Kong, where the weathering has been proved to depths of 70 metres. Here the weathered granite may be dug by a man with a spade, whilst the unweathered granite makes excellent road metal.

5. Another process which may be mistaken in the field for weathering is that of metasomatism. This is a kind of metamorphism which occurs at depth and in which solutions or gases have passed through pores and cracks in the rock under conditions of high temperature and pressure. The resulting changes often alter the rock in a comparable way to weathering near the surface. However an important difference for the engineer is that, whilst the effects of weathering will generally decrease with depth, metasomatism will persist downwards, perhaps for some kilometres' thickness of rock.

Section 4.3. IDENTIFICATION OF HAND SPECIMENS

1. A sample for the purpose of identification of rock type should be as fresh as possible, that is unweathered rock. At most localities a fresh sample may be obtained by breaking away a piece of solid rock outcrop with a hammer, usually with a kilogram head or heavier. The sample should be at least as big as a fist and preferably larger. The freshly broken surfaces are the best for assessment, and each constituent should be separately identified if possible. In this section the most common rock-forming minerals are described, followed by the most common types of rock.

2. One of the simplest field tests is that of mineral hardness. Ten standard minerals have been arranged in order of hardness, forming Moh's Scale from 1 (soft) to 10 (hard), by which the hardness of other substances may be compared (Table 3).

TABLE 3. MOH'S SCALE FOR MEASURING THE HARDNESS OF MINERALS

Standard mineral	Hardness
Talc	1
Gypsum	2
Calcite	3
Fluorite	4
Apatite	5
Orthoclase feldspar	6
Quartz	7
Topaz	8
Corundum (emery, ruby and sapphire are varieties) ...	9
Diamond	10

3. As a rough guide in the field, the hardness of a fingernail is between 2 and 3, the hardness of a penny is between 3 and 4, and the hardness of a penknife is about 6. The list brings out the fact that a rock, however compact it may be, is no harder than the mineral of which it is formed. For example on a road subject to the grinding action of tracked vehicles, a limestone surface (chief mineral calcite) would behave very differently from quartzite (chief mineral quartz), which is harder than steel. Nevertheless some rocks, owing either to the arrangements of the minerals within their mass or to the presence of micro-fissures, have crushing strengths which are little related to the hardness of their main constituents.

Common rock-forming minerals

4. The degrees of weathering stability mentioned below for most minerals will vary; those indicated are for temperate climates, and different behaviour may occur in the polar regions or tropics.

Quartz. Chemical composition SiO_2 (silica), chemically very stable. Hardness 7 (hardest common mineral). Breaks with a conchoidal fracture. Usually colourless.

Feldspar. Chemically complex alumino-silicates of potassium (*Orthoclase* feldspar) or sodium and calcium (*Plagioclase* feldspar). Hardness 6. Breaks along planes. White or grey in colour, less commonly pink. Feldspars weather comparatively easily.

Ferro-Magnesian minerals (includes *Amphiboles* and *Pyroxenes*). Chemically complex calcium and sodium alumino-silicates rich in iron and magnesium, and thus relatively heavy. Hardness 5–6. Break along planes. Colour dark green to black. The commonest amphibole is *Hornblende*, and the commonest pyroxene *Augite*, but they may be grouped together for most engineering purposes. Another ferro-magnesian silicate mineral *Olivine* (confined to basic igneous rocks) is also of dark green to brown colour and of high specific gravity, but has a hardness of 7.

Mica. Chemically complex potassium alumino-silicate (*Muscovite* mica) and potassium-magnesium-iron alumino-silicate (*Biotite* mica). Soft, with hardness 2–3. Relatively stable minerals, but break readily along close parallel planes, forming thin flakes on weathering. Muscovite is colourless, often twinkling in flakes on a rock surface, and biotite is dark green or brown to black.

Chlorite. Chemically a hydrous iron-magnesium alumino-silicate. Soft, with hardness 2–2·5. Breaks readily, forming flakes like mica. Colour green. Very widespread mineral, particularly as a metamorphic or weathering product.

Iron ores. Many compounds of iron occur in variable quantities in rocks. Commonest are the oxides (including the red *Haematite*, Fe_2O_3), the carbonates and the sulphides. The latter includes *Pyrite*, FeS_2, which is a brassy colour, hence its nickname of 'Fool's Gold'. Iron ores are the chief colouring agents in all rock, the ferric iron providing reds and browns and the ferrous iron greens and greys. In some igneous rocks *Magnetite*, Fe_3O_4, occurs as small black grains, and is important as sometimes affecting electrical instruments and compasses.

Calcite. Chemical composition $CaCO_3$. Hardness 3. Has three planes along which it splits equally readily, forming rhomb-shaped crystals. Colour white. Slowly soluble in rainwater containing dissolved CO_2, but otherwise stable.

Clay minerals. A large group of varied minerals, mostly alumino-silicates whose chief common feature is that their crystals are too small to be seen either by the naked eye or a low-powered microscope. Many occur as sheets, which give the characteristic clayey soapy texture. Hardness usually 2–3. Colour usually white, grey or black. (for further discussion *see* Chapter 5).

Common rock types

5. As a rough guide to the identification of rock samples, the chief minerals and the grain size of the more common rocks are now given. It is assumed that it is known whether the rock is igneous or sedimentary; the differences are usually clear-cut, unless the rock is fine-grained. However in this case the engineering properties of the two are usually fairly similar. Although some guide to the colour of the rocks is given, it must be borne in mind that colour is often very variable, depending on quite small quantities of one of the iron oxides which may be present.

6. *Igneous rocks*

Granite. Coarse-grained. Chief minerals quartz, plagioclase feldspar, orthoclase feldspar, biotite mica and muscovite mica. Lesser constituents include iron ores. Overall colour white with dark specks, may be pink or reddish.

Diorite. Coarse-grained. Chief minerals plagioclase feldspar and hornblende. Lesser constituents biotite mica and augite. Colour usually grey or dull green. Granites grade into diorites through granodiorite, the engineering properties of all three are similar when fresh.

Gabbro. Coarse-grained. Chief minerals plagioclase feldspar, augite and olivine. Dark-coloured. Engineering properties similar to granite and diorite, except that gabbro is denser and is very susceptible to weathering.

Rhyolite, Andesite and *Basalt.* All fine-grained, and thus individual minerals cannot usually be seen in hand specimens although their composition corresponds to granite, diorite and gabbro. Colour is important, rhyolite tends to be a light colour, andesite grey and basalt dark grey or black. Dolerite is of the same mineral composition as basalt, but is found only in dykes and sills. Fine-grained igneous rocks sometimes have occasional large crystals, or small rounded cavities filled or partly filled with light-coloured material. The engineering properties of these rocks are similar, except that basalt is the heaviest and rhyolite the lightest in weight.

7. *Sedimentary rocks*

Conglomerate. Coarse-grained, consisting of rounded pebbles 5 mm in diameter or more, cemented in a matrix of finer material. Sometimes the pebbles are all the same size and composition, at other times both size and composition vary considerably. *Breccia* is similar to conglomerate, but the constituent pebbles are angular in shape, rather than rounded. Both have similar engineering properties, but differ markedly from gravel in that they are cemented rather than loosely consolidated.

Sandstone. Medium-grained, from 5 mm grain diameter down to 0·1 mm. Usually quartz grains, more or less cemented together by calcite, silica or iron oxide. Very variable engineering characteristics, from very hard quartzite to scarcely consolidated sand.

Siltstone. Medium to fine-grained from 0·1 to 0·01 mm, in which many grains are visible to the naked eye. Usually a mixture of quartz and clay minerals, with mica sometimes important. Like sandstone, with variable engineering properties depending on the degree of cementation.

Shale and *Mudstone.* Fine-grained, with most constituents finer than 0·01 mm diameter. Main constituent clay minerals, with lesser quartz and mica. Engineering characteristics very variable, depending on the degree of compaction and cementation, and the amount of pore water retained. Shale tends to flake more easily than mudstone.

Limestone. Grains not usually seen. A compact rock, breaking with angular fractures. Composed of calcite, and will effervesce when treated with dilute hydrochloric acid (HCl). Colour varies considerably, chiefly white, but all shades of grey to black or light to dark brown when impure with shale or iron products. *Chalk* is usually a less-consolidated variety of limestone of very fine grain size, but its properties are different (*see* Section 10.5, paragraph 26). *Oolite* is a variety of limestone which includes small (about 1 mm diameter) spherical bodies termed ooliths. The ooliths are usually cemented in compact calcite, and oolites can usually be considered as similar to the softer types of normal limestone for engineering purposes. Oolitic limestone makes the best freestone because of its relative homogeneity, e.g. Bath and Portland stones. *Coral* is also composed of calcite and may be considered as a special case of porous limestone; it is often used in tropical engineering. Its compactness varies considerably due to the many cavities which it usually contains. The tougher varieties of limestone are often useful as aggregates.

Dolostone. Like limestone, but often light brown in colour; it will not effervesce with dilute HCl as the chief mineral constituent is dolomite ($MgCO_3$). Similar properties to tougher limestone for most engineering purposes.

Chert. A very hard non-granular rock, composed of crypto-crystalline silica (SiO_2). Colour varies, but usually brown or black. Very often nodular or interbedded with limestones. *Flint* is a variety of chert found in chalk in Europe. Good aggregate if weathered, but fresh chert may be alkali reactive and is therefore unsuitable for use with Portland cement.

Rock Salt, Gypsum and other evaporites. Since these rocks are soluble in water they are seldom found at the surface except in semi-arid or arid regions, but may be encountered anywhere in boreholes. Since they are chemical evaporation products, the deposits are often crystalline. Specialist advice should be sought if these rocks form substantial parts of sites upon which buildings are to be erected.

8. *Metamorphic rocks*

Slate. Fine-grained, colour usually blue-grey, but may be many other colours depending on minor constituents. It breaks readily along parallel cleavage planes, but from other directions it is tough.

Schist. Medium-grained. Colour variable, with individual minerals usually visible in hand specimen and often including muscovite mica, biotite mica and hornblende. Sometimes with good schistosity, usually tough.

Gneiss. Coarse-grained, with alternating bands of light and dark minerals across the whole outcrop. Colour variable; a large range of minerals possible, including quartz, feldspars, micas and augite. Usually hard and often very tough.

Marble. Metamorphosed limestone, very variable in colour, but often white. Will effervesce on the addition of dilute HCl. Engineering properties similar to compact limestone, but generally tougher.

Quartzite. Term used for any very pure sandstone, whether produced by pressure-welding or by cementation of the silica. Usually a metamorphic rock, but sometimes found in an unmetamorphosed sedimentary succession. Very hard, but may be brittle.

Section 4.4. ROCK PROPERTIES

1. In this section a guide is given to the average physical properties of the commoner rock types. Rock Mechanics are discussed in Section 10.1.

2. Rock strength is an important parameter. It can be measured on a laboratory sample, which should represent as nearly as possible the average state of rock on the engineering site. The Geological Society classification of rock strength is given in Table 4.

TABLE 4. STRENGTH OF ROCKS (GEOLOGICAL SOCIETY CLASSIFICATION)

Uniaxial compressive strength (MN/m^2)	Description
Below 1·25	Very weak
1·25– 5·0	Weak
5·0 –12·5	Moderately weak
12·5 – 50	Moderately strong
50–100	Strong
100–200	Very strong
Over 200	Extremely strong

3. Table 5 shows various properties of the commoner types of rock used by the engineer. Although the table indicates precise limits, it must be remembered that the strength (and many other properties) of a rock is no greater than that of its weakest part, and that most rocks contain planes or zones of weakness or fracture. The figures given are averages only; on a major engineering project, in which any particular property of a rock is very important, tests should be made directly upon specimens from the site. Unlike artificial materials such as metals and plastics, rocks usually lack homogeneity. Several common rock types have been omitted from Table 5 because their properties are so variable, for example the properties of conglomerates vary enormously both on account of their possible range of constituents and also the properties of the binding agent. Even in the table, the figures for the two types of breccia are very tentative.

TABLE 5. PHYSICAL PROPERTIES OF UNWEATHERED INTACT ROCKS

Rocks	Relative density	Porosity (%)	Sorption (%)	Conductivity 10^{-3} cal/cm sec deg C (see Note)	Resistivity (ohm-m)	Seismic vel (km/sec)
(a)	(b)	(c)	(d)	(e)	(f)	(g)
Igneous rocks:						
Andesite	2·2-2·7	2-11	0·1-4·9	4·0-8·5	20-5,000	5·0-6·3
Basalt	2·2-2·8	0·1-9·9	0·1-9·9	4·0-8·6	20-5,000	5·0-6·6
Diorite	2·8-2·9	0·1-4	0·1-0·4	6·0-8·5	500-20,000	5·2-6·6
Gabbro	2·7-3·0	0·2	0·0·3	6·0-9·0	500-20,000	5·4-6·7
Granite	2·5-2·7	0·05-2·8	0·2-1·6	6·2-9·0	500-20,000	4·6-6·0
Rhyolite	2·5-2·7	1-7	0·1-5·6	7·4-8·8	10-5,000	4·5-6·3
Sedimentary rocks:						
Breccia (mainly igneous)	2·5-3·0	0·1-7	—	7·1-8·0	—	—
Breccia (limestone)	2·3-2·5	1-35	—	4·5-6·5	—	—
Chert and Flint	2·6-2·7	1-4	0·1-3	7·0-11·0	—	3·0-7·0
Dolostone	2·5-2·7	0·3-25	0·3-1·2	8·9-13·9	50-10,000	2·8-7·1
Limestone (hard)	2·5-2·7	0·8-27	0·1-1·8	4·7-8·0	200-10^5	1·7-4·2
Limestone (soft), Chalk	2·3-2·5	4-42	0·3-4·1	4·7-6·4	50-10,000	1·0-4·4
Sandstone	1·9-2·6	0·5-42	0·7-13·8	3·5-7·7	20-500	1·4-4·4
Siltstone	2·2-2·5	2·2-24	0·4-6·3	3·0-7·5	20-500	1·5-3·5
Clay and Shale	2·3-2·7	2·9-55	0·2-6·1	2·2-6·9	150-500	2·0-4·5
Coral	2·5-2·7	1-19	0·2-1·0	4·7-6·4	50-5,000	—
Metamorphic rocks:						
Gneiss	2·6-3·2	0·3-2·4	0·1-0·8	4·9-10·4	100-5,000	3·5-7·5
Marble	2·4-2·7	0·1-6	0·1-0·8	4·7-8·0	1,000-10^5	3·8-6·9
Quartzite	2·6-2·7	0·8-7	0·1-0·8	7·4-18·9	500-5,000	5·8-6·3
Slate and Schist	2·6-2·8	0·4-10	0·0·6	4·1-8·9	100-3,000	2·3-5·7

NOTE: 1×10^{-3} cal/cm sec deg C $\simeq$ 0·42 W/m deg C.

REFERENCE LIST—CHAPTER 4

General references

CLARK S P (Jr) Ed, 1966   —Handbook of Physical Constants. *Geol Soc. Am. Mem.* Contains much useful data on rocks.

GASS I G, SMITH P J and WILSON R C L (Eds), 1971   —*Understanding the Earth.* Open University Press, Sussex. An invaluable compilation of articles on most aspects of geology, in particular current theories of earth origin etc.

HOLMES A, 1965   —*Principles of Physical Geology.* Nelson, London. A good simple introductory text-book to geology.

HURLBURT S C, 1957   —*Dana's Manual of Mineralogy.* Sixteenth edition. Wiley & Sons, New York. Deals in detail with all minerals, including data on physical and optical properties, and the classification of minerals and rocks.

STAGG K G and ZIENKIEWICZ O C, 1968   —*Rock Mechanics in Engineering.* Wiley & Sons, London. An introductory, though rather mathematical, approach to rock mechanics, with many references and examples in construction.

CHAPTER 5

LANDFORMS AND SOILS

SECTION 5.1. LANDFORMS

Introduction

1. The relation between landform and geology has been recorded by engineers since William Smith produced the first geological map of Great Britain in 1815, based on his experience from building canals in southern England. A change in landform always reflects some change in subsurface conditions, and may be recorded as such. The change may be in detail, such as a difference in slope which may indicate a spring line, or in the different pattern of landscape produced by the weathering of different rock types. The shape of the landform reflects the history of weathering and erosion of the rocks present, and so the recording of changes in landform distinguishes areas with different geological or geomorphological history. This in itself is useful when preparing an engineering survey (*see* Chapter 9), but an understanding of the factors which shape the landform helps to predict the range of materials and conditions to be found in the different parts of the landscape. This understanding can also explain why differences of great importance to engineers do not always show as different landforms.

2. In considering the factors controlling the development of landscape it is convenient to consider separately the effects of process, geology and time. The main processes include weathering, erosion and deposition, and the distinction between erosional and depositional landforms is particularly significant in the analysis of landform. The effect of time is obviously important in relation to the processes, but in addition a knowledge of the geomorphic history is essential to the understanding of landforms, and the soils with which they are covered. Thus the landscapes of northern latitudes have been formed, or modified, by repeated glaciations, conditions which do not exist in these areas at the present. In some tropical countries extensive areas have remained stable over long periods of time and features such as laterite protected plateaux must be related to landforming processes in a previous geological period.

Weathering

3. The main factor in the evolution of landscape is the weathering of rocks (*see* also Section 3.4). This comprises the physical disintegration of the rock mass and the chemical change of its minerals. Chemical alteration usually predominates with its action assisted by physical weathering, including cracking, which makes new surfaces available for chemical action. Disintegration begins by removal of

the overburden pressure by erosion and the destressed rocks respond by the opening of joints. When preparing deep foundations in a region of recent erosion the opening of joints may require special control, and similar problems are encountered in mining and tunnelling.

4. In igneous and metamorphic rocks the process of weathering leads to the conversion of the original rock minerals into clay minerals. This is largely accomplished by the removal in solution from the rocks of the more soluble salts which then pass into the groundwater and rivers. In arid areas these salts can accumulate in the soil, or in salt pan depressions. The weathered rock which remains still occupies the same volume, as shown by the preservation of mineral structures, and thus is at a lower density. For example, weathered granite from Hong Kong may have a dry density of 1200–1400 kg/m³, whereas the fresh rock would be 2700 kg/m³.

5. The weathering of sedimentary rocks is much simpler, consisting of the removal of cementing material, and the conversion of shale and similar rocks back to clay. With the exception of salt deposits and limestones, the weathering of sedimentary rocks does not lead to any great change in the bulk of the material.

Erosion of slopes

6. The depth of the weathering profile at any point is a result of the balance between the rate of weathering of that particular rock type in that climate and the rate at which material is eroded to form a slope. As the majority of the Earth's surface is formed of erosional slopes, the process of erosion is the dominant factor in determining the shape of the landscape. The main agents of erosion and transportation are water, ice and wind. Of these the most important is the water reaching the surface as rainfall. That part of the rainfall that does not percolate into the ground and become groundwater will flow towards the nearest drainage channel with a velocity of flow depending on slope and on surface roughness. This flow of water steadily erodes the general surface of the ground, but more quickly if the flow is concentrated (e.g. by the effects of differential roughness or vegetation) into gullies cut into that general surface. Further, in those areas in which the geology results in groundwater emerging as springs, erosion of slopes known as spring-sapping will occur.

7. This process of downcutting, whether in temporary gullies or in permanent streams, is the first step in the erosional cycle. The rate of vertical erosion of a stream is controlled by its gradient, flow of water and the amount of material eroded from the slopes of the valley. The stream gradient is controlled by its base level, the lowest point to which the river can erode. This level is ultimately related to sea level, but in many areas temporary barriers, such as lakes or resistant bands of rock, provide a local base level. If such a temporary base is removed, rejuvenation of a landscape can occur owing to the increased downcutting by the streams. A similar effect is caused by a lowering of the sea level.

8. The river itself provides a base level for the slope above it. Even during this first stage of erosion when rivers are cutting downwards, a larger amount of erosion is occurring on the slopes of the valley. Material is either eroded from the surface, or moves down into the valley by mass movement. This may be a slow continuous creep process, normally occurring in the upper, more disturbed layers, or some form of sliding which may also involve deeper layers. These processes leave the slope at its maximum stable inclination, which varies according to the material and its moisture content. In mountainous areas where resistant rocks outcrop, slopes of 30 degrees or more would be expected; in contrast the long term stability of clay slopes with high water table may be less than 10 degrees.

9. As the valley develops from the youthful stage of downcutting to maturity, the river begins to widen the valley floor. The stream develops meanders that undercut both valley sides, but, as the valley widens to form a flood plain, the meander zone will only occupy part of the valley floor. With no further downcutting of the valley the slopes of the valley walls are levelled out, leading to the final stage of erosion, the peneplain, where the landscape is reduced to wide flood plains separated by low divides of deeply weathered material.

10. The material eroded from the valley slopes is transported by the river, either in suspension or as bottom load as long as the stream velocity is sufficient. When the stream is slowed abruptly the transported material is deposited in large quantities. Typical examples of this are the fan deposits which occur where mountain streams come onto a flat plain. The coarsest deposits occur at the foot of the mountains and the finer deposits are carried further away by the slower moving water. Similar large accumulations can occur where rivers enter the sea or lakes; the drop in stream velocity causes the build up of deltas. Variations in stream velocity are also caused by seasonal climatic variations, most erosion and eventual deposition occurring when the stream is in flood.

11. In many cases, because of an upward change of the level of the land relative to the sea during the process, the landscape does not show a simple progression from youth, through maturity to old age. Rejuvenation (extra downcutting) of a valley is caused, with local erosion of landforms, leading to such features as terraces where the older deposits have been eroded. This rejuvenation process is reflected on adjacent coasts in the occurrence of raised beaches. Conversely, in many other areas the drowning of coastal landforms and estuaries is common; this follows when the present general rise of sea level due to the melting of the ice-sheets (*see* Chapter 2) exceeds the effect of regional land movements. It is calculated that sea level has in the last 10 000 years risen 30 metres, although with many minor fluctuations.

Glacial features
12. The landforms produced by glacial erosion and deposition are quite distinctive. The effects of glaciation in highlands are different from those in the lowlands, reflecting the different form of the ice sheets.

13. In mountainous areas there is usually a part of the terrain which remains above the ice where it is weathered by frost action, giving a sharp rocky terrain, typically with very sharp ridge crests. Mountain glaciers usually start from circular basins, known variously as corries, cwms or cirques, where snow accumulates and eventually compacts into ice. The ice flows from the corrie, which usually has a lip at its mouth, to form a mountain glacier. The valleys formed by glaciers typically have a wide flat floor and steep sides, owing to the erosive action of the ice over the whole floor and sides. When the ice melts the valley floor is covered with the moraine deposits which were carried on top of the ice and along the valley floor. The most typical deposits are the lateral moraines at the edge, formed by material eroded from the valley side, and the terminal moraines, which occur as a ridge across the valley marking the end of the glacial advance. Terminal moraines are formed by 'stationary' glaciers, i.e. those in which the rate of melting at the front equals the rate of advance due to ice flow.

14. Where ice extends over the whole area as a sheet the erosional effects produce a general grinding down of the landscape. All loose material is stripped off and evidence of scouring by rocks carried by the glacier may show the direction of ice flow. Away from the centre of ice accumulation, depositional features will also occur, leaving the typical lowland glacial landscape of low relief, confused drainage and many lakes. Towards the edge of the old ice sheets depositional features predominate. Terminal moraines occur as ridges, often as a series of crescents, and not usually more than 50–70 metres high. Many deposits are formed beneath the ice-sheets, including till (boulder clay), which blanket the pre-glacial topography, sometimes to depths of 75 metres or more. The composition of till is variable, depending on the source, but is typically unsorted and unstratified. Drumlins are elongated streamlined hills of till, usually in groups, with the long axis parallel to the direction of ice flow. Other linear ridge deposits, such as eskers and kames, are formed in channels beneath the ice or in contact with the ice and these are often a source of gravel. Extensive deposits are also left by the melt waters from glaciers which rework older materials, so that gravels are found near to the glaciers where the stream velocity was high, while the finer materials were moved farther away.

15. One important aspect of the presence of glaciers and the melt water from them is the effect they can have on the drainage system of an area, even though it is not glaciated itself. The presence of ice, or the deposition of a moraine, may dam the existing drainage system causing a lake to form. Eventually this will find a new outlet which may be in a different direction from the original stream. The large flow of water can create deep valleys in a comparatively short time so that even when the ice retreats the drainage still follows the new course.

Aeolian features

16. Movement of material by wind occurs in areas with little or no vegetation and is typically associated with the hot deserts. Nevertheless, similar landforms can occur in cold areas and also in a coastal environment. Wind is mainly a transporting and sorting agent; the finer silt-size material is carried in suspension

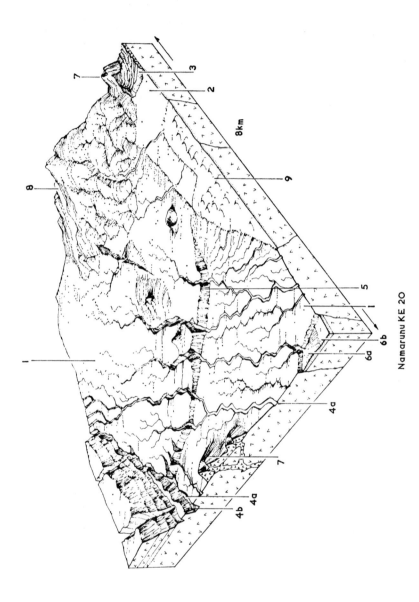

Fig 18. Typical volcanic landforms in Northern Kenya

Namarunu KE 2O

56

NOTES ON FIGURE 18:

Climate: Rainfall 250–500 mm (April)

Facet No.	*Form*
1.	*Lava plateau.* Level to moderately sloping; irregularly undulating and dissected by numerous small shallow branching valleys; extensive.
2.	*Sand covered plain.* Level to very gently sloping; even but locally irregular and broken by small occurrences of lava outcrop; low lying.
3.	*Fresh lava.* Very rough but with gently sloping envelope; up to 500 metres across.
4.	*Valleys.* Winding, narrow (up to 100 metres across) with many branches; may be incised up to 15 metres into lava plateau. Includes: (a) Valley floor up to 60 metres wide. (b) Steep or very steep sides.
5.	*Scarp.* Steep, straight slope; up to 30 metres high; usually straight in plan and occurrences more or less parallel; frequently extend for several km.
6.	*Footslope.* Occurring at bases of scarp and at low end of valleys. Irregular in plan: (a) *Fan.* Upper part of footslope; gently sloping (2–4°); straight; traversed by numerous branching channels; usually 200–500 metres across but some occurrences up to 1 km. (b) *Flats.* Lower part of footslope. Level; even; up to 500 metres wide.
7.	*Volcano.* Circular in plan; up to 100 metres high and 500 metres across; comprising: (a) *Outer slope.* Steep; straight. (b) *Inner slope.* Steep; straight or concave. (c) *Crater floor.* Level or gently concave; usually only about 100 metres across.
8.	*Lava hill mass (from volcano off diagram).* Up to 150 metres high and 4 km across; variable in shape but often deeply dissected into radially disposed steep sided valleys and upstanding ridges; ridge and hill crests are irregular.
9.	*Lodwar land system with Barchan dunes.*

by the wind and may travel great distances before deposition as loess. Such deposits are often formed by wind erosion of fresh glacial deposits; for instance, the brick earth of southern England has been formed in this way. The removal of the fine grained material leaves sand which the wind may move, usually in the form of dunes. If gravel is present, the finer material is removed until a continuous layer of gravel protects the surface. The sand accumulates in sand fields with regular patterns, forming longitudinal dunes or crescentic, barchan dunes. Examples of barchan dunes from the Lodwar land system are shown in Figure 18.

Effect of geology

17. In the preliminary discussion of erosion, the effect of the underlying material has not been considered. In practice the variations in drainage pattern are clearly related to the underlying geology, particularly where the relief is well developed. The simple case of erosion of a uniform clay slope gives rise to a dense fan-like pattern known as dendritic drainage. A more permeable material, such as a loose sand, would give a similar pattern, but with fewer streams. Most materials are not uniform and contain bands of weakness which channel the streams and are shown up in the drainage pattern. These features are related to the rock type and the structures associated with them and their presence may be used to predict the occurrence and extent of the different materials. These two main geological factors, rock type and structure, exert a strong control on the drainage system and on the resultant landform.

18. In sedimentary rocks the varying resistance to erosion of successive beds is of primary importance. If the bedding is near horizontal a hard bed can resist the downcutting of the stream and a waterfall is formed, such as the Niagara Falls. If the beds are folded, or tilted, the outcrop of the resistant layer will form a ridge and the rivers will tend to run parallel to it. The dip of the beds controls the shape of the ridge; where the dip is moderate the typical landform consists of a long regular slope parallel to the dip and a scarp slope where the more resistant bed gives way to the softer material in the valley. The pattern of smooth curves formed by these ridges can be used to deduce the structural arrangement of anticlines and synclines.

19. The direction and frequency of joints in a rock also influence the landform. In the description of weathering it was mentioned that joints provide drainage paths for the water, the major agent in weathering. Thus it follows that a more closely jointed area of rock is more easily weathered and likely to form a valley. The occurrence of tors, large blocks of rock exposed on the surface of a granite landscape, is attributed to variation of joint spacing. Where the joints are more widely spaced the rock is only partially weathered, and later erosion strips the weathered rock, leaving the core stones as boulders on the surface.

20. Jointing is produced by shrinkage of the rock on cooling, by expansion as a result of uplift or by other earth movements, and it thus indicates the history of the rock. Different sequences of rocks are likely, therefore, to have different

Fig 19.   Air photograph illustrating contrasting joint systems

NOTE: The granite (area A) has three main vertical systems at approximately 60 degrees to each other. The sediments (area B) have only two, at right angles, characteristically dividing the sediments into long, parallel ridges. (The white patches at centre of left half of photograph are clouds).

joint patterns which will show in the frequency and direction of valleys (*see* Figure 19). The direction of jointing is often related to the faults in the area as both form in response to the same forces.

21. The lines of weakness induced by both joints and faults are often filled by intrusions which may weather at a different rate from the country rock, thus forming ridges, or sometimes depressions. These features often follow a straight path across the country indicating that the fault is vertical; if the fault were dipping the presence of a valley would deflect its outcrop in the direction of dip, thus producing a sinuous outcrop.

22. Faulting on a large scale is a major landforming process. Large faults are usually complex and often consist of a band of faults, each representing individual movements over a period. The features produced by high angle faults are most conspicuous, producing a relatively straight junction with a scarp. A double fault system producing a depression, such as the East African Rift, is called a graben and an uplifted block is known as a horst. A parallel series of faults which may be found in a high angle thrust zone can produce a repetition of beds, and the resulting topography could be mistaken for a series of dipping strata. Large scale faulting produces linear features, which may erode to form valleys; minor features, such as lateral displacement of streams, are usually temporary. Faults of all kinds may be associated with volcanic activity.

Volcanic landforms

23. Volcanic rocks usually have a more obvious effect on the topography than other rock types as they often occur as subaerial areas of deposition covering parts of the old topography. Where volcanic rocks occur as a bedded sequence the landforms are similar to those found on other sedimentary rocks. The diversity of volcanic landscapes is related to the wide range of properties of volcanic lavas and ashes, and also to the extent of the volcanic region. The presence of local highlands and thermal rises over volcanoes often leads to locally intense rainfall, and this can lead to intense erosion of unconsolidated ashes. Saturation of material deposited on steep slopes also leads to landslides, which may be triggered by movements associated with volcanic areas. Many typical features of a young volcanic landscape are shown in Figure 18. The cones and hill masses are well known volcanic landforms but the lava plateau, with the stepped appearance in the landscape, is a much more extensive landform.

The effect of limestone

24. The features produced by volcanic rocks are mainly associated with their mode of formation, or deposition. The landforms produced by limestones however owe their distinctive character to the unusual weathering characteristics of the rocks. It has been shown that the major factor of weathering is removal of material in solution; but whereas the weathering of a granite might involve the removal of up to 40 per cent of the original rock, over 95 per cent of a limestone

can easily be dissolved. The typical limestone terrain is developed over extensive areas of thick massive well-jointed limestone. The water table lies at depth, and there is little or no surface drainage. The surface is covered with depressions, or small sink holes (dolines), the main drainage of the area being underground via caves. In places the roof of such a cave may break, revealing the stream in a deep gorge. In temperate regions limestone topography can be summarised as hills with many depressions (karst), and little or no surface drainage. In humid tropical areas the effects of weathering accentuate the valleys producing two distinctive forms known as cockpit karst and tower karst. The former consists of conical hills, typically 100–150 metres high, separated by irregular hollows. The latter form develops where the limestone projects above the general ground surface as steep sided cliffs or towers.

Section 5.2. NATURE AND ORIGINS OF SOILS

Definitions

1. *Comparison of engineering and pedological terms.* Engineers usually consider soil to be any unconsolidated deposit which can be shifted by earth-moving equipment without recourse to blasting. In this usage, 'residual soils' are those which are presumed to have developed *in situ* by weathering of surface rocks, from which they inherit some characteristics. The unfortunately chosen term 'transported soil' is often applied to any unconsolidated sediment the properties of which depend mostly on the geological processes of transport, with or without sorting, and deposition; it thus includes such diverse materials as marine clays (e.g. Oxford Clay), tills, river terrace gravels and volcanic ash. Its use is to be deprecated since it invites confusion with older and widely accepted definitions of soil. The unambiguous geological term 'unconsolidated sediment' is greatly to be preferred (*see* also Chapter 1).

To the geologist and pedologist (or soil scientist), soil means that part of the Earth's surface, whether previously transported or not, which has become weathered and otherwise modified *in situ* by solar energy, atmospheric precipitation and organisms, including man. If it has developed to sufficient depth to be of concern to engineers, it is in effect a residual soil on the local substratum.

2. *In pedological terms,* the two-dimension section of a soil seen in a pit is termed the 'soil profile' (Figure 20). The smallest three-dimensional block of soil fully representative of its characteristics is termed a *pedon*; layers formed by processes within the pedon are termed *horizons* and collectively constitute the *solum.* They represent varying degrees of modification of the *parent material,* the geological deposit from which the soil has developed. In soil profile descriptions, the parent material and any subjacent layer that influences soil development (e.g. by restricting drainage) are normally included as soil horizons although not part of the solum (Figure 20). Symbols commonly used in pedological texts to designate horizons are given in Table 6.

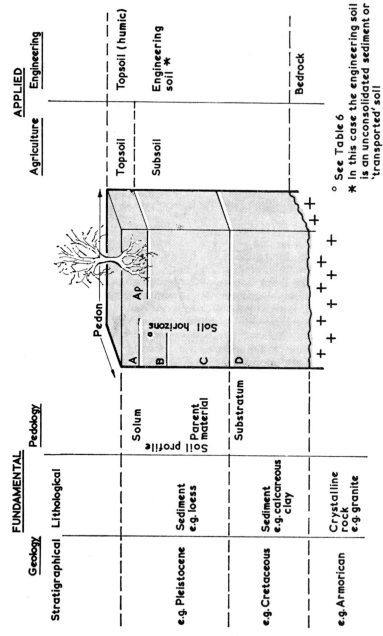

Fig 20. Definitions of surface materials

The diagram shows an idealised block model based on a hypothetical geological situation to illustrate the relationships between different nomenclatures. The stratigraphy and lithologies have been chosen simply as examples and have no general significance.

Note that this is a case where the substratum (or D horizon) would be included in the soil profile description, as the clay would impede percolation and thus affect soil development.

62

TABLE 6. SOIL HORIZON NOMENCLATURE

It is convenient to use symbols designating the dominant pedogenic processes considered to have occurred in each horizon. There is no general agreement on horizon nomenclature and a number of systems are in use. That given below is based on traditional ABCD usage and is suitable for general purposes. Horizons of dual character may be shown by symbols such as BCa, B/Ca, B_{ca}.

Symbol	Common alternatives	Definition
L	A_{00}, O_1	Layer of practically undecomposed litter.
F } H }	A_0, O_2	Organic surface layer with beginning of decomposition of plant remains by fungi and/or soil animals. Organic surface layer in which only a few plant remains are still recognisable.
A		Surface mineral horizon, with maximum biological activity, leaching and weathering, sometimes impoverished in clay and/or sesquioxides.
(A)		Surface horizon of parent material only slightly modified by weathering.
A_h	A_1	Surface horizon containing notable addition of organic matter.
A_e	A_2, E_a	Horizon generally low in organic matter and impoverished in sesquioxides and/or clay ('eluvial' horizon).†
—	E_b	Brown (pale when dry) horizon depleted in clay, typical of sols lessivés.
A_p	S	Rather uniform surface layers produced by cultivation of A (and sometimes other) horizons.
(B)	B, B_w	Subsurface horizon: strong chemical weathering and/or structure development but with no evidence of illuvial colloids.
B		Subsurface horizon with evidence of accumulation of humus, clay or sesquioxides ('illuvial' horizon).†
B_h	B_1	Horizon enriched with illuvial humus.
B_s, B_{fe}	B_2	Horizon enriched with illuvial sesquioxides (B_s) or iron oxides only (B_{fe}).
B_t		Horizon enriched with clay often presumed to be illuvial.
C		Parent material from which the solum is considered to have developed. (In many cases it is impossible to prove that the unweathered material at depth is genuinely that from which the solum has developed).
D	R	Substratum having some important influence on the profile but which is lithologically distinct from the solum.
G	g^*, CG	Horizon in which gleying is a prominent morphological feature.
Ca	ca^*, Ck	Horizon enriched in carbonates.
Cs	cs^*, Cy	Horizon enriched in gypsum.
Sa	sa^*	Horizon enriched in salts more soluble than gypsum.

Those shown with an asterisk are always written as suffixes to other symbols e.g. Cg.

† *See* paragraph 17b.

63

The nature and arrangement of horizons can indicate sub-surface conditions important to engineers such as hydraulic conductivity, the height and degree of fluctuation of the water-table, or chemical conditions affecting the setting of cement or accentuating corrosion problems in iron pipes and piling.

3. Biologists and agriculturalists are concerned with the pedologist's soil down to rooting depth and often distinguish between *topsoil*, the uppermost biologically active layer in which most roots develop and which may be subject to continual mixing by cultivation, and the relatively inert *subsoil* sometimes sharply separated from topsoil at the base of the plough-layer (*see* Figure 20). The *engineering soil* includes the pedological soil and any unconsolidated sediment down to hard bedrock. Topsoil is essentially the same in engineering and agricultural usage.

4. Engineers may need to consider these less familiar views of soil for the following reasons:

a. While the physical properties of most soils are mainly determined by the geological origin of their parent material (e.g. lake bed deposits, beach sands, loess), in others pedological processes are highly significant (e.g. in saline soils, black tropical soils, ferrallites, peats).

b. Many pedological concepts and measurements are relevant to engineering problems (e.g. soil texture, salinity, pH, exchange capacity, hydraulic conductivity).

c. In planning operations in new or remote areas, terrain information may only be obtainable from soil maps and reports produced for non-engineering purposes, such as agriculture, forestry or land-use planning, and written in pedological terminology.

Soil classifications based on particle size

5. Particle size distribution, or mechanical composition, is determined by processes of weathering and soil-formation in 'residual soils', and of sorting and deposition in 'transported soils'. It may be expressed by percentages of different size fractions, which are defined below and which are measured (mechanical analysis) by sieving for the coarser fractions and by settling in water for the finer, or in qualitative verbal terms. There are no universally accepted definitions of the size fractions, those most relevant in the present context are given in Table 7.

6. Results of mechanical analyses are often plotted as summation curves (Figure 21(a)) the forms of which give more information than can be readily visualised from a table of percentages, e.g. the degree of sorting or grading.

7. In the Unified Classification System (UCS) used by engineers (Annex A) soils are classified primarily by the dominant particle size and organic content into gravels (G), sands (S), silts (M), clays (C), organic soils (O) and peats (Pt). Coarse-grained or granular soils (G and S) are further subdivided by grading, and fine-grained or cohesive soils (M, C and O) by liquid limit and plasticity index (paragraph 32). A well-graded soil is one having a good spread of particle sizes.

TABLE 7. SOME IMPORTANT MECHANICAL COMPOSITION SCALES

Pedology and agriculture				Engineering	
International Society of Soil Science scale		BS 1377: 1974. Methods of testing soils for civil engineering purposes (ii)		BS 812 Part I: 1975. Methods of sampling and testing of mineral aggregates sands and fillers (i)	
Fraction	Size range (mm)	Fraction	Size range (mm)	Fraction	Sieve sizes (mm)
(a)	(b)	(c)	(d)	(e)	(f)
Stones or gravel	>2	Cobbles	200–60	Coarse aggregate	75
					63
					50
		Gravel: Coarse	60–20		37·5
		Medium	20–6		28
		Fine	6–2		20
					14
					10
					6·3 (iii)
					5
Coarse sand	2–0·2	Sand: Coarse	2–0·6	Fine aggregate	3·35
Fine sand	0·2–0·02	Medium	0·6–0·2		2·36
		Fine	0·2–0·06 (ii)		1·7
					1·18
Silt	0·02–0·002	Silt: Coarse	0·06–0·02		850μm
		Medium	0·02–0·006		600
		Fine	0·006–0·002		425
					300
					212
					150
Clay	<0·002	Clay	<0·002		75

NOTES: (i) *See* also BS 882, 1201: Part 2: 1973. Aggregate from natural sources for concrete.
(ii) Pedologists sometimes use BS 1377: 1974 and also sometimes divide sand from silt at 0·05 mm following the practice of the US Department of Agriculture.
(iii) Aggregates are specified by percentages passing various sieves, and there is a small percentage overlap between coarse and fine.

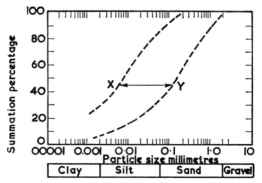

(a) Mechanical composition summation curves (After GILLOTT 1968)

NOTES: 1. A widely used method for representing particle size distributions of sediments and soils. The steepness of a curve is an indication of the degree of sorting in the material.

2. The curves of frost-susceptible soils (paragraph 47) usually fall between those of X and Y.

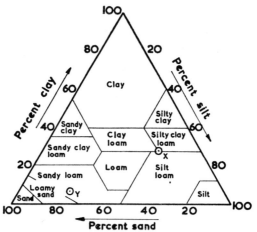

(b) Triangular mechanical composition diagram

(After U.S. DEPT OF AGRICULTURE 1951)

NOTES: 1. If the mechanical composition of a material is expressed in terms of sand (2·0–0·05 mm), silt (0·05–0·002 mm) and clay (<0·002 mm) it may be plotted on the diagram as a single point. This method is more convenient than summation curves when large numbers of soils are to be compared. The points X and Y represent the corresponding soils in (a).

2. By means of the diagram, texture names can be assigned to soils from mechanical analysis; but although the method is widely used, it is potentially misleading for reasons given in the text. (*See* also Soil Survey, 1974).

Fig 21. Mechanical composition diagrams

8. In classifying soils by mechanical composition pedologists usually put most emphasis on the 'fine earth' (< 2 mm); they describe soils broadly as sands, silts, clays and loams (the last-named being soils with a fairly wide range of particle sizes) with a number of subdivisions (e.g. silty clay or sandy clay loam). These are known as *soil textures* and are usually assessed manually in the field. In agricultural usage sands and sandy loams are said to have 'light' textures while clays and clay loams are 'heavy'. Textures are also sometimes defined in terms of mechanical composition by means of a conversion diagram (Figure 21(b)) but this approach can be misleading for practical purposes as soil structure, or fabric, is destroyed and stones, carbonates and organic matter which all contribute to field behaviour are usually removed prior to analysis. Unfortunately it is often difficult to interpret engineering soil classes from pedological textures or mechanical analyses except in rather general terms.

Confusion can arise from differing uses of the terms 'sand', 'silt' and 'clay'. As seen above, each may indicate either the sand, silt and clay fraction or else a soil or geological deposit whose mechanical composition and associated properties, such as cohesion or plasticity, are dominated by that fraction. Such lithological characters may give their names to geological rock formations e.g. Ashdown Sands, Fen Silts, Weald Clay. In addition 'silt' is often loosely used as a term for alluvium while 'clay' can also mean the clay minerals, a group of aluminosilicates with characteristic structures which usually form the bulk of the clay fraction and which are described in paragraph 18.

9. 'Mud' is a widely used but ill-defined term referring to any unconsolidated sediment or soil containing sufficient fines to exhibit plasticity and adhesion to equipment or personnel when moist, particularly after puddling by working or trafficking, and some degree of cohesion when dry. It may be derived from all soils of the Unified Classification System except G and S and from soils of all pedological textures except sand. In geological usage, however, 'mud' also means a fine-grained unconsolidated sediment usually consisting of silicates but sometimes partly or wholly of calcite. On diagenesis it forms mudstone, or shale if bedding planes are well developed.

Weathering and soil-formation

10. The characteristics of soils are best understood by considering their origins. Although it is convenient to describe them separately, weathering and soil-formation (or pedogenesis) take place together at the Earth's surface and cannot be clearly distinguished in practice. Their relationships are shown in Figure 22.

The nature of the pedological soil at any point is controlled by the local interplay of the following factors:

a. Parent rock.

b. Climate (temperature and moisture regimes).

c. Topography and drainage conditions.

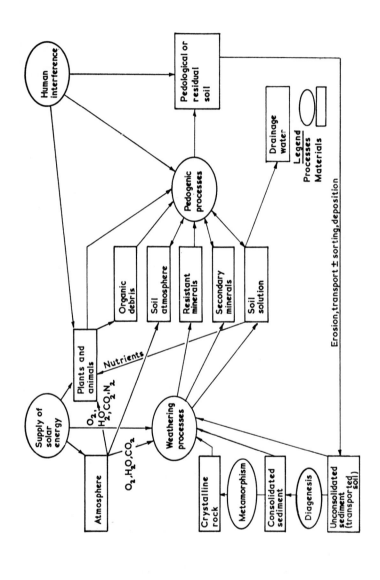

Fig 22.   Relationships between rocks, weathering and soils

NOTE: Sometimes clear distinction between processes and materials is not possible, e.g. plants and animals carry out specific processes as well as providing organic material.

68

d. Age of the land surface.

e. Vegetation and soil population.

f. Human interference.

11. *Rock weathering.* As noted in Section 5.1 physical weathering causes the breakdown of rock into smaller particles with correspondingly larger specific surfaces; this facilitates chemical weathering which produces secondary minerals.

Particularly important are the hydrolysis and oxidation of silicates, normally taking place together under the combined influences of hydrogen ions from the dissociation of water and acids, particularly carbonic, and dissolved atmospheric oxygen. Chemical weathering is inhibited by dry conditions, as in deserts; it is also slowed by low temperatures at high latitudes and altitudes, where although fine-grained rock-flour may be glacially formed, the debris often lack clay minerals, the most important product of chemical weathering.

Silicates vary considerably in their chemical stability. Susceptibility to chemical attack of the commoner rock-forming minerals is in the order; olivine, augite and calcium feldspar > hornblende, biotite and sodium feldspar > potassium feldspar > muscovite > quartz.

12. The essential features of hydrolytic weathering of crystalline rocks are shown in Figure 23. Dissolved hydrogen ions cause breakdown of the surface structure of feldspars and ferromagnesians with the release of:

a. Aluminium in true or colloidal solution.

b. Silicon mainly as silicic acid.

c. Metallic cations in true solution, especially sodium potassium and calcium from feldspars, and magnesium and ferrous iron from ferromagnesians.

Aluminium is relatively immobile and tends to re-precipitate by reacting with silicon from solution to produce amorphous materials such as allophane which may survive, but which are usually precursors to the crystalline, although generally fine-grained, clay minerals.

Silicon is more mobile and that which does not combine with aluminium to form clay minerals escapes in the drainage mainly as silicic acid H_4SiO_4. In tropical areas of high rainfall where there is no impedance to free percolation, clays low in silicon (1:1 minerals, *see* below) and free iron and aluminium oxides (*ferrallites*) are typical end-products; where aluminium oxides are notably concentrated the product is termed *bauxite*. If, on the other hand, removal of silicon is restricted by low rainfall or impeded drainage, clay minerals high in silicon (2:1 and 2:2 minerals) are produced. Occasionally silicon is concentrated by water movement and is precipitated as secondary silica (SiO_2) cementing the matrix into hard *silcrete*.

According to the nature of the rock and the degree of leaching, soluble metallic cations are either lost in the drainage or retained as exchangeable ions on the

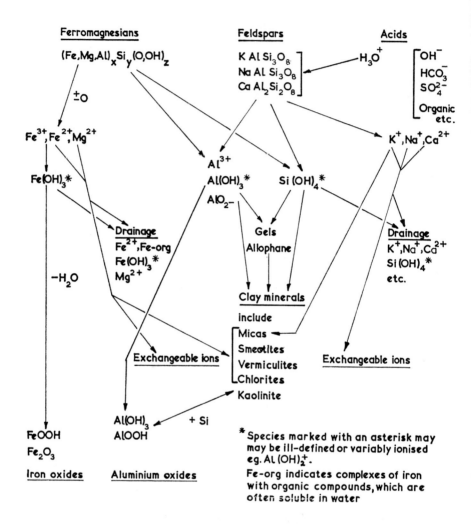

Fig 23.   Schematic representation of silicate weathering

NOTE: To simplify the diagram, the micas muscovite and biotite have been omitted. The former tends to lose some potassium from between the sheets to form clay mica or illite. Biotite is also a ferromagnesian mineral and thus liberates iron, magnesium and silicon as well as potassium; it often changes first to vermiculite.

active surfaces of clay minerals and humus. Potassium ions, because of their particular size, also enter the crystal structure of newly formed illite (paragraph 20) and magnesium is incorporated into some smectites and vermiculites, chlorite and palygorskite. In dry conditions some calcium partially re-precipitates as calcite, or in very arid conditions as gypsum. The former often occurs as small concretions, but in special circumstances it may become concentrated by moving groundwater to form massive sheets of *calcrete* (caliche).

13. In freely drained, well-aerated conditions, ferrous iron (Fe^{2+}) is rapidly oxidised to the ferric state (Fe^{3+}) and mostly precipitates as yellow-brown amorphous ferric oxides or goethite ($FeOOH$), or as red haematite (Fe_2O_3), the latter especially in hot climates. In very acid and/or reducing conditions, particularly in the presence of much organic matter, ferrous iron is stabilised and can move in seeping water as simple salts or as organic complexes.

Recombination of silicon, aluminium and other ions not only takes place at the seat of weathering but in all basins of deposition. For example illite, and sometimes smectites are specially characteristic of marine deposition, and chlorites of diagenesis. Unusual clay minerals high in magnesium such as palygorskite are often associated with playas and saline lakes, and smectites with mbugas.

Together with a few unimportant minerals, quartz (SiO_2) is extremely resistant to chemical weathering and, except in the case of basic crystalline rocks and volcanic glasses from which it is absent, it usually provides most of the coarser fractions of the residue.

14. The weathering of sedimentary rocks involves relatively minor changes as the materials have already passed through at least one previous cycle. Sandstone, conglomerate and breccia mainly disintegrate back into their constituent quartz, iron oxides and rock fragments, although the last may weather further according to mineralogical composition. Shales and mudstones are reduced back to fine-grained quartz and clay minerals, generally with some alteration of the latter by weathering; many older shales and mudstones, for example, contain chlorite which is relatively unstable. Limestones and dolostones mostly pass into solution as calcium and magnesium bicarbonates, leaving a thin residue of clay, iron oxides and quartz with stones of chert or flint. The extent to which mechanical sorting in the original sediment is inherited by the weathered residues depends on the degree of diagenesis and the intensity of weathering.

15. In summary, the solid products of weathered rocks consist of varying proportions of the following:

a. Quartz and other minerals resistant to weathering.

b. Weatherable but as yet undecomposed minerals, e.g. feldspars.

c. Clay minerals and amorphous precursors (paragraph 18).

d. Free iron and aluminium oxides, and sometimes.

e. Calcite.

These constituents are variably distributed among the different size grades (*see* paragraph 8), for example quartz is mainly confined to the sand and silt, and clay minerals to the clay fraction, while iron oxides can occur over the whole range from clay to cobbles.

16. *Soil-formation*

a. *Parent material.* The weathering residue is also the effective parent material of the soil on which the other factors operate. Its mineralogy and chemistry determine its potential for plant growth and its• particle size distribution affects the hydraulic conductivity and permeability to air. Hydraulic conductivity in turn controls the infiltration of rainfall into the surface and its percolation through the soil to the local water-table.

b. *Climatic effects* may arbitrarily be divided into those of temperature and moisture. The former, apart from influencing the evaporation and transpiration of rainfall, controls the rate of all chemical reactions and the production and decay of organic matter. The effectiveness of moisture depends on the percolation rate which is the amount of water free to pass through the soil after evaporation and transpiration from plants has taken place; it determines the degree to which the soil is leached of its more soluble constituents and affects the height of the water-table. The influences of climate may be much modified by local topography and the age of the land surface (*see* below).

c. *Topography* is effective in two main ways. First, the slope controls the extent and direction of water movement in the soil and the amount of run-off. Second, it determines the balance of erosion and deposition in different parts of the landscape. The position is summarised for a deep permeable deposit in Figure 24(a). At **A** erosion is slow, water percolates down to a water-table and a deep freely-drained profile is formed on the weathering residue. At **B** erosion is maximum and the profile is kept shallow; water tends to move laterally as well as downwards. On the concave slope at **C** soils are derived mainly from the hill-wash (colluvium) from **A** and **B**; they are relatively deep but imperfectly drained owing to groundwater at the base of the profile. At **D** soils are again derived from colluvium but are now poorly drained, while at **E** they may be permanently wet so that anaerobic conditions prevail, the decomposition of plant remains is checked and organic matter accumulates. Further, soluble ions leached from the soils at **A** and **B** tend to enrich the lower profiles which normally have higher pH values and contents of exchangeable ions (paragraph 25). Such a sequence of soils, whose inter-relationships are dictated primarily by differences in topography and associated drainage conditions, is termed a *catena*. It may be much more complex if different rocks outcrop down the slope (Figure 24(b)). Once the catena characteristic of a particular area has been recognised (e.g. Figure 24(c)) it is possible in principle to predict the position in the landscape of each constituent soil. It is thus a highly important concept in any form of soil survey.

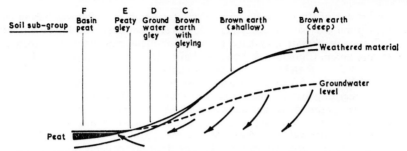

(a) Idealised catena on freely permeable parent material in a temperate humid climate

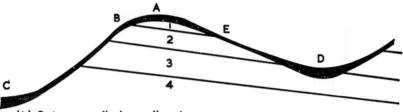

(b) Catena on dipping sediments

At A soils are derived from I, at B from 2 together with downwash from I.

At C the parent material consists of I−4 and may be very variable according to the erosional history of the landscape.

At D the catena is incomplete. If I is more permeable than 2, spring-lines may be expected at the junctions, particularly at E, with consequent modification of soil development and morphology.

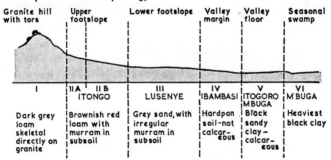

(c) Catena typical of East and Central Africa

Ranges from granite inselberg to mbuga with names and main characteristics of the constituent series. In accordance with usual practice these are given names from type localities

Fig 24.   Soil catenas (Fig (c) after CLARKE 1971)

73

d. *The age of the land surface* influences the depth to which weathering and soil-formation have taken place. On ancient surfaces, the long-continued influence of climate may have become dominant and the mature soils are termed zonal. Climate may, however, have altered greatly during the development of the soil and the concept of zonality is not then appropriate. This is especially so in Africa and Australia where very ancient land surfaces carry soils inheriting features such as lateritic crusts out of context with the present climates. Young or immature soils such as those on sites of recent erosion (e.g. coastal marshes, alluvium, sand dunes, volcanic ash) are shallow with little differentiation from the parent material and are termed *azonal*. *Intrazonal* soils are those whose morphology does not depend primarily on the zonal climate but on special local factors such as a high water-table or calcareous parent material.

e. *Organisms* affecting soil development are plants, animals and microorganisms, which are themselves dependent on the environment. The formation of the pedological soil effectively commences at the time when the surface is first colonised by pioneer species. Thence the soil and its associated population develop in a direction which is determined by the other factors. Specific processes will be noted below.

f. *Human interference* has the following main effects:

(1) Removal of vegetation by felling, grazing or burning, sometimes with consequent disruption of soil structure and acceleration of erosion.

(2) Replacement of forest with grassland which, in a wet climate can induce deterioration in drainage.

(3) Drainage of land with consequent increases in shear strength and aeration, and in some special cases acidity (Section 5.3, paragraph 4).

(4) Irrigation of dry lands; this can raise water-tables so much that soluble salts accumulate by the evaporation of rising groundwater.

Soil-forming processes

17. The main processes of soil formation are:

Erosion and deposition.

Translocation.

Biological transformations, *and*

Structure development.

a. *Erosion and deposition.* These have already been considered in relation to topography (Section 5.1, paragraph 6 and Section 5.2, paragraph 16c).

b. *Translocation.* Materials in solution or colloidal suspension diffuse through the soil in moisture films but more importantly for profile development they are carried by the mass movement of water. Where precipitation exceeds evaporation and transpiration, soluble ions are lost in the drainage water, being replaced by hydrogen and aluminium as the soil becomes more acid. The loss is more or less made good by fresh weathering according to circumstances. As noted under weathering processes, iron and aluminium are in general less mobile (as for example in *brown earths*, Table 10) but under coniferous forest or some heath plants they tend to be translocated as organic complexes from the upper part of the profile (A horizon) and redeposited in the B_S horizon. This process is termed *podzolisation* (Table 10).

In general, horizons from which materials have been leached are termed *eluvial* or *eluviated*; those which have become enriched are *illuvial*.

In a drier climate some of the sparingly soluble salts such as calcite or gypsum may be carried only a limited distance down the profile and reprecipitated as scattered concretions or as a definite illuvial (Ca) horizon (*chernozems, desert soils*) (Table 10). If there is a barrier to percolation, groundwater accumulates and restricts access of air. Iron in this horizon is reduced to the more mobile ferrous state (Fe^{2+}), as inorganic salts or organic complexes with greyish or greenish colours. This process is termed *gleying* and is usefully diagnostic of poor drainage even if free water cannot be seen at the time of observation. Once mobilised, ferrous iron may migrate in groundwater, leaving bleached zones devoid of iron pigments, but being reprecipitated as ferric oxides wherever it encounters more oxidising conditions within the soil or at a surface of seepage, e.g. ochre in streams, and some laterites. In *gley soils* there is usually an upper horizon of alternating oxidation and reduction recognisable by its mottled appearance, the yellow-brown colours of ferric oxide contrasting with greenish or greyish hues. In *ferruginous tropical soils* and *ferrallites* (Table 10) much iron released in weathering is often redeposited in the form of gravelly concretions locally termed *murram*. The word '*laterite*' has been used for two distinct forms of reprecipitated iron. In some circumstances iron oxides form a large part of the soil material which remains soft as long as it is moist; on exposure to the air the iron oxides dehydrate irreversibly to form a strong cement. This material was originally termed laterite, as bricks (Latin *later*: a brick) were made from it by air-drying; it is also known as *plinthite*. The term laterite has also been used for massive accumulations of strongly cemented soil material which have developed at surfaces of seepage or by continued oxidation at the surfaces of permanent water-tables on old peneplains. The ancient laterite crusts (also termed cuirasses and ferricretes) of Africa and Australia are mostly of this type. High groundwater levels can also occur for topographic reasons in arid climates. Water rising by capillarity and evaporating at the surface may then concentrate soluble, especially sodium, salts (*saline soils* Table 10).

A well defined water-table only develops in permeable material; in clay soils where the substratum is virtually impervious the only free water is located at the surface or in the larger pore spaces in the top few centimetres, below which reducing conditions develop if sufficient organic matter is present for bacteriological activity, giving rise to *surface-water gleys*.

c. *Biological activity.* Dead plant or animal material added at or below the surface is attacked by the soil population more or less rapidly. Under favourable conditions, notably when the debris are rich in nitrogen, nutrients such as calcium and phosphorus are in fair supply and where aeration, moisture and temperature are equable, this attack is fairly complete and the debris are converted via complex biological sequences to carbon dioxide, simple salts and a slowly decomposable colloidal residue of *humus* (*mull*). This shows some of the properties of clay minerals (paragraph 12) such as cation exchange (paragraph 25) and is in fact intimately associated with the clay fraction in the topsoil; at greater depths it has little significance. On the other hand, where lack of nutrients and the character of the plant remains inhibit decay, much of the material tends to lie on the surface of the soil as a mat of acid poorly decomposed *mor*. This is characteristic of podzols (Table 10), indicative of acid soil conditions, and may be associated with problems of soil stabilisation with lime or cement (paragraph 43).

Exclusion of air by excessive moisture, especially in cool conditions, also slows the decomposition of plant remains which accumulate as *peat*. *Topographic* or *topogenous* peats are formed in basins (e.g. reed swamp or tropical swamp-forest); *climatic* or *ombrogenous* peats are associated with high rainfalls and humidities especially in cool conditions at high altitudes and latitudes (e.g. hill peats and blanket bogs of western Britain and Ireland, muskegs of Canada). Peats also exhibit the colloidal properties of cation exchange and shrinkage on drying but have low plasticities and bearing capacities. Topographic peats often occur as patches, or as lenses, in alluvium where they may pose unexpected engineering problems.

d. *Structure development.* By structure is meant the arrangement of the clay, silt and sand particles into compound aggregates or *peds*. The size and shape of these depend on the processes in the particular soil which include shrinking and swelling due to moisture changes (paragraph 31), freezing, biological activity and cementing by iron oxides or calcite. The bulk density of a soil horizon depends directly on the nature and extent of structure development.

The term *fabric* is similar in meaning to soil structure but refers particularly to the microscopic arrangement of soil constituents. It may be investigated by optical and electron microscopy.

Clay minerals

18. Clay minerals are a group of alumino-silicates with characteristic structures and properties which usually form the bulk of the clay fraction. Their nature and abundance are of over-riding importance to the engineering properties of cohesive soils, although their effect may be much modified by dilution with non-clay minerals such as calcite. In granular soils they may be important as pore-space fillers or as plastic coatings on coarse particles.

19. *Structures.* All clay minerals are built up from two fundamental units: tetrahedra of one silicon ion (Si^{4+}) and four oxygen ions (O^{2-}) and octahedra of one aluminium (Al^{3+}), or sometimes magnesium (Mg^{2+}), and six hydroxyls

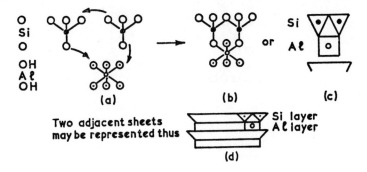

Fig 25.   Structure of 1:1 clay minerals

NOTES: 1. These can be imagined to form by the linking together of silicon-oxygen tetrahedra by sharing corner oxygens; and the substitution of 'downward-pointing' tetrahedral oxygens for hydroxyls in an aluminium-hydroxyl sheet (a) and (b). The actual process in nature is probably not very different.

2. The composite 1:1 sheet so formed is often represented conventionally by diagrams in which tetrahedra are shown as triangles and octahedra as squares or rectangles (c) and (d).

(OH^-) (Figure 25). In the structure of the important clay mineral *kaolinite* these are joined by repetitive replacement of some octahedral hydroxyls by tetrahedral oxygens to form a double sheet.

Since there is one sheet of silicon-oxygen tetrahedra for each sheet of aluminium hydroxide octahedra this is termed a 1:1 mineral. The composite 1:1 sheets are held together by weak surface forces to build up crystallites or micelles of generally small total size. In kaolinite the positive aluminium and silicons exactly balance the negative oxygens and hydroxyls and there is no net charge. A similar mineral common in tropical soils is halloysite, in which the sheets roll up to form tubes or rods. 1:1 minerals are often termed *kandites*.

20. *Micas* are minerals in which, in contrast to kaolinite, there are two silicon layers to every one of aluminium and these minerals are thus termed 2:1. The class includes the macroscopic minerals muscovite and biotite found in crystalline rocks as well as finely divided clay, or hydrous micas, often termed illites and found in soils and sediments.

In further contrast to kaolinite, during the formation of micas some silicons are replaced by aluminiums with a lower charge; the oxygens and hydroxyls are not fully balanced and there is thus a net negative charge on the sheets which is balanced by potassium ions (K^+) residing between them. Potassium is of just

77

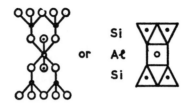

Fig 26.   Structure of 2:1 clay minerals

NOTE: The 2:1 structure is formed in the same way as the 1:1 but substitution of tetrahedral oxygens for octahedral hydroxyls takes place on both sides of the aluminium-hydroxyl sheet.

the right size to fit into spaces between oxygen ions in the surfaces of tetrahedral layers and to 'lock' them rigidly together (Figure 26). Muscovite and biotite have only potassium between the sheets, but illite has water molecules in addition and a less well organised structure; it thus forms small micelles and is confined to clay fractions.

21. In *smectites*, also 2:1 minerals, there is substitution either by aluminium for silicon in the tetrahedral layer (beidellite) or by magnesium for aluminium in the octahedral layer (montmorillonite). (The name montmorillonite is often used incorrectly for the whole smectite group). The negative charge resulting from substitution is balanced in smectites by a variety of cations (M in Figure 28) between the sheets where they are associated with water molecules. The sheets are not rigidly fixed as in micas and the number of water molecules between them depends on the humidity of the environment. Entry or exit of water leads to swelling or shrinking and smectites are therefore often termed 'swelling clays' although, as will be seen later, all clay minerals shrink and swell to some extent.

22. Kaolinite, illite and smectites are the most common clay minerals but a few others are important in particular cases. *Vermiculites* are weathering products of biotite and other micas. They are similar to smectites but have a higher charge on the sheets which, like those in micas, are held together more rigidly by cations, particularly magnesium, and water molecules. *Chlorites* occur particularly in schists also in ancient shales and mudstones; they are further sheet structure minerals, but the spaces between the composite sheets are occupied by extra octahedral sheets of aluminium or magnesium hydroxide and these minerals are thus sometimes referred to as 2:2. *Palygorskite* and *sepiolite*, less common minerals characteristic of saline environments, differ from all the above in having the silicon-oxygen tetrahedra arranged in chains rather than sheets. *Allophane* is a loose term for amorphous co-precipitates of silica and aluminium oxides which have some properties in common with the crystalline clay minerals and it is therefore often listed with them (Table 8). The clay minerals in a soil sample can be determined by x-ray diffraction in combination with other methods such as thermal and chemical analysis and electron microscopy.

TABLE 8. SIMPLIFIED CLASSIFICATION OF CLAY AND RELATED MINERALS

Structure	Class	Examples
(a)	(b)	(c)
Sheet 1:1	Kandites	Kaolinite, halloysite
Sheet 2:1	Micas	Illite Muscovite, biotite (i)
Sheet 2:1	Smectites	Montmorillonite, beidellite
Sheet 2:1	Vermiculites (ii)	
Sheet 2:2	Chlorites (ii)	
Chain		Palygorskite, sepiolite
Structureless		Allophane

NOTES: (i) Muscovite and biotite are macroscopic minerals occurring in crystalline rocks. Finely divided detrital muscovite can occur in some clay fractions but biotite does not owing to its greater susceptibility to weathering.

(ii) Vermiculites and chlorites also occur as macroscopic minerals.

23. *Size and shape.* As already noted, clay minerals occur as small crystallites or micelles with large specific surfaces. Typical dimensions are given in Table 9 from which it will be seen that in order of micelle size the usual order is:

$$\text{kaolinite} > \text{illite} > \text{smectites}$$

The fundamental sheet structure of most clay minerals gives the micelles a platey shape, roughly hexagonal when they are well crystallised but highly irregular when fine-grained or poorly formed (Figure 27). In halloysite, the sheets are rolled up to form tubes or rods, and the chain structure of palygorskite is reflected in its needle-shaped crystallites; the peculiar packing characteristics of these can give rise to geotechnical problems. A summary of clay mineral structures is shown in Figure 28.

Surface charge and exchangeable ions

24. All clay minerals carry surface electrical charges arising from the following sources:

a. Substitution in the crystal structure of Al^{3+} for Si^{4+} or Mg^{2+} for Al^{3+}. (Whereas in the micas this charge is mostly neutralised by rigidly fixed potassium ions except on outside faces, in smectites and vermiculites inter-sheet cations can contribute to the neutralisation of charge while remaining potentially mobile).

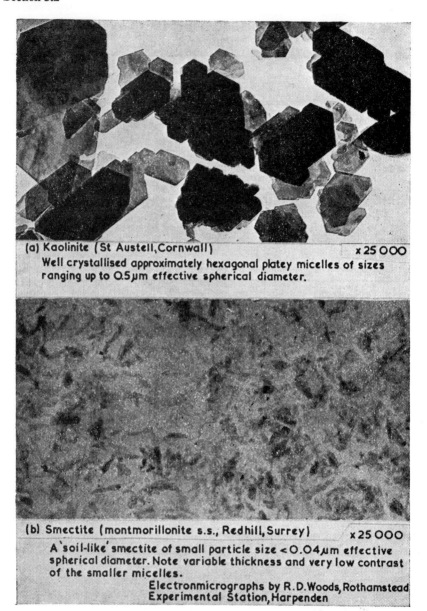

(a) Kaolinite (St Austell,Cornwall)   x25 000

Well crystallised approximately hexagonal platey micelles of sizes ranging up to 0.5μm effective spherical diameter.

(b) Smectite (montmorillonite s.s., Redhill, Surrey)   x25 000

A 'soil-like' smectite of small particle size <0.04μm effective spherical diameter. Note variable thickness and very low contrast of the smaller micelles.

Electronmicrographs by R.D.Woods, Rothamstead
Experimental Station, Harpenden

Fig 27. Electronmicrographs of kaolinite and smectite

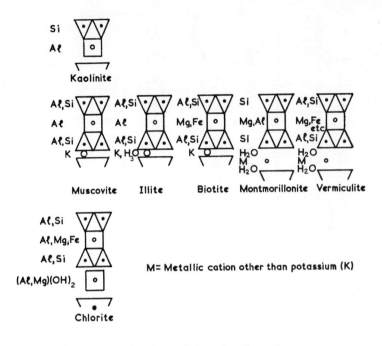

Fig 28.   Summary of clay mineral structures

b. Unsatisfied or broken valency bonds at the edge of the crystal, for example an oxygen ion attached to one silicon instead of the usual two, provides an extra negative charge.

c. Sorption and desorption of hydrogen, hydroxyl and some other ions such as phosphate.

The location of these charges is shown diagrammatically in Figure 29.

25. The total negative charge not neutralised internally is termed the *cation exchange capacity* (c.e.c.) and is usually measured in milligram equivalents (milliequivalents or meq (100 g^{-1}) gram equivalents 100 kg^{-1}) of dry clay.

In kaolinite only **b** and **c** are important; most of the charge therefore develops on the edges and the exchange capacity is small. In other clay minerals **a** contributes most of the charge although **b** and **c** occur at the edges. At low pH, sorption of hydrogen ions suppresses some negative charges at edges and produces extra positive spots; measured cation exchange capacity is thus dependent on pH. Typical values are given in Table 9.

TABLE 9. CHARACTERISTICS OF THE MORE IMPORTANT CLAY MINERALS

Physical properties of clay minerals are very variable, depending on particle size, crystallinity, dominant exchangeable cations etc, and thus on source and subsequent treatment. The figures given are to be regarded as typical but not exact.

Clay mineral	Sheet thickness (Å)(i)	Micelle thickness (Å)(i)	Specific surface (m²g⁻¹)	Cation exchange capacity meq (100 g⁻¹)	Hygroscopicity (ii) (% wt)	Plasticity (iii) PL (%)	LL (%)	PI (%)	Activity and swelling	Main occurrences
(a)	(b)	(c)	(d)	(e)	(f)	(g)			(h)	(i)
Kaolinite	7·1	1000	15	5–15	0·5	Na 26 Ca 36	52 73	26 37	Low	Hydrothermal deposits. Products of strong leaching and weathering especially on acid rocks.
Illite	10·0	200	80	20–40	3	Na 34 Ca 40	61 90	27 50	Medium	Marine clays and shales. Products of intermediate weathering with adequate potassium.
Montmorillonite	Variable	20	800	80–100	17	Na 97 Ca 63	700 177	803 114	High	Some marine clays. Weathered volcanic ashes. Products of restricted weathering (dry conditions or impeded drainage).

NOTES: (i) Ångström (Å) = 10^{-7} mm.

(ii) % weight loss on heating at 105° after equilibrating with atmosphere at 50% relative humidity.

(iii) PL = Plastic Limit. LL = Liquid Limit. PI = Plasticity Index = LL − PL. Figures given are for clays saturated with Na^+ and Ca^{2+} in % dry weight of clay.

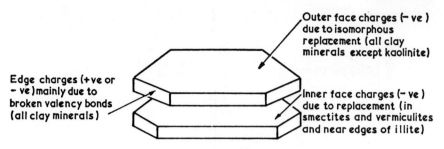

Fig 29.   Location of charges on clay minerals

In general terms, cation exchange capacity decreases in the order:

vermiculites > smectites > illite > chlorites > kaolinite,

but the value for a particular sample is affected by micelle size and degree of crystallinity.

The exchange capacity of allophane is variable but usually higher than that of illite. Soil humus can exhibit values up to 100–300 meq 100 g^{-1}.

26. In natural conditions, the exchange capacity is neutralised by a mixture of cations in equilibrium with those in the pore solution of the soil or sediment. The most important ions are Ca^{2+}, Mg^{2+}, Na^+ and K^+, with H_3O^+ (hydronium) and Al^{3+} in acid conditions, and traces of a great many others such as NH^+_4, Fe^{2+}, Mn^{2+} and Cu^{2+} according to the geological and pedological circumstances in which the clay occurs.

The relative proportions of the cations on the surface may be altered by leaching with water containing a different cation balance. For example in free-draining soils in humid climates there is a tendency for Ca^{2+} to be replaced by H_3O^+ (usually mainly derived from carbonic acid) and to pass out in the drainage waters; the soil becomes more acid and aluminium increasingly soluble, being sorbed as $Al(OH)^+_2$, $Al(OH)^{2+}$ or Al^{3+} according to how low the pH becomes.

27. Ions have differing abilities to replace others, which depend on their size and valency. The usual order of replacing power of the geologically important cations is as follows (although there are complications and the order is sometimes modified):

$$Al^{3+} > H_3O^+ \geqslant Ca^{2+} > Mg^{2+} > K^+ > Na^+$$

Thus in modern salt-marsh sediments Ca^{2+} and Na^+ are often found to be about equally abundant, the high concentration of sodium in sea water compensating for its low replacing power. But when such soils are reclaimed by the building of sea walls, the weakly held sodium is removed by rainwater much more rapidly than the calcium. Sodium may also be readily replaced by calcium in the vicinity of setting concrete.

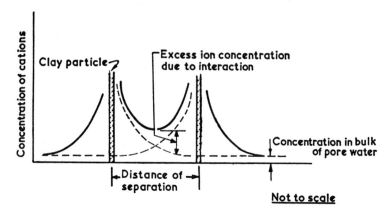

Note interaction of layers giving rise to swelling pressure

Fig 30. Structure of Gouy layers (after YONG and WARKENTIN 1966)

Distribution of ions

28. The exchangeable cations are not all held round the micelle as a single layer, as the surface forces of attraction are opposed by thermal movements in the pore solution and a concentration gradient therefore develops. Likewise, anions, of which the most common in soils are bicarbonate (HCO_3^-), chloride (Cl^-), sulphate (SO_4^{2-}), nitrate (NO_3^-) and phosphate ($H_2PO_4^-$), are repelled by the charged surface and a concentration gradient forms in the opposite direction. Phosphate, fluoride, and some organic anions, may however be held on clay or iron oxide surfaces by special mechanisms and not exchanged in a simple way. The zone round the micelle within which the concentrations of anions and cations differ from those in the bulk of the pore solution is termed the Gouy layer (Figure 30). Its thickness tends to be increased by the following factors:

a. A high surface charge density on the clay micelle (i.e. cation exchange capacity), which depends on its mineral composition and particle size.

b. A high pH, which increases the negative charge and reduces the amount of strongly held H_3O^+ and Al^{3+}.

c. The dominance among the exchangeable cations of weakly held ions, especially Na^+.

d. A low concentration of all electrolytes in the pore solution.

e. The absence of oppositely charged micelles.

f. The presence of active dispersing agents such as some natural organic compounds or the sodium hexametaphosphate often used to disaggregate soil particles in mechanical analyses.

When a combination of the above factors causes the layers to be well developed it prevents the micelles from approaching sufficiently closely for short-range surface forces of attraction to operate. This effect is most marked in active clay minerals such as smectites with large specific surfaces and exchange capacities. Clay minerals in such conditions remain dispersed or *deflocculated* as stable suspensions in water; in soils the micelles are arranged in a random pattern with relatively thick Gouy layers between them.

29. Reduction of layer thickness enables surface attractive forces to dominate. In a clay suspension the micelles *flocculate* or aggregate and settle; in active clays they form loose edge to face (flocculent) structures with much occluded water. In less active clays where repulsive forces are weaker, the micelles are arranged in a semi-oriented face-to-face pattern with much less water.

Mixing of clays with oppositely charged micelles destroys the Gouy layer systems and mutual flocculation occurs; in soils this is probably often a preliminary to the cementing of clay minerals by iron and aluminium oxides which have positively charged micelles. On the other hand some forms of organic matter in soils seem to stabilise clay suspensions allowing them to migrate in percolating water.

Interaction with water

30. So far water has been assumed to surround clay micelles but no consideration has been given to its state. In the water molecule, the two hydrogens lie at an angle of 105 degrees; electron density is lower at the end of the molecule where the hydrogens are closer together and higher at the opposite side. Water is thus a permanent dipole, each end of which is attracted to oppositely charged species, and is capable of forming hydrogen bonds (O—H—O) with other oxygen-containing molecules or crystals. Such bonds between molecules of water in the liquid state are responsible for its high heat of vaporisation, boiling point and surface tension. It is evident that clay mineral micelles with their negatively charged oxygen-containing surfaces, associated cations and large surface areas will retain water molecules strongly, and dry clay minerals in fact take up water hygroscopically from the atmosphere (Table 8). The innermost layers are held very tightly to the surface and the adjacent cations in a rigid pattern analogous to the crystal structure of ice; such water is extremely immobile, and is only fully removed at high temperatures or at pressures approaching those of metamorphism. Further water is held less strongly by osmotic forces in the Gouy layers where the concentration of ions is greater than that in the bulk pore solution. Addition of still more water allows the layers to expand and the clay swells until they are fully developed, exerting a *swelling pressure* if confined. Conversely, application of an increasing overburden pressure on the clay, or an external suction on the water, will compress the Gouy layers as water is removed. The graph (Figure 31) relating moisture contents to equilibrium external suctions is termed the *moisture characteristic*. As would be expected the capacities of the common clay minerals to imbibe water and exert swelling pressures are in the order:

smectites > illite > kaolinite

Dispersed sodium-saturated clays swell more than those flocculated by calcium and/or high salt concentrations. Soil organic matter also swells and shrinks strongly.

31. As clays are dried the volumetric shrinkage is at first equal to the volume of water lost from between the micelles; eventually these come into contact at the *shrinkage limit* (N in Figure 31) and air enters the pores. Although some *residual shrinkage* (NM in Figure 31) takes place on further drying, it is much less marked and no longer related linearly to water loss. Swelling on rewetting is not fully reversible in surface soils, the porosity or voids ratio becoming higher than in the original moist material (*see* drying and wetting cycles in Figure 31 (b)). This may be manifested in semi-permanent cracks or structures of blocky or prismatic soil aggregates which can affect the shear strength of topsoils (paragraph 39).

32. *Plasticity*, the most characteristic property of clay minerals, is the ability of a material to change shape continuously under a stress and to retain the new shape when that stress is removed. It depends directly on the nature and particle size of the clay micelles and on the thickness of the water films between them. At low moisture contents only rigidly fixed water is present, surface attractive forces predominate and the clay is strongly cohesive and non-plastic. With increasing moisture, Gouy layers develop and at the *plastic limit*, cohesion is sufficiently reduced for the plate-shaped micelles (needle-shaped in halloysite and palygorskite) to slide past each other, 'lubricated' by the water films; at the *liquid limit* films are so thick that the clay becomes a slurry with virtually no shear strength (Figure 32). The difference in moisture contents between the plastic and liquid limits (PL and LL) is the *plasticity index* (PI). Typical values for the plastic properties of clay minerals are given in Table 9 but in a particular case these can be substantially modified by micelle size, exchangeable cations, salinity of the intermicellar (pore) solution and surface coatings of, for example, iron oxides. The presence of much organic matter can markedly raise the liquid limit.

Other fines

33. Although the clay and silt fractions of soils are considered together as 'fines' in some systems of soil classification, this can be misleading. So-called 'plastic fines' consist mainly of clay minerals, while 'non-plastic fines' are mostly rounded or angular grains of quartz, iron or aluminium oxides, and sometimes calcite, in the silt grade. In addition to the plasticity, the exchange capacity, water retention, swelling and cohesion are feeble compared with those of clay minerals.

Physical and engineering properties of soils

34. Any soil material results from the local interplay of geological factors (ultimate lithological derivation, modes of transport and deposition, degree of consolidation), and the weathering and soil-forming factors already considered. These determine its mineralogical composition, particle size distribution (and

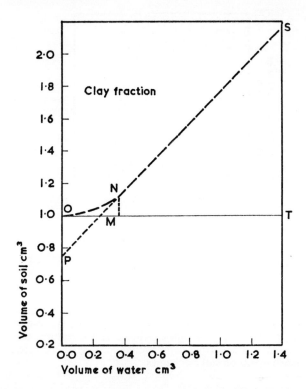

(a) For the clay fraction N is the shrinkage limit, ST is the total shrinkage, NM the residual shrinkage and area NOP the air content

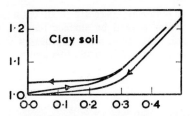

(b) Detail of (a). Note the volume change of the clay soil when taken through cycles of drying and wetting with consequent increase in voids ratio

Fig 31.   Relationships between shrinkage and moisture content
(after BAVER et al 1972)

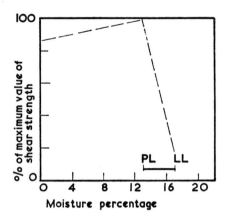

Fig 32. Shear strength of clay soil in relation to moisture content
(after BAVER et al 1972)

Shear strength increases slowly with moisture content up to a maximum at the plastic limit and then falls rapidly as the liquid limit is approached.

its variability as in varves or other bedding), type and amount of organic matter, state of compaction, cementation and contents of water and electrolytes. With these will be associated a characteristic exchange capacity, complement of exchangeable ions, salinity, pH and fabric. All the above features can influence its engineering properties. For further details reference may be made to the books listed and to standard works on soil mechanics.

35. *Retention of soil water*. In cohesive soils dominated by clay minerals, and in peats, water is mainly held by micelle surface forces and in Gouy layers, its removal being accompanied by shrinkage.

On the other hand, in granular materials there are relatively large pores in the framework of grains in which water is held by surface tension, or capillarity. The suction (h cm of water) which must be applied to empty a pore is inversely proportional to the effective radius (r cm) of its neck, through which an air/water meniscus must be drawn:

$$h = \frac{2\gamma}{r}$$

where γ is the surface tension of water.

Thus gravels and coarse sands are easily drained and very fine sands and silts much less so. As there is little or no shrinkage the moisture characteristic can be used to estimate the size distribution of pores between the grains. In most natural soils water is of course held by a combination of the two mechanisms, the relative importance of which depends on the particle size distribution and the modifying effect of structure or fabric on pore sizes.

36. *Movement of soil water.* Water moves in soils in rates and directions according to gradients of pressure or suction, and gravity. The *total soil water potential,* or *head* (φ_A) at a point A is given by

$$\varphi_A = h_A + z_A$$

where h_A is the differential height of water in a tensiometer inserted at A, and z_A is the height of A above an arbitrarily chosen datum.

φ, h and z are most conveniently measured in centimetres.

The rate of movement of water (v) between two points A and B where the potentials are respectively φ_A and φ_B is given by *Darcy's Law*

$$v = K \frac{\varphi_A - \varphi_B}{l}$$

where l is the distance between A and B

and K is the *hydraulic conductivity*, or permeability, of the soil. (This equation is the same as that given in Section 12.3, paragraph 1 but is in the form most commonly used by pedologists).

37. Hydraulic conductivity is a direct function of the size and continuity of pores and therefore depends on particle size distribution, clay mineral type and soil structure or fabric. The value of K for a clay soil may be 5 or 6 orders of magnitude less than that for a sand. It controls the rate at which water can *infiltrate* across the surface and percolate through the soil to lower levels. If water continuously reaches the surface, infiltration falls to a limiting rate equal to K and any surplus either ponds or runs off according to topography, in the latter case possibly initiating erosion if there is inadequate vegetation cover. The subdued relief of clay outcrops is thus related to the low values of K. Hydraulic conductivity is also important in the deformation of soils under load (paragraph 39) and in frost susceptibility (paragraph 46). Hydraulic conductivity is maximum when all pores are full of water (*saturated* soil). In unsaturated soils, the larger pores are empty and moisture films are much thinner and more tortuous; water movement no longer obeys Darcy's equation in its simple form and is essentially a diffusion process down gradients of moisture content and sometimes of temperature. In drier soils a significant part of the movement is in the vapour phase, particularly down temperature gradients. This can be important in hot climates where water vapour may distil into the cooler soil under road bases or other structures, sometimes sufficiently lowering its shear strength to cause failure.

As water drains under gravity from an initially saturated soil to a water-table, the fall in hydraulic conductivity causes movement to slow up until a substantial depth of soil comes to an approximately uniform moisture content known as *field capacity.* This is attained in about 48 hours in granular soils but much more slowly in fine-grained ones. In the absence of further infiltration or transpiration by vegetation, equilibrium would eventually be reached after a period of months, the suction on the soil water at any point then being equal to its height above the water-table. Upward percolation in response to surface evaporation is extremely slow unless the water-table is near the surface (less than about 1·5 metre).

38. *Soil plasticity*. In natural soils the plastic properties of clay minerals are modified or masked according to the amounts of other constituents. A quantitative expression of the influence of clay mineral type on soil plasticity is the *activity*, defined as the ratio of plasticity index of the soil to percentage clay fraction (<2 μm). The influence of common clay minerals on soil activity falls in the order:

palygorskite > smectite > illite > kaolinite

Soil shear strength

39. Shear strength is a complex function, not yet fully understood, of friction between soil particles of all sizes which is dependent on the overburden, and cohesion between clay micelles which is independent of loading. Shear failure may occur by rupture along a shear plane, or by flow, which in cohesive soils is plastic over the range of the plasticity index (Figure 32) and which approximates to viscous above the liquid limit.

The apparent viscosity of a slurry is the ratio of the applied shear stress to the rate of shear.

Shear strength is related to clay mineral type and to exchangeable cations and is strongly influenced by the degree of compaction or bulk density as well as moisture content (Figure 33).

Soil shear properties can be markedly altered by remoulding. In wet granular soils and some flocculated clays, working may increase the voids ratio, moisture films are reduced in thickness and strength rises; such soils are termed *dilatant*, e.g. moist beach sands. More usually shear strength is reduced by working or shock (*rheotropy*) although it may subsequently recover (*rest-hardening* or *ageing*). These changes are caused by temporary breakdown of the system of bonds between clay micelles and their reformation with time; soils containing smectites are specially rheotropic. If the original fabric contains much cementing material, breakdown is often permanent. *Sensitivity* is the ratio of the shear strength of the soil in its original condition to that after remoulding at the same moisture content. Values range from less than 1 in 'insensitive' soils to over 16 in 'quick' clays, such as some recent marine clays in Scandinavia. It has been suggested that these were originally deposited in the flocculated state in salt water, but have since been leached with fresh water so that the original fabric has become metastable and cannot reform when disturbed. Very sensitive soils may be subject to failure due to earth tremor or excessive vibration during construction.

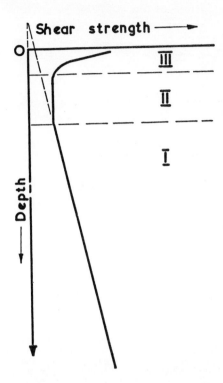

Fig 33.   Changes in shear strength with drying and overburden of London
clay soil (after SKEMPTON and NORTHEY 1952)

Shear strength decreases linearly with reduction in overburden pressure in I, reaches a minimum in the 'uncompressed' soil in II and increases again with drying towards the surface in III. In the upper part of the latter zone, however, pedological processes may greatly modify shear strength.

Other soils which can be sensitive are those containing large amounts of either allophane or fine silt particles.

Shear strength can be measured in the laboratory by shear box and other tests, but practical measurements can be made on site by empirical methods using shear vanes or cone penetrometers.

40. *Soil compressibility.* Compressibility is the relationship between reduction of voids ratio (or increase in bulk density) and applied load (Figure 34 curve ABE). Compression of granular soils depends on the compressibility of the mineral grains, pore water pressures and the extent to which rearrangement of particles is possible. Adjustment to stress takes place relatively rapidly. In

91

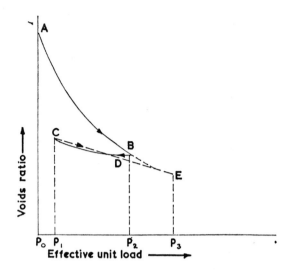

Fig 34. Changes in voids ratio on compression of normally and over-consolidated
clay soil (after TERZAGHI 1955)

On initial compression from p_0 to p_3 the relation between voids ratio and unit load
follows AE (normal consolidation). If on reaching p_2 the load is reduced to p_1 the increase
in voids ratio follows BC. If load reduction is followed by a fresh increase to p_2 the voids
ratio decreases as indicated by the dashed line CD. There is thus no unique relationship
between unit load and voids ratio, the latter being lower in clays which have been pre-
or over-consolidated.

cohesive soils compression is slow and is often termed *consolidation*. The latter is
importantly related to shear strength (Figure 33). In saturated soils the stress is
initially carried by pore water which builds up excess pressure. This is slowly
dissipated by migration into adjacent soil, the rate depending on the hydraulic
conductivity. Compressibility is also affected by the initial arrangement of the
clay minerals; with increasing stress preferred orientation tends to increase and
in the last stages of compression (secondary compression) the load is probably
mainly carried by the swelling pressure of the clay micelles. Compressive
behaviour is strongly dependent on clay mineralogy and exchangeable cations.

Smectites confer high initial and secondary compressibility, but adjustment to
stress may be very slow due to low hydraulic conductivity. These characteristics
are accentuated by the dominance of sodium but this ion is easily replaced and
substantial modification may be expected if ion exchange can occur. Kaolinite is
less compressible and less affected by exchangeable ions; because it is normally
associated with higher hydraulic conductivities, adjustment to stress is usually
more rapid.

41. Increase in the voids ratio of a cohesive deposit on removal of load (e.g. by erosion of overburden or melting of an ice-sheet) is hysteretic (Figure 34 curves BC) due to irreversible rearrangement of clay micelles and their associated water films. Deposits which have been subjected to a compression greater than that due to present overburden are termed *pre-* or *over-consolidated* and exhibit strengths different from those of normally consolidated materials, particularly if smectites or other active clay minerals are present. Pre-consolidated clays can be expected in previously glaciated regions.

42. *Soil stabilisation.* Soil stabilisation is any process which improves the physical properties of a soil (e.g. shear strength, bearing capacity, resistance to erosion, dust formation or frost-heaving). Methods employed are:

a. *Mechanical* (e.g. by increasing compaction or bulk density at optimum moisture content to allow most voids to be filled with fines).

b. *Chemical* and electro-chemical.

c. *Biological* (e.g. by turfing).

Chemical stabilisation may be achieved by relatively expensive organic agents which are either water-repellent, oily or bituminous compounds or cementing resins, or by cheaper inorganic materials such as lime (calcium oxide CaO or hydroxide $Ca(OH)_2$ or cement.

43. Stabilisation by lime involves the formation of new strong bonds between clay minerals and other soil particles and is therefore ineffective in granular soils. Three processes appear to be operative:

a. Rapid replacement of exchangeable cations by calcium with increased intermicellar bonding and a more flocculated fabric.

b. Slower attack on the edges of the clay micelles by alkaline solutions, in which the solubility of silicon is increased, with the possible formation of new silicates similar to those in cement.

c. Crystallisation of calcite to form further inter-particle bridges.

44. Apart from its normal action which is more important in more granular materials, cement probably reacts in similar ways to ordinary lime. The effectiveness of both depends on the cation exchange capacity, smectite-rich soils normally requiring larger additions than those containing kaolinite. Effectiveness can be markedly decreased by excessive acidity, sulphates or organic matter. The presence of these may often be predicted from pedological maps (paragraph 48) but their effects on stabilisation can only be determined by tests on the actual soil material.

45. An electrochemical method of stabilisation involves reduction of water content by *electro-osmosis*. An electrical potential is applied to the soil across two electrodes, to which ions migrate carrying water shells with them.

Soil freezing

46. When air temperatures fall below 0° centigrade soil water may freeze; the lower the moisture content and the higher the corresponding soil water suction the lower will be the freezing point. In high latitudes soils may remain frozen throughout the year as *perennially frozen ground* or *permafrost* which becomes discontinuous as lower latitudes are approached.

Above the permafrost there is an *active layer* which thaws in summer and refreezes in winter. During the thaw absence of drainage through the frozen subsoil can lead to very high moisture contents and correspondingly low shear strengths and bearing capacities. This is a common cause of collapse of road bases in the spring. In more extreme cases solifluction can occur even on slopes of one degree. Permafrost is often preserved by an insulating cover of vegetation or peat, disturbance of which brings about its destruction with detrimental effects on structures. As refreezing of the active layer proceeds downwards from the surface in the autumn, considerable stresses are set up causing heaving, contortions and segregation of stones; these phenomena may be manifested at the surface as stone stripes or *patterned ground* often seen in fossil form in previous periglacial or permafrost regions.

47. *Frost-heaving*, or the formation of mounds or 'blisters' is not caused simply by the expansion of water on freezing but by the development of sub-surface ice lenses usually roughly parallel to the soil surface. These grow by migration of supercooled water, which is favoured by adequate moisture content and hydraulic conductivities. The soils most susceptible to frost-heaving are thus those of intermediate mechanical composition (Figure 21 (a) and (b)) but clay mineralogy and the amount and composition of the pore water are also important. In general, susceptibility falls in the order

kaolinite > illite > smectites

which is probably a reflection of diminishing permeability. Ice lens formation is favoured by high water-tables and low overburden pressures as these also facilitate movement of water towards growing points, but the actual distribution of the ice layers may be controlled by variations in mechanical composition as in varved lacustrine sediments.

Soil surveys

48. Engineering soils surveys are usually site investigations confined to relatively small areas or to the alignments of structures such as roads, railways or watercourses. (For techniques of site investigation *see* Chapter 9). Extensive

surveys may be made for special purposes, for example prediction of seepage from large reservoirs or in agricultural applications such as reclamation, irrigation or drainage.

By contrast, pedological surveys are almost always areal studies to discover, map and explain patterns of soil types in the landscape and attention is normally confined to the pedological profile and its immediate substratum (Figure 20). Similar information is required for agriculture, forestry and land-use planning and these often provide the economic justification for a survey. The report or memoir accompanying a soil map, in addition to description of the soils, frequently contains valuable background information on the district such as the geology (particularly surface geology), physiography, hydrology, climate, natural vegetation, agriculture and other forms of land-use; some pedological memoirs, for example those with recent maps of the United States Soil Survey, also give a limited amount of engineering information. Although soil maps are now being produced in most countries they are still far less common than the geological maps to which they are complementary in predicting site conditions.

Types of soil survey

49. Soil surveys are either *detailed* or *reconnaissance*, and when using a soil map it is important for an engineer to know by which method it has been prepared.

In detailed (or 'field-to-field') surveys soils are observed, in profile pits and by augering, at closely spaced sites selected intuitively by the surveyor or laid out on a grid. Air photographs may be used to facilitate mapping but ground control must be strict. This approach allows mapping units (paragraph 50) to be narrowly defined and soil boundaries to be accurately plotted on large scale maps, e.g. at 1:2 500 to 1:50 000. Such maps are slow and expensive to produce but enable the user to make reliable predictions of soil conditions at chosen sites.

In reconnaissance surveys, the general distribution of soil catenas is assessed by observations at a limited number of selected sites. The surveyor then interpolates and extrapolates from these by means of his background knowledge of the relationships of the local soils with climate, geology, topography, vegetation and land-use; air photographs may often help for this purpose. Mapping units must be defined broadly and the boundaries on the map, which may be at scales from 1:50 000 to 1:1M, can only be approximate. Reconnaissance surveys are relatively cheap and useful for making quick inventories of natural resources or in preparation for detailed survey, but apart from providing background information on the soils in the area, they are of little value to the engineer at specific sites.

Mapping units

50. The choice of mapping units depends on the type, scale, purpose and budget of the survey. There is no full international agreement on the choice and definition

of mapping units. The following is based on British practice but would be understood by any pedologist overseas. The unit most often used in detailed surveys is the *series* which is a group of soils with similar profiles derived from similar parent materials under (presumably) similar conditions of development and usually named after a type locality, e.g. 'Bridgnorth Series'. As the members of a series have similar lithology and drainage conditions they can be assumed to have the same engineering characteristics in the pedological profile and the immediate substratum but not necessarily to the full depth of the engineering soil (Figure 20). In very detailed surveys, series may be split into *soil types* denoting particular textures, e.g. 'Bridgnorth sandy loam', or into *phases* according to practically important differences such as stoniness, slope, erodibility, history of land-use, etc. In some areas, for example on steep slopes or where periglacial disturbance has been significant, series can occur in a very complicated pattern and it may be uneconomic to map them separately. The *soil complex*, defined as consisting of particular series but in an unspecified arrangement, is then often used as the mapping unit.

51. For reconnaissance surveys, particularly at small scales, soil series are too narrowly defined and of too restricted occurrence to be separated on maps and are replaced by higher mapping units such as major groups, catenas or associations. A *major group* consists of soils with similar types and sequence of horizons which have been formed by the dominance of particular processes of soil-development. For example all soils belonging to the major group of *gleys* (Table 10) have a greyish or mottled gley horizon in which iron has been reduced by excess moisture, but this group can be further divided into the *sub-groups* of *ground-water* and *surface-water* gleys and these in turn include many different local soil series on a wide range of parent materials and substrata throughout the world, with correspondingly variable engineering characteristics.

A simple classification of major groups is given in Table 10. For present purposes the catena (defined in paragraph 16 c) is a more useful higher mapping unit. In more intensive surveys catenas may be defined fairly precisely by the sequence of named soil series in the landscape; it is then possible to make reasonably reliable predictions on soil conditions in particular topographic situations (Figure 24(c)). But in less intensive mapping the soils of a catena may only be defined as consisting of particular major groups or sub-groups and predictions about selected sites are then of little value. The term *association* is also used for regional groupings of soil series or of major groups, but the Soil Survey of Scotland use it in a restricted sense to denote a simple catena of series on a single parent material but with differences in drainage conditions.

Integrated surveys and terrain evaluation

52. An efficient and economical method of terrain investigation is the *integrated survey* in which simultaneous studies are made of such fundamental and applied features as geology, geomorphology, soils, vegetation and agricultural potential (*see* also Chapter 8, and MITCHELL 1973).

SECTION 5.3. CHARACTERISTICS OF IMPORTANT SOIL TYPES

Unconsolidated sediments ('Transported soils')

1. These materials are most conveniently arranged according to mode of deposition:

Marine clays.

Marine and beach sands.

Estuarine muds and silts.

Alluvium.

Lacustrine sediments.

Aeolian sands.

Aeolian silts (loess).

Glacial till.

Glacial outwash.

Periglacial deposits.

Pyroclastic (volcanic) deposits.

The upper parts of any of these may be more or less modified by weathering and pedogenic processes. For convenience of reference to pedological texts, the likely texture is given as well as the class in the Unified Classification System (Annex A).

2. *Marine clays:*

a. *Deposition:* slow-settling in relatively deep slow-moving saline water; usually very flocculant with high moisture content when first deposited.

b. *Mineralogical and mechanical composition:* mostly clay minerals (particularly illite) with fine detrital quartz, sometimes calcite and/or sulphides. Clay > silt ≫ sand; no detrital stones but sometimes calcite concentrations up to boulder size in some horizons. Shaley bands may be present and bituminous organic matter can be an important constituent. Marls are clays with much fine calcite. Texture clay or silty clay, class CH.

c. *Characteristics:* high exchange capacity, swelling, plasticity and cohesion; strength dependent on whether or not pre-consolidated. Hydraulic conductivity very low, no clear water-table but surface-water in humid climate in absence of artificial drainage. Topography generally low with low amplitude.

3. *Marine and beach sands:*

a. *Deposition:* from rapidly moving water or on beaches by wave action, where often transitional, to coastal dunes. *Examples:* Lower Greensand, modern beach sands.

b. *Mineralogical and mechanical composition:* mostly detrital quartz, sometimes with iron oxides, calcite etc. Sand $\gg$ silt, clay; stones virtually absent but iron oxide concentrations may occur. Texture sand, class SP but beach sands may pass into shingle; marine sands are usually better graded than beach sands.

c. *Characteristics:* no exchange capacity, plasticity or cohesion. Hydraulic conductivity generally very high although cemented layers may occur; water-table well defined if present. If interbedded with clays etc, may form aquifers and give rise to spring-lines. Topography low to moderately elevated, usually rising above adjacent clays.

4. *Estuarine and coastal muds and silts:*

   a. *Deposition:* in water moving slower than for paragraph 3, faster than for paragraph 2, saline or brackish. *Examples:* estuarine and deltaic marshes and tidal flats world-wide, e.g. The Wash, Dutch polders, mangrove swamps in Malaysia or tropical Africa.

   b. *Mineralogical and mechanical composition:* sand and silt fractions mainly detrital quartz with variable accessories such as muscovite, iron oxides, calcite; often black iron sulphides below surface. Clay minerals variable according to source area. Particle size very variable; silt and very fine sand usually dominant, coarse sand virtually absent. Textures typically silt loam, silty clay loam, fine sandy loam etc, classification ML, possibly CL, with variable grading.

   c. *Characteristics:* exchange capacity varies with amount and type of clay minerals, sodium usually dominant cation; on weathering clay tends to hydrolyse and disperse with rise in pH. (In this context hydrolysis is the partial replacement of Na^+ by H_3O^+ from water or carbonic acid. The effect is to produce a clay still dominated by Na^+ but in an alkaline medium containing sodium carbonate Na_2CO_3 and bicarbonate $NaHCO_3$). However, pH can fall to low values if much sulphide is present to oxidise to sulphuric acid (e.g. reclaimed mangrove swamp). Presence of calcite helps to prevent clay dispersing. Plasticity, cohesion and swelling variable according to clay content and state of dispersion. Hydraulic conductivity intermediate, may fall if clay disperses after reclamation. Topography nearly flat overall but may be much dissected by steep-sided channels.

5. *Alluvial deposits:*

   a. *Deposition:* from water no longer moving fast enough to keep the load in suspension. The main circumstances are:

   (1) From the overflow of modern perennial streams in their floodplains (detritus fine or coarse according to rate of flow).

   (2) From formerly-existing streams, evidenced by valley-side terraces or buried channels (coarse or fine).

   (3) From seasonal torrents as alluvial fans or wadi-fill (coarse or very coarse).

(4) From seasonal sheetwash across broad gently sloping pediments into playas or mbugas (fine).

Examples: fine-grained alluvium of major rivers such as the Nile, alluvial fans and wadi-fills of North Africa, Arabia etc, mbugas of East and Central Africa.

b. *Mineralogical and mechanical composition:* detrital quartz, feldspars, micas, iron oxides, calcite, clay minerals etc, very variable according to terrain being drained. Particle size depends on both source area and speed of water flow, texture ranging from silty clay to sand or gravel; classification and grading likewise variable.

c. *Characteristics:* exchange capacity, plasticity, cohesion and hydraulic conductivity depend mainly on local particle size and cannot be generalised. In flood-plains water-tables generally high and intercalated peats or peaty hollows common; on older higher terraces water-tables are low unless spring-lines present; in wadi bottoms water at very variable depths. Topography: flood-plains nearly flat but slope very gently up to level of stream; terraces roughly level, with slight slope towards sea at the time of their formation; alluvial fans steepening upwards; wadi-fill roughly level with swells and hollows.

6. *Lacustrine sediments:*

a. *Deposition:* in fresh or variably saline water but often with seasonal change in rate of sedimentation giving rise to varves. *Examples:* Triassic sediments of English Midlands, varved clays of Scandinavian Pleistocene lake beds, alluvial and pyroclastic sediments in lakes of the African Rift Valley.

b. *Mineralogical and mechanical composition:* sand grades mostly quartz and other detritals, sometimes organic or inorganic calcite present; gypsum in saline lakes and volcanic glass in active areas. Clay minerals also very variable according to source area; unusual clays such as palygorskite may form in very saline water. Particles mostly fine; textures silty clay, silty clay loam etc, often with large organic component; classification ML, CL, OL, MH, OH.

c. *Characteristics:* exchange capacity variable according to amount and type of clay minerals; pH high in soda-lake deposits; pore waters can be saline. Plasticity, cohesion and swelling variable but not usually very low; poor bearing capacity in locally moist areas. Topography nearly flat with very low amplitude micro-topography.

7. *Aeolian sands:*

a. *Deposition:* from sand grains moving by saltation across open surfaces with little vegetative cover. *Examples:* sand dunes behind beaches, major sand sheets and dune systems in desert areas, reworked glacial sands.

b. *Mineralogical and mechanical composition:* usually nearly all quartz with minor amounts of other detritals, occasionally calcite shell debris may dominate. Sand dominant, silt and clay virtually absent; texture sand, classification SP, with very poor grading.

c. *Characteristics:* exchange capacity negligible, no plasticity, swelling or cohesion; hydraulic conductivity very high. Water-table may be at any depth according to local circumstances but well defined when present. Topography: gently undulating sand sheets or dunes of varying steepness and amplitude; thin sheets may form a blanket over existing topography. Liable to renewed wind erosion if not stabilised by vegetation.

8. *Aeolian silts (loess):*

a. *Deposition:* from dust suspended in air, often brought down by rain and/or trapped by vegetation; much has been reworked by water. Some loess is being formed at the present day around the margins of deserts, but most has probably been derived from the deposits surrounding the ice sheets in glacial times. *Examples:* widespread surface deposit in N. America, NW Europe across USSR to China; thin deposits in southern England.

b. *Mineralogical and mechanical composition:* mostly quartz and other detritals, accessory iron oxides and calcite, clay minerals variable, often dominant. Silt and very fine sand dominant, clay low and coarse sand absent; texture silt, silt loam or very fine sandy loam; classification ML, CL, MH.

c. *Characteristics:* exchange capacity, plasticity and cohesion intermediate swelling very low. Hydraulic conductivity intermediate, drainage locally imperfect. Water-tables poorly defined. Topography: as a thin blanket over existing topography or, if thick, forming subdued undulating terrain; liable to wind or gully erosion in absence of vegetative cover; can sustain almost vertical faces in gullys or artificial cuttings; often subject to settlement.

9. *Glacial till (boulder clay):*

a. *Deposition:* lodgement till deposited as ground moraine below ice sheet; ablation till by 'letting down' of surface debris on melting of ice sheet; all forms typically unbedded but can contain lenses of sand or gravel. *Examples:* widespread in N. America, NW Europe, USSR, etc.

b. *Mineralogical and mechanical composition:* mainly finely comminuted rock from areas traversed by glacier or ice-sheet and thus very variable (e.g. on crystalline rocks in Scandinavia or New England it may consist of finely powdered but unweathered primary minerals, or 'rock flour', without clay minerals while in the English Midlands it is formed from Triassic debris in the west and from reworked marine clays in the east). Mechanical composition also depends on the source area ranging from stiff bouldery clay to gravels low in fines, hence texture and classification cannot be generalised.

c. *Characteristics:* also too variable to generalise. Topography: wide, gently undulating till-plains, cappings on interfluves, valley till complexes, or drumlin systems.

10. *Glacifluvial deposits:*

a. *Deposition:* from fast-flowing melt water inside or below a glacier (e.g. (eskers) or from its snout (end-moraine and outwash).

In contrast to till they show clear but complex bedding. *Examples:* in all areas carrying remains of former or present day glaciers or ice-sheets. May pass into lacustrine or normal alluvial deposits.

b. *Mineralogical and mechanical composition:* stones and sand reflect source area, e.g. granite or basalt debris, flint, etc. Stones and sand dominant, silt less common and clay virtually absent, complex or chaotic bedding and rapid changes in proportion of stones. Texture gravel or sand, classification GP, SP, individual beds poorly graded.

c. *Characteristics:* no exchange capacity, plasticity, cohesion or swelling; hydraulic conductivity very high, water-tables clearly defined if present. Topography: gently undulating on outwash plains, hummocky on end moraines or the ridges of kames and eskers.

11. *Periglacial deposits* (*'head'*):

a. *Deposition:* frost-heaving and solifluxion under freeze–thaw regimes; unbedded or with downhill 'streaking-out' rather than true bedding. *Examples:* in all areas with temperatures regularly fluctuating above and below zero now or in the past, particularly during the Pleistocene glacial periods.

b. *Mineralogical and mechanical composition:* entirely dependent on source material, ranging from coarse unweathered scree to well-weathered soil materials; head often resembles till in its range of particle sizes and lack of sorting; but 'streaking-out' more obviously related to topography, may be well or poorly graded.

c. *Characteristics:* too variable for general remarks on plasticity etc. Often associated with patterned ground and subsurface contortions with rapidly varying lithology, e.g. deep pockets of sand or gravel in clay or chalk. Topography: as thin cappings or absent on interfluves, thickening downslope with concave surface; may interdigitate in complex way with alluvium at valley margin.

12. *Pyroclastic deposits:*

a. *Deposition:* volcanic ash and cinders settling on land or in water bodies. *Examples:* in vicinity of most volcanoes, e.g. African Rift Valley, Sicily, Japan, S. America.

101

b. *Mineralogical and mechanical composition:* volcanic glass, sometimes with recognisable minerals, e.g. augite. Large fragments (cinders) or fine dust (ash) give textures initially gravel or sand to silt respectively, but volcanic material weathers quickly in humid climates to loam or clay; classification: ML to MH weathering to CH.

c. *Characteristics:* exchange capacity, plasticity, cohesion and swelling negligible until material weathered when all may greatly increase. Hydraulic conductivity likewise high or very high initially, tending to fall with weathering. Topography: coarser materials form undulating deposits with thickness increasing towards sources, ash may form thin blankets over existing topography to a considerable distance from the volcano and may merge with lacustrine or mbuga sediments or give rise to surface soils which have no relation to the subjacent rocks. Ash deposits very liable to wind or gulley erosion, and often highly dissected by torrents.

Pedological soils (major groups)

13. Table 10 represents a very brief summary of the main features of the more important soils occurring in the world. Since there is no full international agreement on classification, the soils have been placed in traditional major groups which will at least be recognised by all pedologists and others working with soils. Synonyms, near-equivalents, and related soils for which there is no space, are given to assist with the use of soil reports and memoirs using other systems. However, close correlations must not necessarily be assumed.

REFERENCE LIST—CHAPTER 5

BAVER L D, GARDENER W H and GARDENER W R, 1972 —*Soil Physics* (4th Edn). Wiley, New York.

*BRADY N C, 1974 —*Nature and properties of soils* (8th Edn). Macmillan, New York.

BS 812 Pt 1, 1975 —Methods of sampling and testing of mineral aggregates, sands and fillers. *British Standards Institution, London.*

BS 882, 1201 Pt 2, 1973 —Aggregates from natural sources for concrete. *British Standards Institution, London.*

BS 1377, 1974 —Methods of testing soils for civil engineering purposes. *British Standards Institution, London.*

CLARKE G R, 1971 —*The study of soil in the field.* Oxford University Press.

†FLINT R F, 1971 —*Glacial and quaternary geology.* Wiley, New York. 892 pp. (For Loess *see* pp. 251–266.)

GILLOTT J E, 1968 —*Clays in engineering geology.* Elsevier, Amsterdam.

*GRIM R E, 1962 —*Applied clay mineralogy.* McGraw Hill, New York.

*GRIM R E, 1968 —*Clay mineralogy.* McGraw Hill, New York.

*MITCHELL C W, 1973 —*Terrain evaluation.* Longmans, London.

†SCHULTZ C B and FRYE J C (Eds), 1968 —*Loess and related aeolian deposits of the world.* Univ Nebraska Press, 355 pp.

SKEMPTON A W and NORTHEY R D, 1952 —The sensitivity of clays. *Geotechnique* 3(i), 30–53.

*SOIL SURVEY, 1974 —*Soil Survey Field Handbook.* Technical Monograph 5, Rothamsted.

TERZAGHI K, 1955 —Influence of geological factors on the engineering properties of sediment. *Econ. Geol.* 50 (Anniv Vol 2) 557–668.

US DEPT OF AGRICULTURE 1951 —*Soil Survey Manual* (Agriculture Handbook 18). Department of Agriculture, Washington.

YONG R N and WARKENTIN B P, 1966 —*Introduction to soil behaviour.* Macmillan, New York.

*General references.
†References on Loess.

Section 5.3

TABLE 10. MAJOR GROUPS OF PEDOLOGICAL SOILS

Abbreviations

AlOx	Aluminium oxides and hydroxides (diaspore, gibbsite, etc.)	Ka	Kaolinite
		OM	Organic matter
Ch	Chlorite	P	Precipitation
CM	Clay mineral	p.m.	Parent material
E	Evapotranspiration	Sm	Smectite
FeOx	Iron oxides or hydroxides (goethite, haematite, etc.), amorphous oxides	Vm	Vermiculite
Il	Illite		

For horizon nomenclature in column (g) see Table 5

NOTE: Numbers are assigned to Major Groups for the clarity of the table but they have no other significance

Major groups	Synonym or approximate equivalent	Parent material	Climate and typical occurrence	Hydrology and leaching
(a)	(b)	(c)	(d)	(e)
I Immature soils	Raw soils, skeletal soils, lithosols, regosols, sols minéraux bruts, entisols.	Any. Especially recent deposits e.g. dune sands, volcanic ash, alluvium, colluvium.	Any. World-wide on freshly exposed materials or at sites of continued erosion.	Variable according to site. Run-off may be main cause of erosion and hence immaturity.
II Brown earths	(i) Brown forest soils, sols bruns acides, inceptisols, braunerde. (ii) Brown earths with textural B horizon, sols bruns lessivés, grey-brown podzolics, grey wooded soils, inceptisols.	Almost any, including decalcified calcareous rocks. Not favoured by very siliceous rocks e.g. quartzites, dune sands, in leaching climate.	P > E temperate, oceanic to subcontinental. N. America, W. Europe, USSR, China, N. Zealand, mountains in low latitudes.	Free drainage, moderate leaching, often seasonal. Carbonates removed.
III Podzols	Spodosols.	Especially siliceous rocks e.g. rhyolites, quartzites, sandstones, gravels.	P > E or ≫ E especially cool— N. America, N. Europe, USSR; in mountainous tracts generally, but also on suitable p.m. in humid tropics e.g. coastal sands in Malaya.	Free drainage, strong leaching.
IV Chernozems	Black earths, sols isohumiques, mollisols.	Especially base-rich unconsolidated sediments e.g. loess, alluvium.	P ∼ E continental, cold winters, hot summers. Ukraine, S.E. Europe, Mid-West USA and Canada, Argentine, Australia.	Free drainage. Leaching restricted due to climate.

Weathering	Profile and colour of surface soil	Vegetation or land-use and humus type	Remarks
(f)	(g)	(h)	(i)
Feeble or physical only (but p.m. may inherit weathering from earlier cycle).	C or (A)C Colour according to p.m.	None or pioneer plants e.g. lichens, marram, sand-sedge. Humus: negligible.	Shallow soils differing little from p.m. Development hindered by youthfulness of surface (recent p.m. or erosion), cold (arctic and montane soils) or drought (desert soils). *Rankers* are soils with feeble development but a little humus accumulation.
Moderate, variable depth. CM: 1:1 or 2:1 according to local factors. Moderate amounts of FeOx.	(i) A(B)C (ii) A (B) B_t C Shades of brown, paler in drier climates. (Soils may inherit red colours from p.m.).	Deciduous, mixed forest or ever-green originally, now often arable or grassland. Humus: mull.	Less strong leaching and/or more base-rich p.m. than podzols. Not normally very deep. Moderately acid to neutral. (ii) have lower clay-rich B_t horizons. Analogous, usually paler coloured, varieties in drier climates (*cinnamon soils*, etc).
Strong, variable depth. CM: especially Ka, Vm. Fe, Al mobile from A to B horizon.	Ideally $A_0A_hA_eB_hB_{fe}C$ but not all may be seen. Dark humic surface on pale or purplish grey A_e horizon.	Coniferous forest or heath plants (heather, bil-berry, etc). Humus: mor, often with thick litter.	Favoured by base-poor p.m. and/or strong leaching. Depth variable, usually greatest on sands. Acid to strongly acid, may pose cement stabilisation problems. B horizon may form thick cemented 'pans'. In wetter conditions may pass to *gley podzols or peaty podzols*.
Restricted, CM: Il, Ch, Sm. No free oxides; secondary calcite present.	A C_{ca} C Very dark brown to black.	Steppe grassland or cultivated. Humus: deep dark mull.	Deep soils with friable well-structured. A horizons and good trafficking properties but which may become powdery and liable to wind erosion after cultivation. Neutral reaction, with calcite concretions or flecks in C_{ca} horizon. With increasing moisture pass into *prairie* and *sub-arid brown soils*; with increasing dryness into *chestnut soils*, *sierozems* and *desert soils*. May occur in catenas with saline soils.

TABLE 10. MAJOR GROUPS OF PEDOLOGICAL SOILS (*continued*)

Abbreviations

AlOx	Aluminium oxides and hydroxides (diaspore, gibbsite, etc.)	Ka	Kaolinite
Ch	Chlorite	OM	Organic matter
CM	Clay mineral	P	Precipitation
E	Evapotranspiration	p.m.	Parent material
FeOx	Iron oxides or hydroxides (goethite, haematite, etc.), amorphous oxides	Sm	Smectite
		Vm	Vermiculite
Il	Illite		

For horizon nomenclature in column (g) see Table 5

NOTE: Numbers are assigned to Major Groups for the clarity of the table but they have no other significance

Major groups	Synonym or approximate equivalent	Parent material	Climate and typical occurrence	Hydrology and leaching
(a)	(b)	(c)	(d)	(e)
V Gleys	Hydromorphic soils, meadow soils, aquepts, aquents: (i) Groundwater gleys. (ii) Surface-water gleys.	Any but: (i) essentially permeable. (ii) essentially impermeable, especially clays.	(i) Dependent on groundwater only. (ii) Normally $P > E$ Worldwide.	Drainage and leaching impeded by: (i) high water-table. (ii) substratum.
VI Organic soils	Peats, bog soils, sols tourbeaux, histosols: (i) Topographic or basin peats. (ii) Climatic peats (hill, raised-bog or blanket peats).	Plant remains: (i) often includes alluvium and/or colluvium. (ii) pure organic debris.	(i) Any except warm dry, but favoured by cool moist. (ii) $P \gg E$ temperate to cold (high latitudes, or high altitudes in any latitude).	(i) Permanent water-logging by groundwater. (ii) Continually moist surface due to high rainfall, low evaporation.
VII Halo-morphic soils	(i) Saline, white alkali, szik, sodic soils, solonchaks. (ii) Alkali, black alkali, non-saline sodic soils, solonetz. (iii) Degraded alkali soils, solods (also written soloths, soloti).	Especially unconsolidated basin sediments (alluvium, lake deposits, volcanic ash, fresh marine sediments) W. USA, S.E. Europe, Central Asia, Middle East, Pakistan, S. Africa, Australia, E. China.	$P < E$ temperate continental to hot. With increased leaching, can change in order: (i) →(ii) →(iii).	(i) High water-table with upward movement of salts. (ii), (iii) More leaching than in (i) due to higher site, climatic change or irrigation.

Weathering	Profile and colour of surface soil	Vegetation or land-use and humus type	Remarks
(f)	(g)	(h)	(i)
Restricted, generally less than free-drained members of the same catena and 2:1 CM more abundant.	AC_gC or $A(B)_gC$ Brown or grey-brown surface on grey or mottled gley horizon; p.m. may show no signs of gleying.	Moisture-tolerant forest or grass-land or semi-marsh, often modified by agricultural drainage. Humus: usually mull but passing into peaty mor in gley podzols, or into peat in peaty podzols.	(i) In low-lying sections of many catenas, shear strength and trafficking properties depend on p.m. and on height of water-table. Seams of peat or running sands may be present. pH often higher than in other soils of the catena. Reducing, corrosive conditions likely in gleyed horizon. Similar soils with sulphides in coastal marshes (polders, mangrove swamps) which may become intensely acid on drainage. Pass into *peaty gleys* in very wet conditions. (ii) Mainly in low-lying sites but also on small elevations and hilltop sites on very impermeable clays. Trafficking good to extremely poor according to season; may shrink and swell strongly.
Not applicable.	No pedological profile; layers represent succession of vegetation. After reclamation may become $A_o\,B_o\,C_o$ Brown or black.	(i) Aquatic or semi-terrestrial plants. OM often ∼ 80%. (ii) Bog plants adapted to moist acid conditions OM may be >99%. Humus: peat.	(i) Occupy very low water-logged sites. Very variable depths of organic debris usually over gleyed substratum. Well to poorly humified according to nature of groundwater. Neutral to strongly acid. Swampy in natural state; shrink strongly on drying but still low shear strength and poor trafficking. May be powdery and erodible when dry. (ii) Similar to (i) in many respects but generally more fibrous and acid.
Restricted in present cycle, usually 2:1 CM, little FeOx. Silica mobile in (ii).	(i) A_{sa} or A_{psa} C_{sa} (ii) $A(B)\,B_t\,C$ (iii) $A\,B_t\,C$ C_{sa} horizons may be gleyed. Salts occur as surface crusts or flecks or in solution below the surface.	Sparse vegetation tolerant of salt (i) or alkaline conditions (ii). Humus: mull, very low.	(i) Generally occupy low sites where salts may accumulate in dry conditions due to saline p.m. (e.g. old lake beds), sea spray, rising groundwater etc. Salt crust may be hard and trafficking good until it is broken. May be damp patches of low bearing capacity between hard surfaces. Neutral reaction but salts may be highly corrosive. (ii) Develop from (i) by removal of salts and hydrolysis of colloids; become strongly alkaline with deflocculated clay, hard when dry, slimy when wet. May be dense clay-rich B_t horizon. (iii) Similar to (ii) but with acid surface.

107

TABLE 10. MAJOR GROUPS OF PEDOLOGICAL SOILS (continued)

Abbreviations

AlOx	Aluminium oxides and hydroxides (diaspore, gibbsite, etc.)	Ka	Kaolinite
Ch	Chlorite	OM	Organic matter
CM	Clay mineral	P	Precipitation
E	Evapotranspiration	p.m.	Parent material
FeOx	Iron oxides or hydroxides (goethite, haematite, etc.), amorphous oxides	Sm	Smectite
		Vm	Vermiculite
Il	Illite		

For horizon nomenclature in column (g) see Table 5

NOTE: Numbers are assigned to Major Groups for the clarity of the table but they have no other significance

Major groups	Synonym or approximate equivalent	Parent material	Climate and typical occurrence	Hydrology and leaching
(a)	(b)	(c)	(d)	(e)
VIII	Ferruginous tropical soils	Tropical red and yellow loams, latosols, alfisols, ultisols.	Any but tend to be replaced by ferrallites on very basic rocks.	$P > E$ tropical or subtropical monsoon. S. America, Africa, E. Asia, Australia.	Free drainage, strong leaching but may be seasonal desiccation at surface and/or water-logging at depth.
IX	Ferrallites	Red earths, krasnozems, lateritic soils, latosols, oxisols, ultisols.	Any, especially basic rocks.	$P \gg E$ equatorial; soils sometimes relict from earlier hot wet climates. S. America, Africa, S. Asia, Australia.	Free drainage, intense leaching at surface, waterlogging often almost permanent at depth.
X	Black tropical soils	Margalitic soils, black cotton soils, regurs, 'mbuga', 'vlei' or or 'dambo' soils, vertisols.	Basic igneous rocks especially basalt, tuffs; fine-grained basin sediments.	$P > E$ tropical monsoon or equatorial. S. America, Africa, S. Asia, Australia.	Restricted drainage at surface, impermeable at depth. In wet season surface flooded and whole profile impermeable.
XI	Calcareous soils	Calcimorphic soils, rendolls: (i) Rendzinas. (ii) Red and brown calcareous soils.	Limestones, dolostones, unconsolidated calcareous sediments e.g. tills, loess, shell sands, alluvium.	Any. Worldwide.	Free drainage or slightly impeded; leaching dependent on climate.

Weathering	Profile and colour of surface soil	Vegetation or land-use and humus type	Remarks
(f)	*(g)*	*(h)*	*(i)*
Strong and deep, CM: Ka; Il, Vm subordinate or absent, Sm absent. Allophane common, FeOx abundant but not AlOx.	A(B)C or A(B) B_t C There may be gleying at depth. Red to deep yellow.	Tree or bush savanna, sometimes forest, often under cultivation for which they are better suited than ferrallites. Humus: pale-coloured mull.	A and (B) horizons deeper and more ferruginous than in soils in temperate regions, but weathering less than in ferrallites. Many examples are relict from earlier climates and have frequently been truncated by erosion. B_t horizon may be well developed and surface pallid from loss of FeOx. Red Mediterranean Soils are sub-tropical variants.
Very strong and very deep. CM: Ka; 2:1 absent. Abundant FeOx and AlOx.	A B C. Red.	Forest but relict soils may be under thorn or savanna. Shifting or settled cultivation but soils inherently very poor. Humus: mull.	On mantles of very deeply weathered p.m. on old surfaces, often relict from past climates. High clay, low silt but matrix is often very strongly aggregated by FeOx and AlOx into 'pseudo-sand' or gravelly concretions ('murram'). May contain incipient laterite, with bleached zone at depth, or may be associated with exposed laterite sheets.
Restricted. CM: Sm and Il. No free FeOx or AlOx.	AC or A(B) C or A(B)$_{sa}$ C. Black.	Grassland or specialised savanna. Humus: mull.	Occur regularly in catenas with ferruginous tropical soils in depressions ('mbugas', 'dambos', 'vleis'). Deep uniform black A horizon with very high clay content often mostly Sm; strong swelling and shrinkage, offering very unfavourable foundations and very poor trafficking when wet. May develop strong micro-relief ('gilgai'). Neutral reaction; may be strongly calcareous and/or saline at depth.
(i) Weak (mainly solution of calcite). (ii) Moderate, some FeOx formed. CM: initially as inherited from limestone, thereafter may alter according to degree of weathering.	(i) AC. (ii) A(B) C or A(B) B_t C. (i) Grey to black. (ii) Brown to red.	Most vegetation types depending on climate. Humus: mull.	(i) Very shallow soils differing only from immature soils by greater solution of calcite and accumulation of humus. (ii) Deeper soils tending to approach the zonal soil of the region (brown earths, ferruginous tropical soils etc) and often similar to them in profile characters but always with free clacium carbonate. Red Mediterranean variants often termed 'terra rossa'.

CHAPTER 6

GEOPHYSICAL METHODS

SECTION 6.1. INTRODUCTION

1. Geophysics is the study of the physical processes and properties of the Earth. Applied Geophysics in particular is concerned with the determination of the nature and shape of unseen rock bodies by measurement of physical characteristics at or near the Earth's surface.

Since geophysical methods are by their nature indirect, it follows that interpretation of results entails some uncertainty; therefore engineers should not expect geophysical predictions to be exact. However geophysical techniques are invaluable in feasibility studies and site investigation work, over both large and small areas. In particular geophysics provides partial or continuous data which can be quickly integrated with a knowledge of the geology of an area provided by surface or borehole data, often at a smaller cost and effort than that of sinking additional boreholes and when so integrated the data can be interpreted with greater confidence.

2. Geophysical data are often complex, and interpretation difficult; it is therefore important that frequent consultation takes place between engineer, geologist and geophysicist, so that all are aware of the engineer's requirements on the one hand, and of the limitations of the methods and results on the other. Two general types of information can be obtained from geophysical measurements:

a. *In situ* values of the physical properties of the rock mass (e.g. seismic velocity, resistivity value) which can be correlated with other parameters or characteristics of more immediate engineering significance (e.g. porosity, degree of rock fracture).

b. Delineation of structural features, such as boundaries between different lithological types, faults, depths to bedrock, etc.

In some instances known boundaries might be traced between boreholes or outcrops; in first stage reconnaissance surveys unknown boundaries might be indicated and subsequently identified in boreholes.

3. The most useful geophysical techniques for engineering application are seismic refraction, resistivity, borehole logging and borehole-to-borehole methods, and marine continuous profiling reflection techniques: gravity,

110

magnetic and electromagnetic methods are also sometimes used. Some examples of situations in which geophysics would probably be used are:

a. Determination of a bedrock profile beneath cover of unconsolidated material (e.g. alluvium), using seismic and resistivity techniques, along a highway route or beneath a dam site.

b. Detection of faults and zones of weathered or fractured rock beneath cover of drift, using seismic, resistivity, and possibly electromagnetic techniques, along tunnel lines.

c. Determination of geological structure beneath the sea bed, or at a proposed estuary crossing, using continuous seismic reflection profiling to outline thicknesses of deposits and the positions and attitudes of faults.

d. Detection and approximate outlining of near-surface cavities, filled sink holes etc., using resistivity, occasionally gravity, and, for deeper cavities, borehole-to-borehole acoustic velocity measurements.

e. Mapping of groundwater salinity variations using resistivity measurements after 'calibration' of aquifer resistivity against porewater salinity in selected boreholes.

f. The outlining of large buried channels and sedimentary basins in regional hydrogeological studies, using regional gravity and seismic refraction or reflection measurements.

As a general guide accuracy greater than ± 10 per cent cannot be expected for depth determination from seismic and resistivity surveys; the reliability depends on the uniformity of the ground, and the degree of borehole control. In shallow investigations of less than 10 metres depth the accuracy may be less.

TABLE 11. **RELATIONSHIP OF ROCK TYPES TO RESISTIVITY AND SEISMIC VELOCITY (after PATERSON and MEIDAV 1965)**

Increasing seismic velocity ↑			
'Basic' igneous rocks		Unweathered limestones, metamorphic rocks, and 'acid' igneous rocks	
Calcareous and indurated shales			
Shales and consolidated clays		Sandstones, consolidated gravel, weathered rock	
Clay and clayey fill	Sand and bouldery fill		Loose boulders

Increased resistivity →

111

4. There can be definite advantages in the integration of different techniques. In the case of surface seismic and resistivity surveys, discrepancies between the interpretations may indicate the presence of a 'blind zone' or hidden layer and show where drilling should be carried out. Combinations of techniques are also used to obtain the best estimates of rock properties. For example, a combination of neutron, sonic and formation density logs gives the best value of porosity, and most closely identifies the lithology in boreholes (SCHLUMBERGER 1966). Combined resistivity and seismic velocity measurements on the surface have been used to estimate bulk strength and rock type in the field (Table 11). The table shows general relationship of rock types and their consolidation states to various resistivity and seismic velocity combinations. The relative positions of the boundaries in the table may shift in different areas. Furthermore in an ideal investigation there might be several alternating stages involving geophysical methods, borehole investigations and combined interpretation and re-interpretation as the knowledge of ground conditions becomes more refined.

SECTION 6.2. PHYSICAL PROPERTIES OF ROCKS AND SOILS

Rock resistivities

1. *Definition:* If a conductor carries a current with parallel lines of flow over a cross-sectional area A square metres, then the resistance R ohms between two equipotential surfaces L metres apart is given by

$$R = \rho L/A \qquad \text{Eqn 6(1)}$$

where ρ is the resistivity of the material in ohm-metres.

2. A wide range of resistivities is exhibited by rocks and minerals from highly resistive igneous rocks ($>10^5$ ohm m) to very conductive sulphide deposits (<0.01 ohm m). A comprehensive list of values is given by SLICHTER and TELKES (1942). The range for some common rocks and soils likely to be encountered in engineering projects is shown in Figure 35. The most highly conductive materials such as graphite and sulphide deposits are representative of electronic conductors in which electricity is conducted by the flow of electrons as in metallic conductors. In the great majority of rocks, however, electricity is conducted by ions in the pore water which acts as an electrolyte, and the mineral constituents of the rock normally make a negligible contribution to the conductivity of the rock except where certain clay minerals are present. The resistivity of the rock is thus mainly determined by the resistivity of the pore water (ρ_w) the porosity (φ) and degree of saturation (S). In sedimentary rocks, a relationship of the following type can be valid provided that clay minerals are not present;

$$\rho = a\rho_w\varphi^{-m}S^{-n} \quad \text{(KELLER 1966)} \qquad \text{Eqn 6(2)}$$

where ρ_w = resistivity of the pore water, φ = fractional pore volume, S = fraction of pore volume filled with water, a, m and n are arbitrary parameters.

112

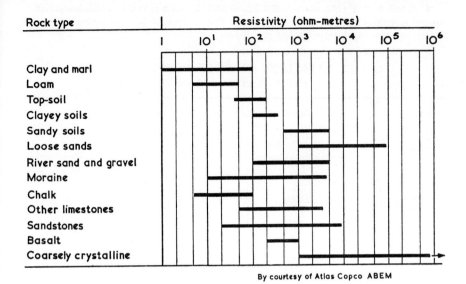

By courtesy of Atlas Copco ABEM

Fig 35.   Resistivity values for natural materials

m ranges from 1·0 for unconsolidated sand to 2·5 for well cemented granular rocks. n is approximately 2·0 for water saturation greater than 30 per cent of pore space.

3. Where clay minerals such as montmorillonite or vermiculite occur, e.g. as minor constituents in sands, the effective pore water resistivity is greatly reduced by ion exchange between the pore water and the clay minerals. As there is no practicable method of quantitatively determining the effect in the field, Equation 6(2) is of little practical use in these circumstances. In areas where conductive clay minerals are not present, a form of Equation 6(1) can sometimes be applied to obtain approximate values for groundwater salinity (proportional to $1/\rho_w$) when porosity is known and can be assumed to be reasonably constant in the area (*see* example from Holland, Section 6.4, paragraph 2). For saturated rock below the water-table, $S = 1$ and Equation 6(2) simplifies to:

$$\rho = a\rho_w \varphi^{-m} \qquad \text{Eqn 6(3)}$$

Assuming fairly constant values of φ, m and a in the area, one can theoretically map variations in ρ_w from values of ρ obtained in surface resistivity measurements, provided that preliminary measurements have established values for φ, m and a.

4. Some typical rock resistivities based on Equation 6(3) for particular ground-water salinities are given in Table 12. The values of ρ_w are taken from the Schlumberger log interpretation handbook.

TABLE 12. VARIATION OF ROCK RESISTIVITY WITH
PORE-WATER SALINITY

Water salinity ppm of NaCl	ρ_w ohm m (24°C)	Values of ρ for water saturated rock of 20% porosity		
		Sand(1) $\rho=3\rho_w\varphi^{-1}$ ohm m	Sandstone(2) $\rho=1\cdot5\rho_w\varphi^{-1\cdot7}$ ohm m	Limestone(2) $\rho=2\cdot2\rho_w\varphi^{-1\cdot7}$ ohm m
(a)	(b)	(c)	(d)	(e)
100	48·0	720	1104	1632
500	10·0	150	230	340
1000	5·0	75	115	170
2000	2·7	40·5	61	92
4000	1·2	18	27·6	40·0

(1) BREUSSE and DUPRAT (1964).
(2) KELLER (1966).

5. Alternatively, in areas where groundwater salinity is reasonably constant, variations in porosity or 'effective porosity' due to rock fracture can be mapped with the resistivity method. The method is particularly sensitive to weathering in igneous and metamorphic rocks by virtue of the contrast between the highly resistive fresh rock and the low resistivity clay minerals produced in the weathering process.

Seismic velocities

6. Seismic waves generated artificially by explosive or other sources are of three main types: longitudinal or compressional (P) waves in which the direction of particle motion is the same as the direction of wave propagation, transverse or shear (S) waves in which the direction of particle motion is perpendicular to the direction of wave propagation, and surface waves of which the most common is the Rayleigh wave travelling along the ground surface and causing particles in its path to follow an elliptical motion in a vertical plane. The velocities of propagation of these waves are denoted V_P, V_S and V_R respectively and $V_P > V_S > V_R$. For an ideal isotropic material, V_P and V_S depend on Young's modulus (E), bulk density (γ) and Poisson's Ratio (ν) according to the following relationship:

$$V_P = \sqrt{\frac{E}{\gamma} \frac{1-\nu}{(1-2\nu)(1+\nu)}} \qquad \text{Eqn 6(4)}$$

$$V_S = \sqrt{\frac{E}{\gamma} \frac{1}{2(1+\nu)}} \qquad \text{Eqn 6(5)}$$

7. In theory, therefore, from determinations of V_P, V_S and γ it is possible to calculate E and v for an ideal, unfractured, homogeneous and isotropic rock and this forms the basis of the 3–D seismic borehole logging device. In actual situations in rock, this ideal state is not normally encountered so that values of Young's modulus so determined will differ considerably from the deformation modulus determined statically by plate-loading or other similar tests. However, it is sometimes possible to have a direct relationship between dynamic E and static E so that the seismic method can be used to map variations in deformation modulus, at a dam site for example, once this 'calibration' has been carried out (KNILL and PRICE 1972; STAGG and ZIENKIEWICZ 1968).

8. The values of V_P as measured in the field can be used in an empirical way by correlation against rock fracture spacing or Deere's R Q D (Rock Quality Designation) (*see* Chapter 9). If laboratory velocity values for intact rock cores are determined (V_{lab}), the ratio $\dfrac{V_P}{V_{lab}}$ termed 'fracture index' by KNILL (1970) gives an empirical assessment of degree of rock fracture. The recording of shear wave velocities in field and laboratory entails difficulties of identification and measurement because the slower travelling shear waves appear as later arrivals on the field seismic record. Nevertheless, because shear waves are not transmitted by water, V_S is a more sensitive indicator of fracturing beneath the water-table. Examples are given later of the use of velocities in rock quality assessment.

9. The theory of propagation of seismic waves in sediments is complicated by the fact that one is dealing with a multi-phase material which does not necessarily behave elastically and in which there is complex interaction between particles which make up the sediment. A useful working equation is that due to WYLIE et al (1956) in which the compressional wave velocity V_P is related to the sediment porosity (φ) for a normally consolidated sediment:

$$\frac{1}{V_P} = \frac{\varphi}{V_P \text{ (fluid)}} + \frac{1 - \varphi}{V_P \text{ (lab)}} \qquad \text{Eqn 6(6)}$$

where V_P (fluid) is the velocity in the pore fluid (water for saturated sediment, air for a dry sediment) and V_P (lab) is the compressional wave in intact rock as measured in the laboratory. An example of the use of this equation in relation to fracturing in chalk is given in Section 6.4, paragraph 3. Examples of field seismic wave velocities (V_P) are shown in Figure 36.

Geophysical logging relationships

10. Similar relationships apply to the results of geophysical logs obtained with acoustic or sonic and resistivity tools in boreholes. In addition, special radioactive tools which are in standard use in the oil industry give promise of useful adaptation to engineering requirements. In particular, the natural gamma

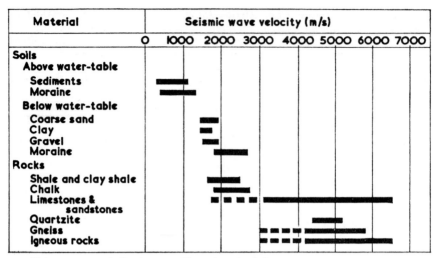

Material	Seismic wave velocity (m/s)
	O 1000 2000 3000 4000 5000 6000 7000

Note: ■ ■ ■ ■ indicates weathered material

Fig 36.   Seismic wave velocities (after ABEM 1972)

log (correlating with clay content), the gamma-gamma and neutron probes which give values of rock density and moisture content, together with the cross-plots between these and the sonic and resistivity log results enable both porosity values and an estimate of lithology to be made (*see* Section 6.3, paragraph 28).

Rock properties from other techniques

11. Gravitational and magnetic methods depend on density and magnetisation of rocks respectively. Because of the inherent ambiguity of interpretation caused by the nature of the potential fields, their use is confined to studies of geological structure on a large scale. Electromagnetic techniques rely principally on variation in ground resistivity and are particularly used in the search for conductive mineral deposits. They can be used in suitable circumstances for the location of faults which behave as narrow steeply dipping conductive sheets.

SECTION 6.3. **METHODS, TECHNIQUES AND INTERPRETATION**

Electrical resistivity

1. *Methods of field measurement* (*see* KUNETZ 1966, and GRIFFITHS and KING 1965).

To measure the electrical resistivity of the ground, current from a generator or battery is fed into the ground surface via two current electrodes C_1 and C_2 and the potential difference between two other, potential electrodes P_1 and P_2,

116

in the vicinity is measured. From the known current input, the measured potential difference and the distances between the various electrodes, a value for the 'apparent resistivity', ρα is obtained. ρα is the resistivity that a completely homogeneous half space of infinite extent would have to have in order to produce the measured values of potential difference ΔV, and the current I for the electrode configuration used. At a distance r from a current electrode set in a homogeneous

half space of resistivity ρα, the potential is $V = \dfrac{I\rho\alpha}{2\pi r}$. By the principle of super-

position, one can calculate the value of ΔV between P_1 and P_2 due to current I entering at C_1 and leaving from C_2 (Figure 37).

$$\Delta V = \frac{I\rho\alpha}{2\pi}\left[\frac{1}{C_1P_1} - \frac{1}{C_1P_2} - \frac{1}{C_2P_1} + \frac{1}{C_2P_2}\right] \qquad \text{Eqn 6(7)}$$

Hence ρα is obtained. In practice ρα includes contributions from all the variations in resistivity in the neighbourhood of the electrodes.

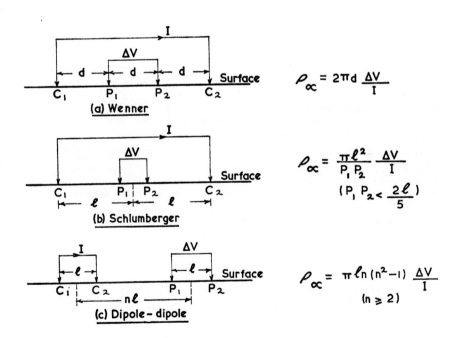

Fig 37.   Resistivity electrode configurations

117

2. Lateral variations can be explored by traversing a configuration of fixed geometry, i.e. the constant separation traverse. Variations in depth are investigated by expanding the separation between the electrodes while keeping the reference point of the system stationary, i.e. the expanding electrode measurement. Although constant separation traverses are occasionally useful (for example, to outline channels of highly resistive gravels contained in more conductive alluvium), they are difficult to interpret quantitatively. The expanding electrode technique can give useful quantitative data where nearly horizontal strata are concerned. An illustration of the effects of vertical discontinuities at different depths on constant separation traverses with various values of current electrode separation (C_1C_2) is given in Figure 38.

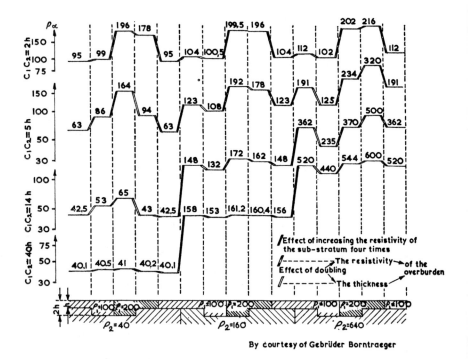

By courtesy of Gebrüder Borntraeger

Fig 38. Variation of the apparent resistivity due to vertical contacts in the sub-stratum and the overburden for different electrode separations (from KUNETZ 1966)

In expanding electrode surveys variations in the resistivity of the surface layer can change the measured values of $\rho\alpha$ and might be confused with resistivity change at depth. If suspected, these lateral variations can often be detected and sometimes eliminated from field curves by carrying out two or three measurements at the same point with different orientations of the line of electrodes. Lateral variations should be suspected if the recorded ρα against d curves are not smooth.

3. Three electrode configurations are illustrated in Figure 37. Of these, the Wenner configuration is the most commonly used by British geophysicists. It has the advantages of ease of operation from a central point, and that the potential difference measured is a maximum for a given current and that more theoretical curves are available for interpretation of expanding electrode measurements. The Schlumberger configuration is widely used in Europe; it is easier and quicker to operate in the field, but has disadvantages in the plotting of expanding electrode results. The dipole–dipole configuration is now widely used in induced polarisation surveys in which measurements of change of apparent resistivity with frequency are made. In these it is essential to separate the potential wires from the current wires. The Wenner configuration will be assumed hereafter in which the electrodes are equally spaced in a straight line, distance d apart, and the apparent resistivity, derived from Equation 6(7) is:

$$\rho\alpha = 2\pi d \Delta V/I \qquad\qquad \text{Eqn 6(8)}$$

With a constant separation traverse (d fixed), variations in ρα over an area of interest can be mapped and give information on vertical features such as faults and on variations in depth to a particular bed.

4. By increasing d, one obtains greater effective penetration into the ground and a plot of ρα against d can be interpreted in terms of the resistivities and thicknesses of the (assumed) horizontal layers beneath the point of measurement. The actual depth of penetration also depends on the resistivities of the various layers and, particularly where relatively conductive layers are encountered, on the frequency of the applied current. It should never be assumed that the depth of exploration equals the separation d. On the contrary, theoretical curves illustrate the point that in order to explore to a depth h, one normally has to expand to a value of d equal to at least two or three times the value of h (*see* MOONEY and WETZEL 1956).

5. *Practical aspects of measurement.* Either AC or DC equipment can be used for both expanding electrode and constant separation measurements. The theoretical principles are strictly applicable only in the case of direct current. However, in most circumstances the errors introduced by the use of AC are either negligible or can be avoided, and there are practical advantages in its use. The main ones are the elimination of polarisation at the potential electrodes, and of natural earth potentials, which can only be avoided in DC measurements by the use of special non-polarising electrodes, and of additional circuitry to balance out natural earth potentials. AC measurements suffer from severe limitations in highly conductive ground. There are two main effects; 'coupling' between

current and potential wires increases as the resistivity of the surface layer decreases, leading to error in the measured potential differences.

Also the use of AC over a highly conductive stratum is often impossible because of 'skin-effect' in which current flow becomes concentrated at the surface of the conductor, thereby increasing its effective resistance. This effect becomes more marked at bigger separations so that where great depth of exploration is required in strata of very low resistivity (of the order of 1 ohm m), DC measurements become essential. Ideally and under most operating conditions, a low frequency (3–4 Hz) square wave or 'chopped DC' is best as this overcomes most of the problems associated with both AC and DC measurements.

6. There is a variety of instruments on the market. One instrument employs a battery-powered oscillator to give low power current of 110 Hz. Resistance is measured directly. Its low power and relatively high frequency mean that its maximum separation in the Wenner configuration is limited in most practical cases to little more than 100 metres. Another instrument, readily available in Britain, is much slower to operate but has a lower operating frequency (4 Hz) and measures ranges of lower resistance, which enable it to work with larger separations. Because of its very high input impedance, it is also able to operate where high electrode 'stake' resistances are encountered. The effect of these contact resistances between the electrodes and the ground is to limit the amount of current which enters the ground and hence makes the potential differences between the potential electrodes very small and very difficult to measure. Stake resistances can be considerably reduced by the liberal application of salt water and by the use of multiple connected stakes or aluminium foil which greatly increases the area of contact but is time consuming in preparation. Since field curves are plotted logarithmically, suitable increments in electrodes spacing should be chosen so that approximately equal distances separate the plotted points on the graphs.

Interpretation of field results

7. The only sound basis for the quantitative interpretation of expanding electrode measurements is the use of appropriate theoretical curves. Sets of theoretical curves have been prepared for a number of cases involving 2, 3 and 4 layers of varying resistivities for both Wenner and Schlumberger configurations (MOONEY and WETZEL (1956), MOONEY et al (1966), ORELLANA and MOONEY (1966), and VAN NOSTRAND and COOK (1966)). Comparison between theoretical and field curves involves the direct superposition of one on the other, both being plotted on double logarithmic paper. Theoretical curves are

plotted for ratios $\dfrac{\rho a}{\rho_1}$ where ρ_1 is the resistivity of the uppermost layer against

ratios d/h (Wenner) or l/2h (Schlumberger), where h is the depth to the lowest

layer. Field curves are plots of the apparent resistivity (ρa) against d $\left(\text{or } \dfrac{1}{2}\right)$.

The field curve is superimposed on the theoretical curve which most closely matches it. For a perfect match the curves are identical and the origin of the theoretical curve corresponds to the point x = log h, y = log ρ_1 on the field curve. Hence, one reads off ρ_1 and h directly and the resistivities and depths of the other layers follow from the ratios of resistivity and depth assigned to the theoretical curves. Great care is needed in the choice of appropriate theoretical curve and frequently more than one curve can be fitted to a particular field curve. In such cases, the choice may have to be decided by the most reasonable interpretation in the light of other information (Figure 39).

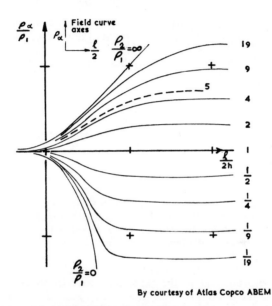

By courtesy of Atlas Copco ABEM

**Note: Axes shown refer to theoretical curves
Scale of field curve is not shown as this is normally plotted
on logarithmic paper**

Fig 39.   Two-layer master curves (partly after Compagnie Générale de Géophysique) For curves in the upper half $\rho_2 > \rho_1$ and for curves in the lower half $\rho_2 < \rho_1$. A measured resistivity curve is shown in coincidence position.

8. *Sources of error and limitations in resistivity methods.* Because of the nature of the potential field measured at the ground surface, ambiguities in the physical interpretation of resistivity data can arise. In some instances, different theoretical curves can be found to fit the same field curve, and the alternative solution may represent quite different sequences of resistivity layers. The remedy lies in obtaining a large number of field curves from the area in question and in having one or

more 'calibration' boreholes. It is often possible to establish the normal resistivity sequence at sites where interpretations are unique and then to discard abnormal interpretations at ambiguous sites. Another type of ambiguity occurs where the lowest layer is highly conductive, and the effect of a fairly thin layer of intermediate resistivity is effectively masked. For example a four-layer system with a thin third layer whose resistivities are of ratios $1 : \frac{1}{3} : \frac{1}{5} : \frac{1}{100}$ would be indistinguishable from another of ratios $1 : \frac{1}{3} : \frac{1}{20} : \frac{1}{100}$. Errors can also arise in correlating resistivity layers with lithological layers. This can commonly occur above the water-table where changes in water saturation may cause resistivity changes as large as those caused by lithological changes (Equation 6(1)). This type of error is best avoided by resistivity borehole logging in a 'calibration' borehole.

Seismic refraction surveys

9. The most useful information from seismic surveys conducted at the ground surface on the disposition of strata down to shallow depths is likely to be obtained from refraction shooting with multi-geophone spreads. A prerequisite for the successful application of the refraction technique is that the sediments and rocks beneath the survey area exhibit an increase in velocity with depth.

10. An energy source, normally consisting of a sledgehammer, falling weight or an explosion, is used to generate sound waves which are recorded at each of the geophones after travelling by various paths through the ground. In most surveys designed to elucidate geological structures, for depths of 100 metres or less, only the first arrival on each trace of the recording paper is timed. To illustrate the transmission of sound in layered media, it is helpful to consider the ray paths which are perpendicular to the expanding wave fronts generated by an explosive source on the ground surface. In an actual refraction survey, observations are made of the first arrivals at a series of receiving geophone positions D_1, D_2, D_3, etc (Figure 40). For short distances between shot point and detector, the direct wave travelling with velocity V_0 in the upper medium arrives first. As the distance is increased, the refracted wave travelling immediately beneath the interface of the second medium with velocity V_1 will 'catch up' with the direct wave and arrive first.

These arrivals are caused by the headwave, which is generated at the interface by the critically refracted wave, such that $\sin i_c = \frac{V_0}{V_1}$. Plotting of the results as arrival time against distance from shotpoint permits the construction of a travel-time curve consisting of the straight line segments or a curve passing through points T_{D1}, T_{D2}, T_{D3}. (Figure 40).

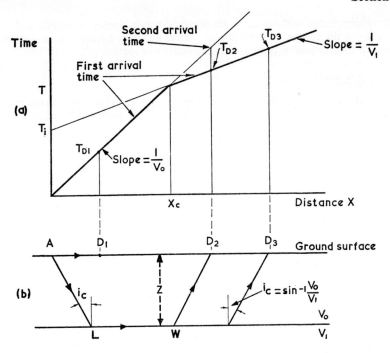

Fig 40.   Theoretical ray paths of least time and time-distance curve for two layers separated by horizontal interface

Practical aspects of refraction surveys

11. A variety of seismic survey instruments is commercially available, ranging from simple hammer source instruments with single geophone and single channel recorder, which only records interval time, to multi-channel (12 or 24 geophones) recorders with full record traces. Multi-channel instruments can be used with any type of source, particularly explosive, which is not economic for single channel recording. Falling weight sources are being used increasingly for shallow surveys, producing greater energy input and a better controlled impact than the hammer, while avoiding the restrictions applied to the use of explosives. For a single channel recorder the traverse is done twice, once in each direction.

12. In multi-channel recording, geophones are laid out in a straight line across the area of interest. Common intervals are 2–5 metres for more distant geophones, depending on the detail required, the depths to layers of interest and the thickness of any weathered or dry surface layers. For any given geophone spread, shot points will normally be sited at intervals along the spread, such as the mid-point, at each end and at a distance off each end. This procedure enables reasonable interpretation to be made in situations of dipping or undulating horizons.

123

13. Auger holes will normally be required for explosive to enable sufficient tamping for the charge. In drift deposits shots at about 1 metre depth consisting of 150 to 1500 grams of gelignite are normally adequate.

Interpretation of refraction surveys

14. For a single horizontal interface, an explosion at a shot point on the ground surface will give rise to many ray paths which will be partly reflected and partly refracted at the interface, such that angle of incidence (i_1) equals angle of reflection, and angle of refraction (i_2) obeys Snell's law of refraction:

$$\frac{\sin i_1}{V_0} = \frac{\sin i_2}{V_1} \qquad \text{Eqn 6(9)}$$

15. The velocity V_0 of the upper medium is given by the slope of the first portion of the curve; V_1 of the lower medium by the second portion (*see* Figure 40(a)). Interpretation of the depth to the refracting layer can be carried out from either the 'intercept time' T_i, or the critical distance X_c according to the following equations:

$$\text{Thickness of upper layer} \quad Z_0 = \frac{T_i}{2} \sqrt{\frac{V_1 V_0}{V_1^2 - V_0^2}} \qquad \text{Eqn 6(10)}$$

$$Z_0 = \frac{X_c}{2} \sqrt{\frac{V_1 - V_0}{V_1 + V_0}} \qquad \text{Eqn 6(11)}$$

and since

$$\sqrt{\frac{V_1}{V_1^2 - V_0^2}} = 1/\cos i_c$$

$$Z_0 = \frac{T_i}{2 \cos i_c} V_0 \qquad \text{Eqn 6(12)}$$

Instrument handbooks normally explain the basic interpretation.

16. For multi-layer cases, a similar procedure is adopted but the computation is more complex. For example, for two horizontal layers of thicknesses Z_0, Z_1:

$$T_{i_1} = \frac{2 Z_0 \cos i_{1_c}}{V_0} \qquad \text{Eqn 6(13)}$$

$$T_{i_2} = \frac{2 Z_0 \cos i_{0_2}}{V_0} + \frac{2 Z_1 \cos i_{2_c}}{V_1} \qquad \text{Eqn 6(14)}$$

where T_{i_1} is the 'intercept time' for the signal arrival from the first interface and T_{i_2} is the intercept time for the second interface. (The suffix 'c' indicates a critical angle, the suffix 1 and 2 relating to the first and second layer. i_{0_2} indicates the inclination of the path through the top layer for critical refraction at the second interface, i.e. the base of the second layer).

Dipping interface and apparent velocities

17. The apparent velocity of the headwave from a single horizontal refracting interface is given by $Va = \dfrac{\Delta x}{\Delta t}$ where Δt is time difference between corresponding first arrivals at D_1 and D_2. Hence, $Va = V_0/\sin i_c = V_1$ which corresponds with the concept of the headwave being propagated at the V_1 boundary.

For an interface dipping at angle α to the horizontal, the apparent velocities of the headwave are modified to:

$$V_{\text{down dip}} = V_0/\sin (i_c + \alpha) \qquad \text{Eqn 6(15)}$$

$$V_{\text{up dip}} = V_0/\sin (i_c - \alpha) \qquad \text{Eqn 6(16)}$$

For dipping beds, reversed refraction spreads are shot to obtain apparent velocities up and down dip directly from the appropriate segments on the travel-time curves. Equations 6(15) and 6(16) enable i_c and α to be calculated. As the intercept times at the two ends (A and B) of the spread will be different, calculation is made of the perpendicular depths h_A and h_B to the refractor beneath each shot point:

$$T_{iA} = \frac{2h_A \cos i_c}{V_1} \qquad \text{Eqn 6(17)}$$

$$T_{iB} = \frac{2h_B \cos i_c}{V_1} \qquad \text{Eqn 6(18)}$$

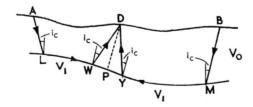

Fig 41.   Ray paths of critically refracted rays. (Used in the isolation of the time-depth to the refractor)

18. Clearly the foregoing computations relate to the simplest physical conditions, comprising defined isotropic layers with all surfaces planar and parallel. In practice, interpretations are required from more complicated rock structures entailing, for example, irregular interfaces (Figure 41) for which special techniques are needed (HAWKINS 1961), such as the use of several traverses on different azimuths.

19. *Limitations of the seismic refraction technique*

a. In loose sediments, the compressional wave velocity above the water-table may be lower than that in air. The first arrivals will thus be air waves and will lead to erroneous depth estimates of a refractor. Below the water-table, a velocity corresponding to the velocity of sound in water (1950 m/s) may be recorded if the material is loosely compacted. Such a value could give an erroneous impression of the degree of consolidation or firmness of the ground.

b. Seismic layers do not necessarily correspond to lithological layers and velocity contrasts related to sonic impedance may not be sufficient to allow mechanically different layers to be distinguished. 'Calibration' drilling should always be undertaken to identify such layers.

c. *Hidden layer.* Where there are three layers, with velocities V_0, V_1, V_2 in descending sequence $V_2 > V_1 > V_0$, the thickness of the second layer being small, the refracted wave from the third layer overtakes the refracted wave from the second layer before the latter can be recorded as a first arrival. By ignoring the layer which is too thin to give rise to first arrivals on the record, the interpreter can obtain serious errors in depth computation to the third layer. Its presence should be allowed for, by comparison with adjacent areas where it is detected from borehole records, or from other geophysical data. A minimum thickness for such a hidden layer can be computed by analysis of intercept time in relation to the velocities.

d. *Blind zone.* For a three layered system such as that illustrated by Figure 42, i.e. $V_2 > V_0 > V_1$, the intermediate layer is not detectable by seismic refraction since the refracted rays never return directly to the surface from its upper interface. Failure to recognise this situation leads to a computed depth to the next layer which is too large. Figure 42 also illustrates a circumstance in which the thickness of the third layer may be over estimated. Interpretation in such a case requires a knowledge of the thickness and approximate velocity of this 'blind zone' from other evidence.

e. Erroneous depths can be obtained, particularly in shallow problems, where there is uncertainty about the true velocity in the superficial or drift layers. Glacial deposits are highly variable and likely to be orthotropic, i.e. with different horizontal and vertical velocities. These velocity variations may be determined by the use of several short traversee.

Seismic reflection surveys

20. Continuous seismic reflection profiling is now widely used as a technique for subaqueous engineering studies, for maritime and offshore works and in

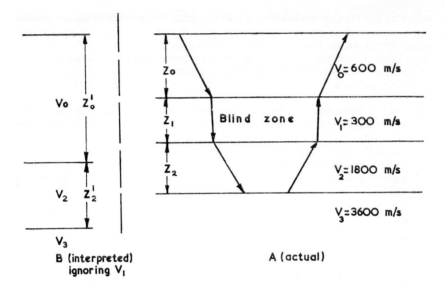

Fig 42.   Effect of blind zone. Velocity inversion in an intermediate layer, in A,
results in over-estimates of layer thicknesses in B

rivers, lakes and estuaries. In land survey, reflections cannot normally be recognised on the record from depths of less than about 100 metres because the high energy shear and surface wave arrivals completely obliterate the relatively small amplitude reflections. In marine survey, continuous seismic profiling techniques successfully exploit the use of reflections from shallow depths because the water cannot support any shear or high velocity surface wave energy. The method of reflection profiling is similar in concept to echo-sounding in that a piezoelectric or other type of sound source transmits short pulses of sound down to the sea bed and below, reflections from the various sub-bottom layers being recorded by a pressure sensitive hydrophone or hydrophone array. Sources include the sparker (an underwater high voltage spark) with predominant frequency in the 1–2 kHz range, used for reflections down to several hundred metres, pinger (piezoelectric) (5 kHz), used for greater resolution in the shallow sub-bottom layers, and others such as the boomer, air gun and gas gun, with mechanical sources of sound.

21. The records (Figure 50) have the appearance of geological cross-sections with the vertical scale a time scale. In computing depths, therefore, the velocities through the water and the strata of the ground need to be known. These can be determined from sonic logging of nearby boreholes or by refraction spreads in the area, the latter only for near isotropic velocities. In shallow water, problems can arise because of multiple reflections between the sea-bed, river bottom or other prominent reflector and the water surface. Interpretation difficulties arise

wherever the acoustic impedance of differing rock or sediment types are similar. For example, where the very stiff/hard glacial tills of the North Sea have similar sonic properties to the underlying solid rocks, the interface cannot be distinguished. The great merits of the method spring from the immediate visual presentation which is obtained rapidly and economically and from the fact that, unlike refraction techniques, it does not rely upon velocity increasing with depth.

22. When embarking on an investigation by reflection profiling, it is often advisable to test more than one type of source. Moreover, the circuitry between receiver and recorder is capable of adjustment and 'tuning' may well be optimised by trial-and-error. Theoretically, the reflected fraction of sound pulse transversely striking a boundary between layers of properties E_1, γ_1, v_1 and E_2, γ_2, v_2 (Young's modulus, bulk density and Poissons Ratio (*see* Section 6.2, paragraph 10)) will be:

$$\frac{\sqrt{A_1\gamma_1} - \sqrt{A_2\gamma_2}}{\sqrt{A_1\gamma_1} + \sqrt{A_2\gamma_2}} \qquad \text{Eqn 6(19)}$$

where A_1 and A_2 are $\sqrt{\dfrac{E_1(1 - v_1)}{\gamma_1(1 - 2v_1)(1 + v_1)}}$ and $\sqrt{\dfrac{E_2(1 - v_2)}{\gamma_2(1 - 2v_2)(1 + v_2)}}$ respectively.

The layers need to be thick in relation to the wave length of sound wave in the ground, but on the other hand the longest waves have the greatest degree of penetration. It is possible to obtain records on magnetic tape and, with signal processing such as filtering or cross-correlation, to gain the greatest degree of information of a specific feature.

23. One specialised technique which has been very successfully used in reflection surveys on land for engineering work is the 'Vibroseis' system. In contrast to the conventional 'single-shot' source for seismic exploration, the Vibroseis system employed a truck mounted vibratory source which introduces a repetitive pulse of controlled varying frequency to the ground (controlled linear frequency sweep). The returning signal from each reflector in the ground is also a frequency swept signal. A signal processing system is used to recover the reflected pulses by cross-correlation between the outgoing and reflected pulses. The Vibroseis system is thus similar in principle to the conventional seismic reflection method, but uses a more sophisticated source and signal processing system. However, it has several operational and practical advantages:

a. No explosive charges are needed and hence it can be used in built-up areas.

b. The cross-correlation technique helps to eliminate the 'noise' problem.

c. The sweep frequencies can be optimally selected to suit the geological conditions of the area.

d. The energy of the source can be increased by using multiple vibrators.

Borehole logging techniques

24. On large engineering schemes, borehole logging techniques should be seriously considered as giving considerable supplementary information. The results can be used to correlate between boreholes, to assist in the choice of sampling intervals, to identify horizons such as silts which might have been lost completely in the drilling process, and to give quantitative information about rock properties and lithologies in the borehole (*see* Table 13).

25. The techniques fall into three categories: electrical, sonic and radioactive, described in the SCHLUMBERGER (1966) publications: Log Interpretation—Principles and Log Interpretation—Charts. The electrical resistivity devices employ various electrode configurations down the hole and measure continuously the resistivity variation in the wall of the borehole as the instrument is raised from the bottom of the hole to the top. In the normal resistivity device which has

TABLE 13. ENGINEERING USES OF BOREHOLE GEOPHYSICAL LOGS

(NOTE: Only Logs 2C, 4A, 4B and 4C can be used in cased holes)

Serial No.	Type of log	Parameters measured	Application
(a)	(b)	(c)	(d)
1	*Borehole geometry* A. Caliper log	Continuous record of borehole diameter	Correction of other logs correlation with lithology
	B. Verticality test	Point determination of inclination and azimuth of borehole	Correction of inter-borehole measurements and geological log measurements of ground movement
	C. Television inspection	Visual examination and photography of borehole wall	Location of voids. Mapping of discontinuities
2	*Sonic logs:* A. Sonic or continuous velocity log (CVL) B. 3–D Sonic log	Continuous record of (P wave) velocity in the borehole wall Continuous record of both P and S wave velocity in the borehole wall	Determination of rock quality, degree of fracturing, porosity As for 2A and determination of dynamic moduli in homogeneous material (with density log)
	C. Inter-borehole sonic log	Record of apparent velocity between adjacent boreholes at discrete depths	Location of cavities and determination of rock quality between boreholes

(a)	(b)	(c)	(d)
3	*Electrical logs:* A. Single point resistivity log	*Continuous records of:* Apparent resistivity between single electrode in borehole and ground surface	Lithostratigraphical correlation
	B. Normal resistivity log C. Induction (resistivity) log D. Laterolog (resistivity) log	Apparent resistivity of borehole wall measured by electrode array Focused beyond borehole invaded zone	Ditto and rock quality/ porosity/groundwater salinity estimation
	E. Microlaterolog (resistivity)	Resistivity of invaded zone	Estimation of formation permeability
	F. Spontaneous Potential (SP) log	Natural potential differences in borehole wall	In conjunction with resistivity, lithological identification and correlation
4	*Radioactive logs:* A. Gamma log	Rapid sampling with continuous record of natural gamma ray emission	Estimation of shale/clay content of rock
	B. Gamma–gamma or Formation density log (FDL)	Gamma ray intensity after bombardment by gamma ray source	Determination of electron density and hence rock density
	C. Neutron log	Quantity of hydrogen in rock around borehole i.e. water and hydrocarbons	Determination of porosity and shale content

two potential and one current electrode in the down-hole sonde, thin resistive beds are recorded with resistivity values which are much too low, while thin conductive beds show values which are too high because of the influence of the adjacent thicker beds. The laterolog 7 employs seven electrodes in an array which focuses the current into the borehole wall as a disc-shaped sheet thus enabling better values for true resistivities to be recorded. Spontaneous Potential (SP) logs are often recorded at the same time as resistivity logs and the two enable a semi-quantitative assessment of lithology and hydrogeological characteristics to be made.

26. The sonic logging device is a transmitter receiver system with transmitter(s)/receiver(s) mounted at fixed positions, a known distance apart, on the sonde. In the more advanced devices, two transmitters and four receivers are used to eliminate the effects of tilt of the sonde and variations in the borehole diameter. The transmitters emit short high frequency pulses several times per second and differences in travel time between receivers are recorded to obtain velocities (of the refracted wave). Porosity of the formation (φ_s) as measured by sonic logging

can be obtained, knowing the matrix and fluid velocities, from the equation by WYLLIE et al (1956):

$$\varphi s = \frac{(\Delta t_{\,log} - \Delta t_{\,matrix})}{(\Delta t_{\,fluid} - \Delta t_{\,matrix})} \qquad \text{Eqn 6(20)}$$

In the 3–D sonic log, one transmitter and one receiver are used at a time and the form of the record enables compressional, shear, and 'tube' (surface wave down borehole wall) waves to be recorded. When taken together with the results of the density log, these recorded velocities allow the theoretical dynamic elastic moduli to be determined. The limitations of this have already been outlined (Section 6.2, paragraph 7).

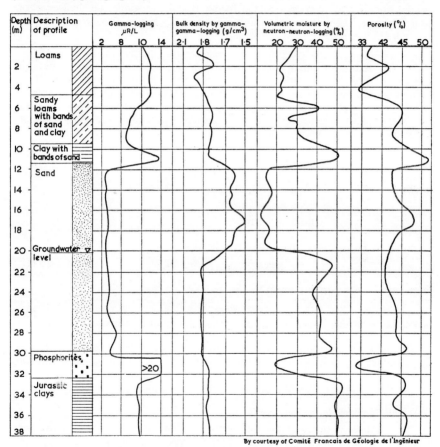

By courtesy of Comité Francais de Géologie de l'Ingénieur

Fig 43.   Borehole log example—radioactive porosity logs
(FERONSKY 1970)

131

27. Radioactive logs include the gamma-ray (or natural gamma), the gamma-gamma (or formation density) and the neutron logs. The natural gamma log records gamma radiation from elements such as Potassium 40, uranium and thorium isotopes which tend to concentrate in clays and shales. Hence, the natural gamma log reflects clay or shale content. The gamma–gamma log uses a source of gamma rays which are emitted into the borehole walls and lose energy on collision with electrons in the formations. A receiver/detector records the returning gamma-ray intensity, a high count indicating low electron density and hence low formation density. The neutron log employs a source of fast neutrons, e.g. radium-beryllium; the neutrons are slowed down and finally absorbed by hydrogen atoms in the formation. The neutron log is thus primarily a function of the quantity of hydrogen in the media round the sonde, i.e. hydrocarbons and water and therefore mainly measure porosity but is also affected by clay content (Figure 43).

28. The natural gamma and neutron logs can be used in cased holes with reduced precision; the other logs can only be used in uncased holes. These techniques are most effectively used in combination so that the variables of lithology and porosity can be separated and estimated. For example, in shaly sands, the shale percentage (P) affects the apparent porosity value (φ_{FDL}) obtained from the formation density log:

$$\varphi_{FDL} = \varphi_{true} + P\,\frac{(\varphi g - \varphi\,\text{shale})}{(\varphi g - \varphi\,\text{fluid})} \qquad \text{Eqn 6(21)}$$

where
$$\varphi g = \text{grain density} = 2 \cdot 65\ \text{t/m}^3$$
$$\varphi\,\text{shale} = 2 \cdot 45\ \text{t/m}^3$$
$$\varphi\,\text{fluid} = 1\ \text{t/m}^3$$

For a water saturated shaly sand this becomes:

$$\varphi_{FDL} = \varphi_{true} + 0 \cdot 12\ P \qquad \text{Eqn 6(22)}$$

Similarly for the porosity given by the neutron log (φ_N):

$$\varphi_N = \varphi_{true} + 0 \cdot 5\ P, \qquad \text{Eqn 6(23)}$$

since the 'hydrogen index' of shale is generally about 50 per cent.

The neutron-formation density cross plot is drawn up from these relationships and enables φ_{true} and P to be determined from the recorded values of porosity φ_{FDL} and φ_N.

Borehole to borehole measurements

29. Borehole to borehole and surface to borehole acoustic or seismic measurements have been made in various different ways. The 'velocity scanning' technique (McCANN et al 1975) is perhaps the most successful to date as it employs a fairly high power (up to 1000 joules) source in the form of a sparker in

132

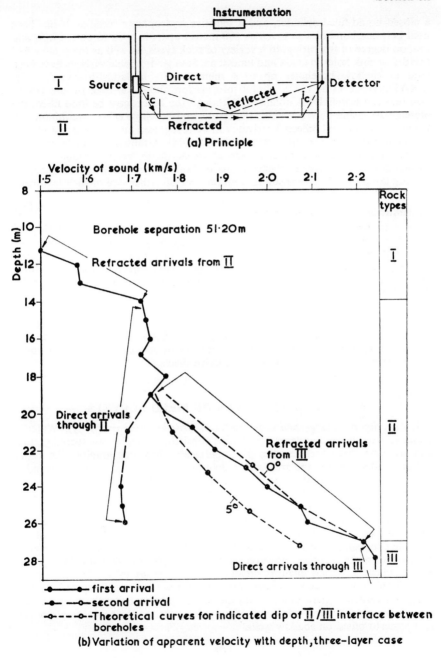

Fig 44.   Inter-borehole scanning technique (after McCANN et al 1975)

a closed liquid-filled cylinder, and a sensitive hydrophone receiver. It has been used between boreholes as much as 100 metres apart in rock to obtain information on degree of fracturing, in a variety of rock types, as well as the presence of cavities or sink holes in chalk and limestone. Smaller 1–2 joule variants have been used in shallow boreholes up to 6 metres apart in landslip investigations (CRATCHLEY et al 1972). Useful information about the rock or soil condition between boreholes is obtained. Although the holes must be fluid filled, the presence of casing does not affect the results and borehole effects are minimised. Direct, refracted and reflected arrivals can be recognised on the records and thus can significantly help in interpretation (Figure 44). Although normally operated so that transmitter and receiver are at equal depth below ground surface, particular features such as master joints or fault planes can be investigated by varying the positions of transmitter and receiver on either side of such features.

Gravitational and magnetic surveys

30. Gravitational and magnetic surveys are used predominantly for geological mapping and for mineral prospecting. Each may play a part in defining major structural features at an early stage in planning the largest engineering projects. The existence of high gravity anomalies, giving rise to appreciable local departures of the geoid (the surface of gravitational equipotential) from a true ellipsoid, may call for corrections to optical surveys with long sight lines.

31. Electromagnetic techniques depend on variations in ground resistivity and are used principally in the search for conductive mineral deposits. The methods also have application for the location of steeply inclined fault zones where these can behave as narrow conductive sheets.

Section 6.4. EXAMPLES OF FIELD SURVEYS

1. *Location of underground water basins in Nigeria (using resistivity)*. Water is obtained from limited underground basins consisting of highly weathered granite and granite-gneiss which overlies the impermeable bed-rock granite. The initial object of all surveys in the area was to determine the position of the water-table and to locate the deepest parts of the basin. In practice, the problem turned out to be a 3-layer one in terms of resistivity, with (a) laterite forming a highly resistive surface layer (up to 1000 ohm m), (b) the weathered layer (50–100 ohm m) and (c) bedrock of effectively infinite resistivity. No water-table was detectable and no attempt could be made to determine groundwater salinity in view of the variable clay content of the weathered layer. The top of the bedrock was determined with some accuracy, however. A good example is from Kura Town, near Kano, where a buried channel greater than 60 metres in depth was located immediately outside the town walls. Four or five boreholes subsequently drilled along the centre of the channel confirmed the depths to bedrock and produced high yields (14 000 litres per hour) of potable water for the town supply which had hitherto relied on shallow wells.

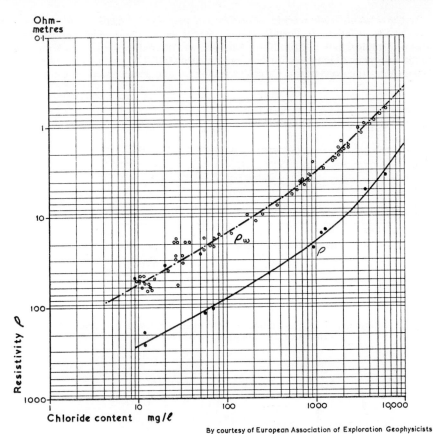

By courtesy of European Association of Exploration Geophysicists

Fig 45. Graph of resistivity against chloride content of groundwater in Northern Holland. Resistivity ρ of the sandbeds, ρ_w of pore water (VOLKER and DIJKSTRA 1955)

2. *Determination of groundwater salinity beneath the Zuider Zee, Holland (using resistivity)* (VOLKER and DIJKSTRA, 1955). An extensive series of measurements was carried out using DC equipment over the reclaimed areas of the Zuider Zee and from boats and pontoons over the then remaining parts. Because of the insignificant lateral variations and the apparently constant porosity, a direct empirical relationship was obtained between the interpreted resistivity of the aquifer and the groundwater salinity measured in an adjacent borehole (Figure 45). From a few control boreholes, the whole area was mapped in terms of salinity of the aquifer.

Section 6.4

3. *Correlations with seismic velocities and mapping of chalk 'grades' at the CERN Site, Mundford, Norfolk* (GRAINGER et al 1973). The almost level area of Middle Chalk at Mundford, Norfolk, was offered by the United Kingdom to the European Committee for Nuclear Research (CERN) as a possible site for a 300 GeV proton accelerator. Extensive geological and geotechnical investigations

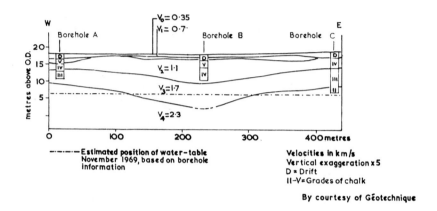

Fig 46. Interpreted velocities compared with chalk grades in boreholes, CERN site Mundford, Norfolk (GRAINGER et al 1973)

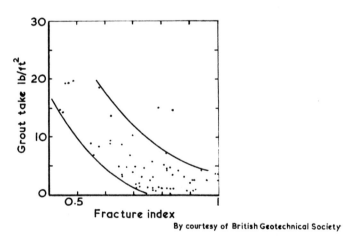

Fig 47. Relationship between fracture index and curtain grout take (KNILL 1970)

were carried out by WARD et al (1968) to demonstrate that the load-deformation characteristics of the chalk at the proposed site met with the stringent requirements for the machine. A visual classification of the chalk into grades, based on characteristics observed in large diameter boreholes over the site was quantified by means of a tank loading test and three plate loading tests (BURLAND and LORD 1969), the aim being to try to assess the load-deformation properties of the chalk from a visual inspection. Detailed investigation could not be carried out over the whole of the site by conventional means. Subsequent seismic refraction measurements with twelve channel recorder and falling weight

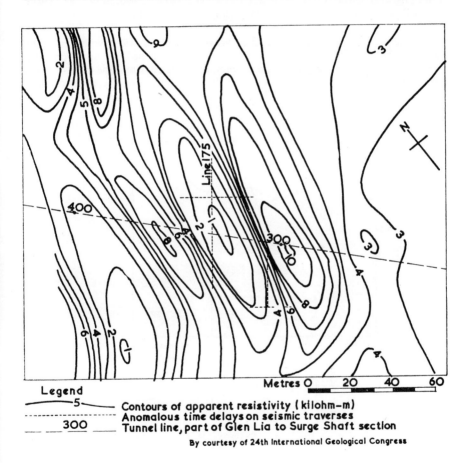

Legend

—— 5 —— Contours of apparent resistivity (kilohm–m)
----------- Anomalous time delays on seismic traverses
— — 300 — — Tunnel line, part of Glen Lia to Surge Shaft section

By courtesy of 24th International Geological Congress

Fig 48.   Plan showing resistivity contours over part of the low pressure tunnel, Foyers, Scotland (CRATCHLEY et al 1972)

137

source established a reasonable correlation between the chalk grades and the seismic velocities as shown in Figure 46, the grades ranging from V, highly weathered chalk with evidence of bedding and jointing destroyed, to I, hard brittle chalk with widely-spaced closed joints.

Corresponding velocities are respectively 0·7 and 2·3 km/sec. The water-table had negligible effect on the velocities in grades II and III chalk. In grades IV and V chalk, however, the velocities are raised significantly by the presence of water from 0·7 or 1·1 km/sec (unsaturated) to 1·95 km/sec (saturated). There is thus a need for care in interpretation where the water-table is known to be high. These velocities are consistent with the Wyllie equation (Eqn6(6)).

4. *Fracture index at British dam sites.* Work by KNILL (1970) with a hammer seismograph at various British dam sites has established an approximately linear relationship between the state of fracture of the rock and the ratio of compressional wave velocity measured in the field (i.e. including the effect of fractures) to that measured in the laboratory (fracture index). The work has also indicated some correlation between fracture index and grout take (Figure 47).

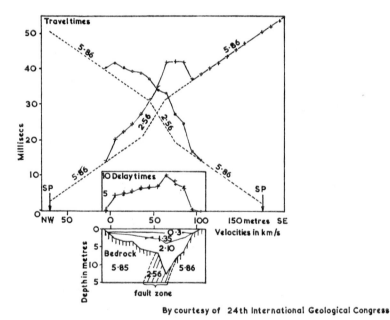

By courtesy of 24th International Geological Congress

Fig 49.   Seismic interpretation across the Glen Lia fault, Foyers, Scotland
(CRATCHLEY et al 1972)

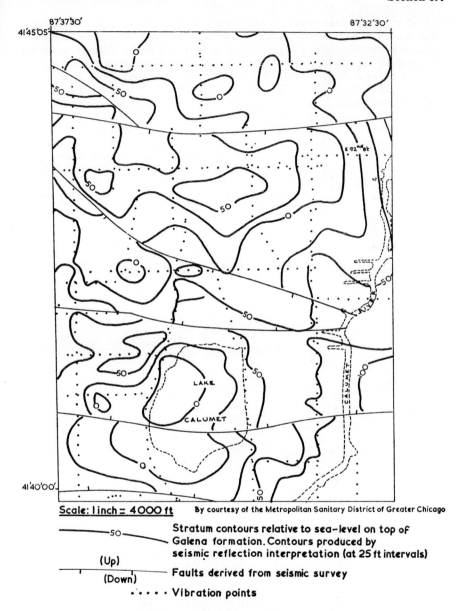

Scale: 1 inch = 4000 ft   By courtesy of the Metropolitan Sanitary District of Greater Chicago

———————50——————— Stratum contours relative to sea-level on top of
                    Galena formation. Contours produced by
                    seismic reflection interpretation (at 25 ft intervals)

(Up)
———————————————— Faults derived from seismic survey
(Down)

· · · · · Vibration points

Fig 50. Interpretation of Vibroseis survey, Chicago deep sewer tunnel
         (after MOSSMAN and HEIM 1972)

139

5. Detection of faults—Foyers low pressure tunnel, Scotland. Resistivity mapping with a Schlumberger rectangle technique (Figure 48) and with seismic refraction measurements using explosive sources and twelve channel recording has enabled fault traces to be mapped and some information on the precise position of the faults beneath drift to be obtained (Figure 49). These faults occur in and adjacent to the Foyers granite (CRATCHLEY et al 1972). In the resistivity technique the current electrodes were fixed 600 metres apart, while the potential electrodes were traversed over the area shown to map variations in apparent resistivity. The trace of a major fault, later penetrated by the tunnel, is indicated by the low resistivity zone which crosses the tunnel line at the 330 metre point (Figure 48) with corroboration from seismic traverses which reveal anomalous time delays, caused by infill along the top of the fault planes. Across the major Glen Lia fault at Foyers, velocities in the near surface sediments were obtained from short refraction spreads not shown on the diagram (Figure 49).

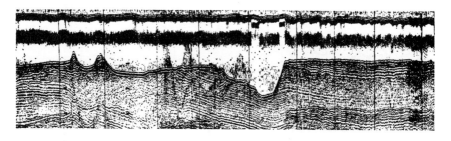

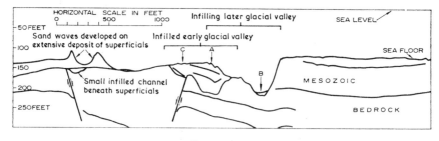

By courtesy of the Institution of Mining and Metallurgy

Fig 51.   Hydrosonde profile made across part of the Straits of Dover
(after SARGENT 1966)

Note the extensive deposit of sands and gravels upon which sand waves are developed. At 'A' is seen a late glacial valley infilled during the main interglacial, and at 'B' is a valley excavated during the last glaciation. Note also the small deposit of sediment at the bottom of the valley.

6. *Vibroseis reflection survey for sewer tunnels in Chicago.* An example of the effective use of the Vibroseis technique is given by MOSSMAN and HEIM (1972) in an account of a seismic survey for the Metropolitan Sanitary District of Greater Chicago. Tunnels were planned down to a depth of 200 metres and Vibroseis reflection technique was used to obtain detailed structure contours and fault traces on various reflecting surfaces including the Galena dolomite (Figure 50), within which it was proposed to place the deeper tunnels. The number of interpreted data points averaged $7\frac{1}{2}$ per square mile at a cost which would have provided an average of one borehole per 9 square miles. As a result thirty faults were disclosed by this survey, most of which were unknown previously.

7. *Continuous marine profiling.* Continuous marine profiling in connection with the investigation for the proposed channel tunnel has been extensively carried out over the past few years (SARGENT 1966). The example shown in Figure 51 indicates the type of detail which can be delineated with this technique.

SECTION 6.5. **PRESENTATION OF GEOPHYSICAL INFORMATION**

1. There is no generally accepted international set of geophysical symbols for use on maps and there seems little point in establishing a comprehensive set of definitive symbols or specific types of measurements. Three types of symbols may be required in the presentation of geophysical information. These are:

a. Symbols to indicate the site and type of a particular measurement (e.g. an expanding electrode resistivity measurement).

b. Symbols to show measured quantities, and derived physical properties (e.g. resistivity contours, gravity contours).

c. Symbols to show geological features interpreted from geophysical measurements (e.g. the line of a fault determined geophysically).

Symbols under **b.** would be standard contour lines as this type of information would have to be shown on a separate map or overlay.

2. *Symbols for sites of measurements.* These symbols would be incorporated on the type of map which shows positions of boreholes, trenches, etc.

130 ————————————————Expanding electrode resistivity measurement
▲ ————Constant separation resistivity traverse
⊙S₁ ————Single seismic refraction spread
⊙S₃ ⊙S₄———— Reversed seismic refraction traverse
○ GL ————————————↑——Borehole, geophysically logged
⊙²⁰¹ ————————————Position of gravity or magnetic station

141

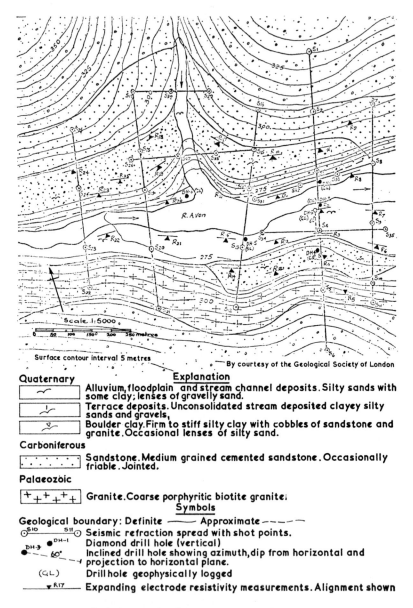

By courtesy of the Geological Society of London

Explanation

Quaternary

~	Alluvium, floodplain and stream channel deposits. Silty sands with some clay; lenses of gravelly sand.
ı	Terrace deposits. Unconsolidated stream deposited clayey silty sands and gravels,
▽	Boulder clay. Firm to stiff silty clay with cobbles of sandstone and granite. Occasional lenses of silty sand.

Carboniferous

. . . . .	Sandstone. Medium grained cemented sandstone. Occasionally friable. Jointed.

Palaeozoic

+ + + + +	Granite. Coarse porphyritic biotite granite.

Symbols

Geological boundary: Definite ——— Approximate – – – –

⊙—^{S10}———^{S11}—⊙ Seismic refraction spread with shot points.

●$^{DH-1}$ Diamond drill hole (vertical)

●$^{DH-3}$—$^{60°}$—ᴵ Inclined drill hole showing azimuth, dip from horizontal and projection to horizontal plane.

(GL) Drill hole geophysically logged

———▾^{R17}——— Expanding electrode resistivity measurements. Alignment shown

Fig 52.   Hypothetical example of the presentation of geophysical data—surface geology and geophysical data points

142

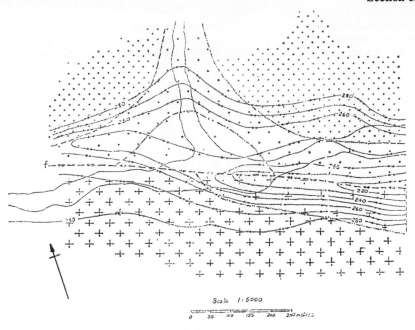

Scale 1:5000

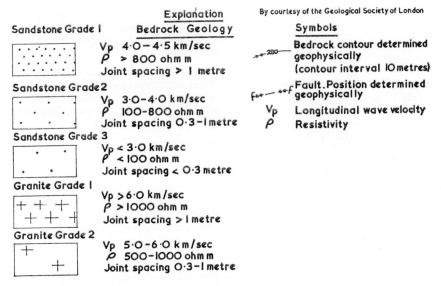

| Explanation | By courtesy of the Geological Society of London |

| | **Bedrock Geology** | **Symbols** |

Sandstone Grade 1

Vp 4·0 – 4·5 km/sec
ρ > 800 ohm m
Joint spacing > 1 metre

Sandstone Grade 2

Vp 3·0 – 4·0 km/sec
ρ 100 – 800 ohm m
Joint spacing 0·3 – 1 metre

Sandstone Grade 3

Vp < 3·0 km/sec
ρ < 100 ohm m
Joint spacing < 0·3 metre

Granite Grade 1

Vp > 6·0 km/sec
ρ > 1000 ohm m
Joint spacing > 1 metre

Granite Grade 2

Vp 5·0 – 6·0 km/sec
ρ 500 – 1000 ohm m
Joint spacing 0·3 – 1 metre

Symbols:

—280— Bedrock contour determined geophysically (contour interval 10 metres)

f— —f Fault. Position determined geophysically

Vp Longitudinal wave velocity
ρ Resistivity

Fig 53. Geophysical interpretation of Figure 52

143

3. *Symbols of interpreted features.* Symbols are required for contours on bedrock or particular marker horizons, faults and boundaries determined primarily by geophysical methods. These should be close to the normal geological symbols but distinctive.

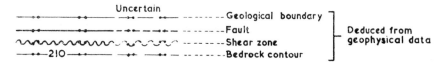

Figures 52 and 53 are hypothetical illustrations of the use of these symbols and of the sort of information which can be derived from geophysical surveys

REFERENCE LIST—CHAPTER 6

BREUSSE J J and DUPRAT A, 1964 —Contribution of geophysics to sea water intrusion surveys. *Terres et Eaux, No. 42,* pp. 2–11. Eng. trans. pp. 9–11.

BURLAND J B and LORD J A, 1969 —The load-deformation behaviour of Middle Chalk at Mundford, Norfolk: a comparison between full-scale performance and *in situ* and laboratory measurements. Proceedings of conference on *in situ* investigation in Soils and Rocks. *Br. Geotechnical Society,* pp. 3–16.

CRATCHLEY C R, GRAINGER P, McCANN D M and SMITH D I, 1972 —Some applications of geophysical techniques in engineering geology with special reference to the Foyers hydroelectric scheme. *24th I.G.C. Montreal,* Section 13, pp. 163–175.

FERRONSKY V I, 1970 —Nuclear Techniques in Engineering Geology. *Proc. of 1st Int. Congr. of Eng. Geology,* Paris, pp. 719–731.

GRAINGER P, McCANN D M and GALLOIS R W, 1973 —The application of the Seismic Refraction Technique to the Study of the fracturing of the Middle Chalk at Mundford, Norfolk. *Géotechnique* 23 No. 2 pp. 219–232.

*GRIFFITHS D H and KING R F, 1969 —*Applied Geophysics for Engineers and Geologists.* Pergamon Press, pp. 223.

HAWKINS L V, 1961 —The reciprocal method of routine shallow seismic refraction investigations. *Geophysics,* 26, 6, pp. 806–819.

KELLER G V, 1966 —Electrical properties of rocks and minerals pp. 537–577 in S P Clarke Jr (Ed). *Handbook of Physical Constants*, Geol. Soc. of America, Memoir 97.

KNILL J L, 1970 —The application of seismic methods in the prediction of grout take in rock. Proc. of Conference on *in situ* investigation in soils and rocks. *Br. Geotechnical Society*, pp. 93–100.

KNILL J L and PRICE D G, 1972 —Seismic evaluation of rock masses. *24th IGC, Montreal*, Section 13 pp. 176–182.

*KUNETZ G, 1966 —Principles of direct current resistivity prospecting. *Geoexploration Monographs* Series 1, No. 1. 103 pp. Gebrüder Borntraeger, Berlin.

McCANN D M, GRAINGER P and McCANN C, 1975 —Inter-borehole acoustic measurements and their use in engineering geology. *Geophys. prosp.* Vol. 23, No. 1 pp. 50–69.

MOONEY H M, ORELLANA E, PICKETT H and TORNHEIM L, 1966 —A resistivity computation method for layered earth models. *Geophysics*, Vol. 31, No. 1, pp. 192–203.

MOONEY H M and WETZEL W W, 1956 —*The potentials about a point electrode and apparent resistivity curves for a two- three- and four-layered earth*. University of Minnesota, Minneapolis.

MOSSMAN R W and HEIM G E, 1972 —Seismic exploration applied to underground excavation problems. *Proc. N. American Rapid Excavation and Tunnelling Conference*, Chicago, Vol. 1, pp. 169–192.

*MUSGRAVE A W (Ed) 1967 —Seismic refraction prospecting. *Society of Exploration Geophysicists*, Tulsa, pp. 604.

ORELLANA E and MOONEY H M, 1966 —Master tables and curves for vertical electrical sounding over layered structures. *Interciencia*, Madrid.

PATERSON N R and MEIDAV T, 1965 —Geophysical methods in highway engineering, *48th Ann. Convention of the Canadian Good Roads Assoc.*, Saskatchewan.

SARGENT G E G, 1966 —Review of acoustic equipments for studying submarine sediments. Trans. Section B of *IMM* Vol. 75, B113–119.

SCHLUMBERGER, 1966 —*Log interpretation—principles* ⎱ Paris
Log interpretation—charts ⎰

SCOTT J H, LEE F T, —The relationship of geophysical measure-
CARROLL R D and ments to engineering and construction
ROBINSON C S, 1968 parameters in the Straight Creek tunnel
pilot bore, Colorado. *Int. J. Rock. Mech.
Min. Sci.*, Vol. 5, 1–30.

SLICHTER L B and TELKES M, —Electrical properties of rocks and
1942 minerals, pp. 299–319 in F Birch (Ed)
Handbook of physical constants, Geol.
Soc. of America Special Paper 36.

*STAGG K G and —*Rock Mechanics in Engineering Practice*,
ZIENKIEWICZ O C (Ed) 1968 Wiley, London.

*VAN NOSTRAND G and —Interpretation of resistivity data,
COOK K L, 1966 *U.S.G.S. professional paper* 499, pp. 310.

VOLKER A and DIJKSTRA J, —Détermination des Salinités des eaux
1955 dans le sous-sol du Zuiderzee par
prospection géophysique. *Geophysical
Prospecting*, Vol. 3, No. 2, pp. 111–125.

WARD W H, BURLAND J B and —Geotechnical assessment of a site at
GALLOIS R W, 1968 Mundford, Norfolk, for a large proton
accelerator. *Geotechnique*, 18, pp. 399–431.

WEST G and DUMBLETON M J, —An assessment of geophysics in site
1975 investigations for roads in Britain. Transport and Road Research Laboratory
Report 680.

WYLLIE M R J, GREGORY A R —Elastic wave velocities in heterogeneous
and GARDNER L W, 1956 and porous media. *Geophysics*, 21, 41–70.

Recommended for general reading

CHAPTER 7

GEOLOGICAL MAPS AND OTHER INFORMATION SOURCES

SECTION 7.1. GEOLOGICAL MAPS

Introduction

1. Small scale geological maps exist for almost all the land surface of the world. Large scale geological maps with supporting detailed information are available for most of the British Isles, north-west Europe and the urbanised or mineralogically important areas elsewhere. It can be assumed that for any projected engineering site some geological information already exists. For most effective use it needs to be acquired, assessed and applied during the initial planning and design of a project. Sought too late, it may provide an explanation rather than a solution for practical difficulties. This chapter outlines the type of information that may be available, where to get it and how to use it.

2. A geological map is the primary source of information. An engineer needs to obtain the most useful map(s), and written amplification which is relevant to his particular investigation, for from these he can assess whether specialist geological advice will be necessary, and predict features for observation and detailed study during site investigation. Yet to be of maximum use, the map must be the right type, reliable, and correctly interpreted.

Types of geological map

3. For many mineral exploration and other geological purposes, superficial and unconsolidated deposits such as alluvium have to be ignored unless they are very thick and they may even be omitted from geological maps, which are then said to present the 'Solid' geology only. In British practice these are called 'solid' geological maps and although outcrops of alluvium, till, etc may be plotted in outline they are not coloured. Such maps may, therefore, mislead users concerned with the nature of the immediate ground surface. Geological maps of many countries show all these superficial deposits, but in doing so many important boundaries of underlying bedrock formations may be obscured or rendered more difficult to interpret.

4. This problem becomes acute in areas that are heavily and irregularly covered with glacial till, as is the case in most of the British Isles. In parts of the United Kingdom and in a few similar areas it has been found necessary therefore to produce two maps for each sheet area, Solid and Drift, superficial deposits being

shown in full on the latter. The example given in Section 7.2, paragraph 18 *et seq* illustrates this difference, and the possible uses of each.

Map explanations

5. Colours or symbols are normally allocated to named formations of sand, clay, limestone, etc, but in many cases these rock or sediment units are not homogeneous and it is necessary for instance to refer to a limestone formation of which a considerable proportion of the material is either sand or clay. The detail it is possible to portray is determined by the scale of the map, but the problem exists at all scales.

6. In Germany and some other parts of Europe this problem is avoided by mapping rock units on the basis of their geological age rather than their composition. These may be called chronostratigraphic maps; they require the intermediary of specialist geological knowledge to interpret.

7. *Appreciation.* It is thus essential to assess the nature of any map and its relevance to the extraction of the type of information required. It is unusual for more than one type of map to be available for any one area outside the UK.

Conventional symbols

8. Geological maps normally carry a legend as key to the features represented on them by colour or symbol. Important features usually included are:

a. *Units mapped.* Relative disposition of different rock units will be indicated by colour or symbol (or both). Rocks differentiated by composition will probably also differ in engineering properties, whereas rocks differentiated by age need not necessarily do so. Some of the symbols recommended by the Geological Society of London for use on uncoloured maps and sections in their report (GEOL SOC 1972) are shown in Figures 54 and 55. Many other symbols may be encountered, including those shown in CP 2001 (1957). It is important, therefore, when reading engineering geology maps or plans to refer to the legend. When preparing new plans, it is recommended that the report of the Geological Society Working Party (GEOL SOC 1972) should be followed.

b. *Boundaries.* Observed boundaries between units are normally indicated by a solid line, inferred boundaries by a dashed or dotted line. The latter are more common since soil, vegetation and buildings frequently mask the junction. The proportion of observed to inferred boundary shown on a map will give some indication of its reliability.

c. *Faults.* The presence, pattern, magnitude or probable occurrence of faults have engineering significance, for not only are the rocks adjacent to faults displaced relative to one another, but their geotechnical properties are modified by the fracturing associated with the fault line.

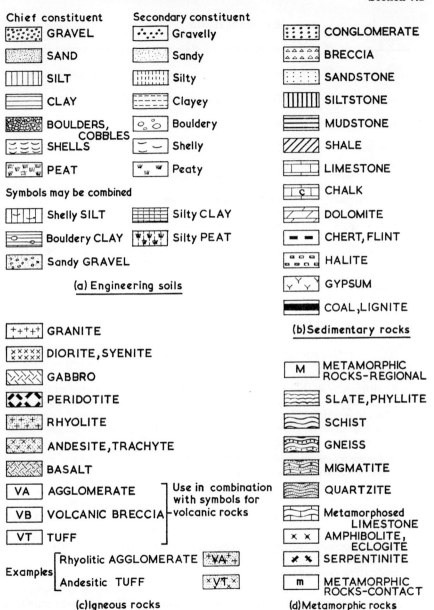

Chief constituent | Secondary constituent

GRAVEL | Gravelly | CONGLOMERATE
SAND | Sandy | BRECCIA
SILT | Silty | SANDSTONE
CLAY | Clayey | SILTSTONE
BOULDERS, COBBLES | Bouldery | MUDSTONE
SHELLS | Shelly | SHALE
PEAT | Peaty | LIMESTONE

Symbols may be combined | | CHALK

Shelly SILT | Silty CLAY | DOLOMITE
Bouldery CLAY | Silty PEAT | CHERT, FLINT
Sandy GRAVEL | | HALITE
| | GYPSUM
| | COAL, LIGNITE

(a) Engineering soils

(b) Sedimentary rocks

GRANITE
DIORITE, SYENITE
GABBRO | | M | METAMORPHIC ROCKS-REGIONAL
PERIDOTITE | | SLATE, PHYLLITE
RHYOLITE | | SCHIST
ANDESITE, TRACHYTE | | GNEISS
BASALT | | MIGMATITE
VA | AGGLOMERATE | Use in combination with symbols for volcanic rocks | QUARTZITE
VB | VOLCANIC BRECCIA | | Metamorphosed LIMESTONE
VT | TUFF | | x x | AMPHIBOLITE, ECLOGITE
| | | SERPENTINITE
Examples | Rhyolitic AGGLOMERATE VA | m | METAMORPHIC ROCKS-CONTACT
| Andesitic TUFF VT

(c) Igneous rocks

(d) Metamorphic rocks

Fig 54.   Recommended symbols for engineering soils and rocks
(GEOL SOC 1972)

149

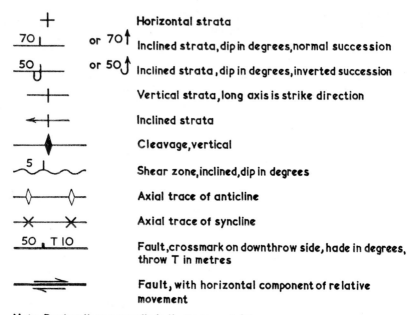

		Horizontal strata
70	or 70	Inclined strata, dip in degrees, normal succession
50	or 50	Inclined strata, dip in degrees, inverted succession
		Vertical strata, long axis is strike direction
		Inclined strata
		Cleavage, vertical
5		Shear zone, inclined, dip in degrees
		Axial trace of anticline
		Axial trace of syncline
50 T 10		Fault, crossmark on downthrow side, hade in degrees, throw T in metres
		Fault, with horizontal component of relative movement

Note. Broken lines normally indicate uncertainty

Fig 55. Recommended structural symbols (GEOL SOC 1972)

d. *Attitude of rocks.* An estimate of the sub-surface disposition of the rocks can be made from measurements of dip and strike shown on the map together with the surface outcrop pattern. Geological maps of many countries are drawn from aerial photographs and in these cases the structures shown may not have been confirmed by field surveys.

e. *Localities of geological significance.* These are places where significant fossils or minerals have been found.

f. *Localities of engineering significance.* Quarries, mines, shafts, adits, boreholes and wells are marked on some maps. Either one can see for oneself the disposition of rocks below land surface at these points, or seek detailed geological information obtained during their construction.

g. *Physiography and land use.* Geological maps are usually printed upon a topographical base map. Relation of physiography and land use to the underlying geology may be obvious and of engineering significance.

h. *Thickness of strata.* Rock units may vary quite widely in thickness over the area depicted on a map. The limits of variation will normally be indicated by figures on the legend and some indication of significant variation can be obtained from cross-sections when these are included to illustrate the map.

9. Inspection of a geological map and reference to its key should therefore indicate to the engineer the disposition of rock types of different properties; regions where boundaries need confirmation; places where faulting may cause construction problems; whether the geological structure of the region is complex and needs specialist interpretation; whether detailed geological information might already be held by government, mining or engineering concerns; if geology and land use have obvious relationship and engineering implications; and whether the strata are sufficiently variable to require mapping on a larger scale.

Scale

10. It is important to obtain maps of an appropriate scale. Small scale maps (of the order of 1:500 000) exist for much of the world. They are easily obtainable and may have some use for broad civil engineering planning. Their use is limited by the lack of detail shown and often a variation in reliability of information in different parts of the map.

11. Published map sheets in technologically advanced countries are commonly of medium scale; 1:50 000. They tend to be of more uniform, reliable quality and may have an associated written text in amplification. Such medium scale maps are normally summary maps produced from a number of larger scale maps.

12. Larger scale maps are seldom published, but geological mapping is frequently carried out in the scale of 1:10 000. Such maps may therefore be available in manuscript for reference or photocopying. Maps of even larger scale are sometimes prepared for areas of particular detailed interest, such as coal-fields. Most site investigations require geological maps of at least 1:10 000 scale either at the feasibility or detailed planning stage and plans of much larger scale e.g. 1:1 000 or 1:500 may be required at pre-contract or pre-construction stage depending on the complexity of either the geology or the project.

Reliability

13. A geological map is an interpretation of a number of observations. It will contain and be based upon facts, but it should be clearly understood that it is not in itself to be relied upon as entirely factual. Its reliability can be assessed by reference to:

   a. *Literature.* Duration and method of field work, reference to earlier geological work and its detail will indicate the quantity of work done on the area.

   b. *Authority of publication.* A map may be published by a government department, academic institution or commercial firm. Its prestige may indicate quality of map work.

   c. *Scale.* Generally speaking, the larger the scale, the higher the degree of reliability.

151

d. *Reliability of base topographical map.* If the base map is old and inaccurate, positions of geological data marked upon it may also be unreliable·

e. *Geological detail.* Lack of detail indicates general rather than precise survey.

14. The map should be assessed in the field by walking over the ground to confirm or determine as far as possible the number and nature of significant rock types present, the sequence of the strata, the extent of each soil and rock (especially the precise position and nature of their boundaries) and, where appropriate, the geological structure of the region.

15. Use of satellite photographs for geological mapping. The Earth Resources Technology Satellite (ERTS) provides multi-spectral scanner images (MSS) of the Earth's surface. These pictures are taken in three different spectral bands; green, red and infra-red, to give a false colour composite and have been used for environmental monitoring, land use and general resources inventory. If the area was cloud-free at the time of photography the transparencies show good detail. Providing that the interpretation is done by a geologist familiar with the ground, the photographs will assist in:

   a. Regional geological mapping.

   b. Recognising regional changes in major sedimentary units.

   c. Recognising major structural features and lithological units.

   d. Distinguishing igneous intrusive bodies.

In some cases the photographs might be used to construct an uncontrolled photomosaic from which a regional geological map could be prepared; this could be an economical method of providing geological maps where none exist, or supplementing old or incomplete data.

Section 7.2. MAP INTERPRETATION

1. Once a reliable geological map of suitable type and scale is available, it needs interpretation. Where the geology is complex, this is best undertaken by a qualified geologist. An outline guide to the interpretation of simple geological structures is given below. Further information, including the procedure by which geological maps are constructed, may be found in EARLE 1965, BLYTH and DE FREITAS 1974, HIMUS and SWEETING 1968, and LAHEE 1961.

Basic principles

2. The map is a two-dimensional representation of a three-dimensional structure. Its interpretation involves an ability to visualise the disposition of rocks in three dimensions. In the block diagram used below, the top surface shows the rocks as they would appear on a geological map, the other faces of the block give the subsurface interpretation.

3. Generally sedimentary rocks can be considered to have formed as a succession of near horizontal sheets or beds, their upper and lower surfaces approximately parallel. This pile of sheets seldom remains horizontal, for stresses in the Earth's crust may cause it to be folded, faulted and intruded by igneous rocks. These processes together with the effect of variation in sedimentation pattern of the beds and their subsequent erosion determine the surface outcrop of bedrock as it appears on a geological map. Principal effects are outlined below, initially assuming that erosion has worn down the topography to a flat horizontal surface and that the sedimentary pile has not been overturned.

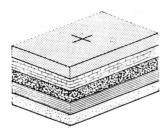

Fig 56. Horizontal beds (as tabulated in key on most geological maps)

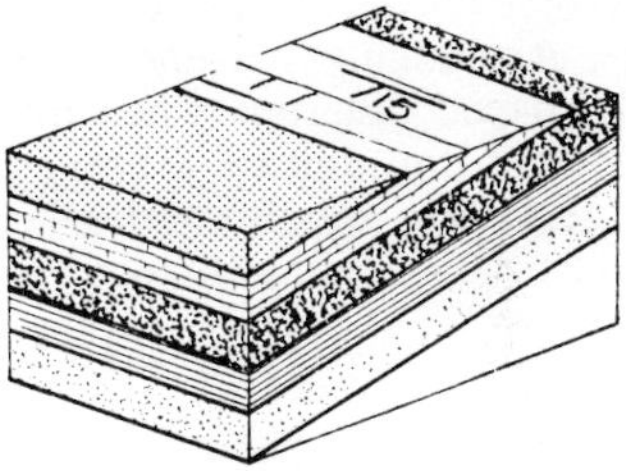

Fig 57. Inclined beds (note indication direction of and amount of dip)

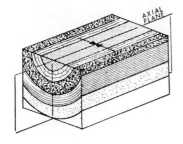
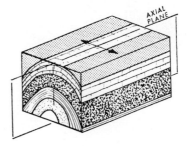

Fig 58. Syncline (note symbol over youngest bed exposed)

Fig 59. Anticline (note symbol over oldest beds exposed)

Folds

4. The key to a geological map will indicate the rock units mapped, oldest at the bottom and youngest at the top, as might be expected if the beds were not affected by folding (Figure 56). In areas where the rocks have been tilted (Figure 57), the direction and amount by which they dip will usually be indicated by symbol. Two symbols are commonly used to indicate dip (Figure 55), the direction of cross-mark or arrow indicating the direction of dip and figures showing the amount in degrees. In the absence of symbols, the direction of dip can be

153

estimated as being perpendicular to the outcrop boundary at a level surface of any two conformable rock units, in the direction of the younger beds, or as described at paragraph 15.

5. Beds folded down into a syncline can be recognised by symbol (Figure 55), by dip symbols pointing towards each other, or as a young bed founded by older beds on each side (Figure 58). Conversely, beds folded up into an anticline can be recognised by symbol (Figure 55), by dip symbols pointing away from each other, or as a bed bounded by younger beds on each side (Figure 59).

6. Where the axis of a fold is inclined rather than horizontal, the fold is said to plunge. A plunging syncline (Figure 60) and a plunging anticline (Figure 61) give U-shaped outcrops when mapped on a flat surface. Concentric outcrops with inward dip indicate a basin (Figure 62), those with outward dip a dome. Close repetition of alternating anticlines and synclines with parallel axes is termed isoclinal folding (Figure 63). A monocline (Figure 64) is a local steepening of an otherwise uniformly dipping or horizontal series of beds, and is composed

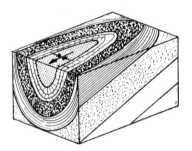

Fig 60.   Plunging syncline

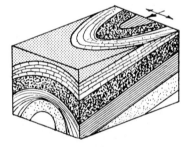

Fig 61.   Plunging anticline

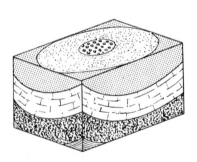

Fig 62.   Basin

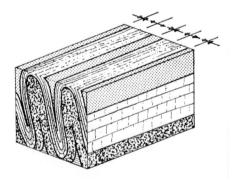

Fig 63.   Isoclinal folding

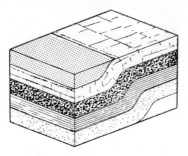

Fig 64.   Monocline

of an anticlinal bend above, followed by a synclinal bend at a lower level. This term has, however, also been used by some American authors to describe a series of inclined beds, all dipping in the same direction.

Faults

7. Faults are fracture planes through rocks. Three main types occur (*see* Section 3.3):

    a. *Normal faults*—vertical movement only.

    b. *Thrust faults*—chiefly horizontal movement of one block over another.

    c. *Wrench faults*—horizontal movement of one block alongside the other.

8. Normal faults are by far the most common. When they occur parallel to the strike of dipping beds, their effect is to cause omission (Figure 65) or repetition (Figure 66) of rock units at the mapped surface. When they occur parallel to the direction of dip, the effect is to cause a lateral displacement of the outcrops on either side of the fault (Figure 67). Wrench faults produce the same appearance,

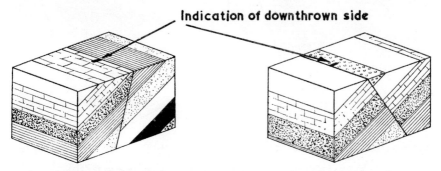

Fig 65.   Omission of bed at surface (fault parallel with strike and hading with dip)

Fig 66.   Repetition of bed at surface (fault parallel with strike and hading against dip)

155

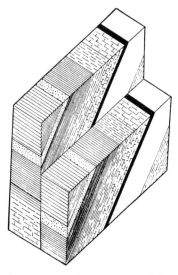

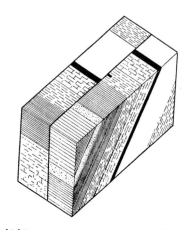

(a)Position of beds after faulting   **(b)Denudation to common level**

Fig 67.   Lateral displacement of bed at surface due to fault parallel
with dip (note indication of downthrown side)

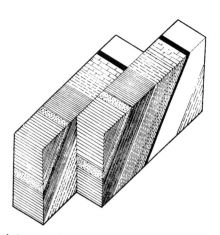

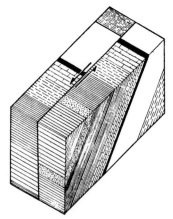

(a) Position of beds after faulting   **(b)Trimmed section of (a)**

Note: Horizontal section is the same as Fig 67(b)

Fig 68.   Lateral displacement of bed at surface due to wrench or tear fault (note
direction of movement indicated)

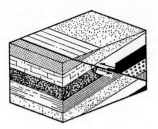

Fig 69. Thrust or reverse fault (note
direction of movement indicated)

but by movement in a lateral rather than vertical sense (Figure 68). Thrust faults
are usually low-angle and may indicate considerable displacement of rocks,
those above the thrust plane being thus very different from those beneath
(Figure 69).

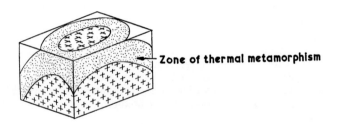

Fig 70. Batholith or boss

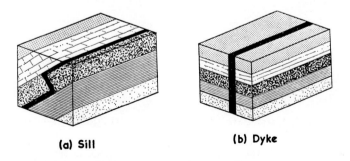

(a) Sill   (b) Dyke

Fig 71. Minor igneous intrusion

157

9. *Plutons.* Plutons are masses of generally coarsely crystalline rock formed at depth in the Earth's crust (Section 3.1). Their surface outcrop is irregular and subsurface structure may be difficult to determine from the map alone (*see* Figure 70). However, in practice most bodies of coarsely crystalline rock can be regarded as batholiths.

10. *Igneous rock intrusions.* Minor igneous rock intrusions (Figure 71) generally form sheetlike masses of less coarsely crystalline rock, emplaced nearer the surface of the Earth's crust (Chapter 3). A sill tends to follow the bedding planes of the intruded rocks. A dyke cuts across them, usually almost vertically.

11. Igneous rocks are always distinctively marked on geological maps by colour or ornament, and their presence therefore made obvious. Subsurface interpretation is, however, more speculative than with sedimentary rocks, since they may be intruded along irregular lines of weakness. Minor intrusions may be too small to mark adequately on printed geological maps, and they are easily obscured by soil and vegetation. If they are present at all, therefore, an engineer should anticipate finding more than are shown on the geological map.

Variation in sedimentation pattern

12. If the rate and type of sedimentation varied across an area when its sedimentary rocks were formed (a normal occurrence), then interpretation of the subsurface type, thickness and disposition of rocks is very speculative from map data alone (cf Figure 72, where there is no surface indication of a subsurface limestone unit). Borehole information is therefore required. The map key often indicates the order of thickness variation and any lateral change in rock type, enabling the magnitude of the problem to be predicted.

13. Breaks in sedimentation are frequent, giving rise to unconformities (Chapter 3). Rocks may change very significantly in type and orientation across such boundaries (Figure 73).

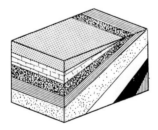
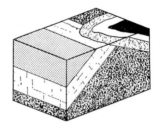

Fig 72. Non-uniform rate of sedimentation (variation of thickness of beds)

Fig 73. Break in sedimentation (unconformity)

Effect of topography on outcrop pattern

14. For simplicity, outcrop patterns have been illustrated in block diagrams as they would occur on level ground. In practice, ground is seldom level, and outcrop pattern is controlled both by the shape of the rock unit and by topography. Thus a map could show a strip of young beds bordered by older

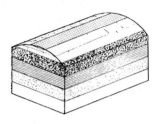 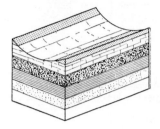

Fig 74. Effect of hills on outcrop (compare Figure 59)   Fig 75. Effect of valleys on out-crop (compare Figure 58)

strata. On level ground, the interpretation would be of beds folded into a syncline (Figure 58). In hilly country, exactly the same outcrop pattern could indicate horizontal beds (Figure 74). Similarly, the effects of anticlines and valleys can be confused (Figure 59 cf Figure 75).

15. Direction and amount of dip of inclined strata can be calculated from their outcrop pattern relative to the topography, most easily where V-shaped outcrops are related to river valleys (*see* Figure 76). It is often useful to prepare contour maps of specific rock interfaces from data provided by geological maps and bore-holes and also from geophysical surveys. These are called structure contour maps.

Construction of geological sections

16. To avoid confusion, a vertical section is usually provided with geological maps, drawn between selected points to indicate the key structure. It will seldom cross the precise area of engineering interest. An engineer may therefore need to acquire the skill of constructing geological sections from maps (*see* EARLE 1965) or employ a geologist to do this for him.

Conclusion

17. Engineering implications of such map interpretation are discussed in Chapters 9 and 11. Two examples follow here to illustrate geological data potentially available to an engineer.

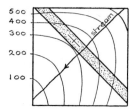

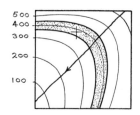

(a) Outcrop crossing valley contours without deflection indicates vertical strata

(b) Outcrop which parallels topographic contours indicates horizontal strata

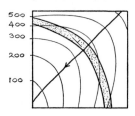

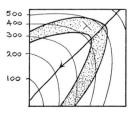

(c) Outcrop which forms a blunt V pointing up a valley indicates strata dipping upstream

(d) Outcrop which forms a narrow V pointing up a valley indicates strata dipping downstream but at an angle smaller than the valley gradient

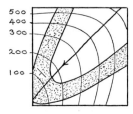

(e) Outcrop which forms a V pointing downstream across a valley indicates strata dipping downstream but at an angle greater than that of the valley gradient

Fig 76. Outcrop patterns related to topography

EXAMPLE 1: **BRITISH ISLES**

18. The area of Holy Island, Co. Durham, illustrates the geological data commonly available in a developed western country. A range of relevant maps has been published by a government agency (in this case by the Institute of Geological Sciences). Its regional setting can therefore be appreciated from the 1:1 584 000 and 1:625 000 scale geological maps, and more detail is shown on the 1:63 360 maps, of which both 'solid' and 'drift' editions are available. Parts of these two are reproduced as Maps 1 and 2. They are the largest scale geological maps of the area available for purchase and will therefore form the initial basis of interpretation. However, since the Institute has prepared field maps on the scale of 1:10 560 (Map 3) and coal mines have been developed in at least part of the area, the existence of more detailed maps and information can be anticipated.

19. From the 'drift' map (Map 2) it can be seen that boulder clay (pale blue on map) together with sand and gravel (pink on map) occur at much of the ground surface, especially in the south-west. The thickness of these beds is not given, it can be expected to vary widely, and may need to be found by augering at any particular site. However, since these drift deposits cover such large areas and even the railway cuttings do not penetrate to bedrock, their thickness is likely to be substantial, measured in terms of metres.

20. On the 'solid' map (Map 1), older sedimentary rocks (lettered d^{1a}) occur near the centre, younger ones to the east and west. Ground level can be considered as almost uniformly level, so a fold in the form of an anticline can be inferred. The outward pointing dip symbols confirm this. The oldest beds are restricted to south, so the anticline must plunge approximately northwards. A section across the area from Doddington to Holy Island illustrates the structure (Figure 77(a)). Faults, symbolised to indicate that they are normal faults, cut across many of the rocks. A large outcrop of quartz dolerite (qD) generally but not entirely follows the bedding, and must therefore have been intruded as a sill. Smaller, linear outcrops of quartz dolerite (as on Holy Island itself) indicate the presence of dykes.

21. *Site selection.* The 'drift' map clearly indicates the limited areas where hard bedrock occurs at the surface. Till (Boulder Clay) is widespread: sites on it may encounter the problems associated with clay soils. Sand and gravel patches would give better drainage. Peat would give its particular problems.

22. *Foundation and excavation techniques.* The widespread surface occurrence of till with numerous, sometimes large, boulders of hard rock, could make mechanical excavation difficult. Any clay would necessitate low angle slopes for cuttings.

23. *Construction materials.* Igneous rocks generally make the best construction materials. Quartz dolerite is the only type found in this area, and

161

although in itself an excellent material, e.g. for road metal, it occurs in two different situations: as a sill and as dykes. The sill could be most easily worked. The drift map shows where the rock is exposed at the surface without over-burden, the solid map can be used to interpret its dip between the sedimentary strata. In the dykes, the narrow linear outcrop implies that the rock is nearly vertical. Quarrying problems in the two situations are therefore very different (cf quarries shown on Map 3). Some beds of limestone might also be hard enough to provide good road stone. They would need more detailed study to ascertain this, and quarries would have to be located where the dip of strata and topography combined to reduce overburden to a minimum (deduceable from 1:10 560 map).

24. *Water supply.* In general, sandstones and certain limestones are permeable to water. In this area, the Fell Sandstone is a good water-bearing horizon. The solid map shows where it outcrops and how it dips, permitting the depth to the sandstone to be calculated and boreholes sited accordingly. The same procedure applies to units within the limestones. Water may also be located in the sand dune and raised beach areas.

25. *Trafficability.* The boulder clay might be expected to give rise to clay soil problems. Peat and some alluvium (locally widespread) would generally be too soft to cross without prepared roadways.

26. Detailed information to amplify these deductions could be sought from the sources indicated in Table 14.

EXAMPLE 2: **THAILAND**

27. The area west of Bangkok, Thailand, illustrates the geological information that might be available in a developing territory overseas. The only map of the area published to date is small scale 1:1 000 000 (Map 4). Rock units are only broadly differentiated, on the basis of presumed age. There is little supporting literature.

28. Clearly the region west of Ratchaburi and Kanchanaburi is underlain by old (Palaeozoic: Devonian to Carboniferous) rocks, principally sediments or metamorphosed sediments, with large granite batholiths intruded. The sub-surface geotechnical properties of these rocks will differ greatly from those of the alluvium (coloured yellow on map) of the Bangkok Plain, and from the younger, unmetamorphosed (Mesozoic: Triassic to Cretaceous) sandstones and shales of the Khorat Plateau (coloured blue to green). The map is thus useful in broad planning, but illustrates neither the extent and variation of surface 'drift' material nor the details of bedrock type and structure given for the previous example.

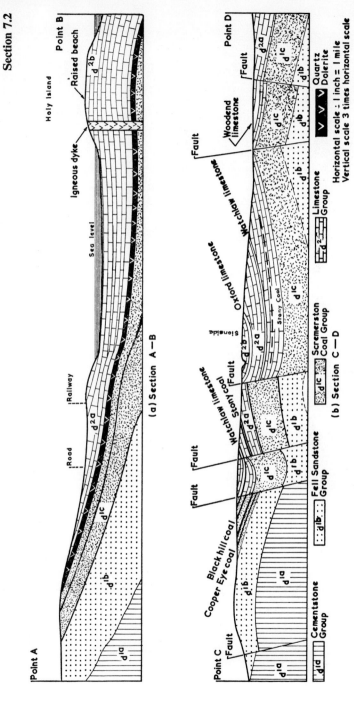

(a) Section A—B

(b) Section C—D

Fig 77.   Cross-sections of Holy Island map (Map 1)

29. In practice, the map was used to guide the general deployment of a well-drilling team in 1969. Selection of areas of operation required more detailed maps than were then available, and larger scale (1:50 000) maps were quickly compiled from:

a. Unpublished data made available at the Department of Mineral Resources, Bangkok.

b. Unpublished maps and reports prepared by students during training in geological field mapping, made available in the Geology Department, Chulalongkorn University, Bangkok.

c. Interpretation by a geologist of air photographs of the region.

d. Detailed ground survey of areas judged significant by a geologist.

SECTION 7.3. SPECIAL GEOLOGICAL MAPS

Engineering geology maps

1. A series is available which covers northern Germany at scale 1:250 000 but maps of this type are not yet in widespread production elsewhere. They are simplified maps in which the details and stratigraphy are omitted as far as possible, and physical characteristics of the rocks and soils, and their uses, are stressed. The legends of such maps are prepared so as to furnish those facts and inferences that are the most important to a civil engineer. They are a valuable aid in planning and are being prepared for regions outside Germany. Small scale maps may, however, be over-simplified. The preparation of maps and plans in terms of engineering geology is described in a Geological Society report (GEOL SOC 1972). Such geotechnical maps can be produced by an engineering geologist from the standard geological maps, and DUMBLETON and WEST 1970 (p. 24) list references which describe the procedure.

Hydrogeological maps

2. The purpose of hydrogeological maps is to enable various areas to be distinguished according to their hydrological character in relation to the geology. At present only limited areas have been so mapped, either to show the structure and distribution of the various water bearing strata, and/or to show the depth, and sometimes the seasonal variation, of the water table. Future maps should indicate, on a regional basis, such items as the extent of the principal groundwater bodies, the scarcity of groundwater elsewhere, the known or possible occurrence of artesian basins, areas of saline groundwater and the potability of groundwater. They should also show, according to scale, information of a local character, such as the locations of boreholes, wells and other works, contours of the groundwater table, the direction of flow and variations in quality of water. In general, any information leading to a better understanding of occurrence, movement, quantity and quality of groundwater, should be shown with sufficient geology to lead to a proper understanding of the hydrogeological conditions.

Section 7.3

Resource maps

3. These maps indicate the occurrence and distribution of economic rocks (e.g. gravel, limestone, minerals, coal, oil etc). They are often on a small scale and generalised, but may record useful (if rapidly outdated) production information.

Soil maps

4. Soil maps generally deal with the top 1 to 1·5 metres of material at the land surface. They indicate what kinds of soils are present, where they are located and to some extent what use they can best serve. The classification adopted varies with intended use. Most soil maps are prepared for agricultural purposes, and although there is no agreed worldwide system which will provide for the precise classification of all the varieties of physical and chemical composition existing in the soils of the world, discrimination between units is generally on the basis of geographical association, parent material, texture, subsoil characteristics and drainage class. Some but not all such criteria have engineering significance. Engineering soil maps, especially those designed for military use, discriminate rather between soils according to their permeability, stability under stress, bearing capacity and important variations in these last two properties with changing moisture content. Agricultural soil maps are in more general production than those for engineering purposes, but the latter can to some extent be translated from the former, especially where a geological 'drift' map is available. Areas of poor drainage and difficult trafficability are readily distinguished. These may be emphasised on military cross country movement or 'Going' maps.

Landform maps

5. Land use and land form are interrelated. Some maps are available which show present land utilisation. Other maps indicate the relative value of land for agricultural use. Where such maps exist, they may serve as a preliminary indication of land values and engineering characteristics (*see* DUMBLETON and WEST 1970 (p. 10) for information concerning England and Wales). Landform is itself a reflection of the type of climate and the nature of the geology (Chapter 8). Patterns of landforms can be mapped as land systems, subdivided into facets and elements, which broadly relate to the geology of the ground and more particularly to its agricultural or engineering properties. The advantage of such maps is that they can be easily prepared from air photographs; the disadvantage is their small scale (usually 1:500 000 to 1:50 000). The technique has so far developed primarily in East Africa.

Subsidiary map types

6. All the above categories of maps are potentially useful to the engineer and should be sought where appropriate. There are also more specialised maps sometimes obtainable through geological sources which may meet a particular need. Examples of some of these are:

a. Geophysical maps: such as Aeromagnetic and Gravity Anomaly maps.

b. Geochemical and mineral maps: which indicate the distribution of specified substances.

c. Structural and tectonic maps: which indicate the geological structure of an area but not necessarily the rock types.

d. Single feature maps: to emphasise the distribution of a single rock type or rock property.

Section 7.4. GEOLOGICAL INFORMATION

1. Although geological information is available both in the form of maps and in written texts, to obtain the most useful map(s) and written amplification, both published and unpublished data may have to be acquired.

Published data

2. Most countries now publish geological maps with supporting literature. This basic literature, which is generally readily available and intelligible to a non-specialist, may take the form of a memoir, dealing with the geology of one map sheet or area (or with one aspect of the geology of several map sheets); a book broadly describing the geology of a country or significantly large region; or simply a brief summary of the geology of an area or information printed on the reverse side of a map.

3. Supplementary literature, more difficult to obtain or assess, but often incorporating maps, exists as papers in the bulletins of various institutions and in scientific journals. Where the basic literature is inadequate, this supplementary literature must be used. However, handling problems are caused by the sheer magnitude of published data (some 30 000 geological papers are published annually), and sifting and extracting relevant information may thus take time and necessitate the use of a geologist.

Unpublished data

4. Although a wealth of geological information is available in published form, much that is detailed and therefore potentially useful is never published in full, e.g.:

a. Specialist reports.

b. Borehole logs.

c. Field notebooks, maps and detailed records from which publications have been summarised.

d. University theses and dissertations.

Such information, because of its detail, may be of the greatest practical use for engineering purposes, yet problems may be encountered in locating it and securing permission for its release.

Sources of geological information

5. The main sources of geological information in England are listed in Table 14.

TABLE 14. SOURCES OF GEOLOGICAL INFORMATION IN ENGLAND

Serial No.	Place	Information available	Remarks
(a)	(b)	(c)	(d)
1	A. GENERAL INFORMATION: Institute of Geological Sciences Exhibition Road South Kensington London SW7 2DE (Regional offices in Leeds, Edinburgh and Belfast)	Maps and literature, world-wide coverage.	Constituent body of Natural Environment Research Council. Specialist advisory service is also available through staff members. Library of the Overseas Division contains information on most developing countries. Visits should be made by appointment.
2	National Reference Library of Geology Geological Museum Exhibition Road South Kensington London SW7 2DE	6 in to 1 mile maps of Britain. Many publications and maps of countries world-wide. Comprehensive filing system.	Available for copying. Also much unpublished data such as large scale maps and borehole logs. Access free. (Part of the Institute of Geological Sciences.)
3	Libraries of British Museum (Natural History) Cromwell Road South Kensington London SW7 5BD	All British and many foreign publications (including geological maps) relevant to the work of the museum.	Access free.
4	Library of Geological Society of London Burlington House Piccadilly London W1V 0JU	Information on Britain and overseas countries, also contain an engineering geology section.	Consultation fee must be paid by non-members.
5	B. SPECIALISED INFORMATION: National Coal Board Hobart House Grosvenor Place London SW1X 7AE (Also area and regional offices)	Detailed information about coalmining areas.	

6	Nature Conservancy (Geology and Physiography Section) Foxhold House, Thornford Road Headley nr Newbury Berks	Geology of conserved areas, sites of special scientific interest, some coastlines, new roads.	Constituent body of Natural Environment Research Council.
7	Institute of Hydrology Maclean Building Crowmarsh Gifford Wallingford Berks (Also Regional Water Authorities)	Hydrogeological data.	Constituent body of Natural Environment Research Council.
8	Soil Survey of Great Britain Rothamstead Experimental Station Harpenden Herts (Also at Aberdeen, Scotland)	Pedological soil surveys.	Constituent body of Agricultural Research Council.
9	Department of the Environment Lambeth Bridge House London SE1 7SB	Data on some resources and sites, UK and overseas.	Minerals Division deals with geological aspects of planning and environment, also sand/gravel supplies.
10	Transport and Road Research Laboratory Crowthorne Berks G11 6AU	Geological information relevant to road construction.	Under Department of the Environment.
11	The Water Resources Board Reading Bridge House Reading RG1 8PS	Hydrogeological data.	Under Department of the Environment.
12	The Building Research Station Garston nr Watford Herts	Some geotechnical data relevant to building construction.	Under Department of the Environment.
13	The Civil Engineering Laboratory Cardington Bedfordshire	Data relevant to foundations and to slope stability.	Under Department of the Environment.

Section 7.4

Books

6. The bibliographies of the Geological Society of America have been issued since 1933 and are available at most University or National Reference Libraries. They list all the geological information published about any country for a given year and in its present form have separate location and author indices. The value of the bibliographies lies in the detail of their references, but many of the publications listed may be difficult to obtain other than from the largest national lending libraries, e.g. The British Library, Boston Spa, Yorks.

7. Many countries have an established Geological Society which publishes papers primarily related to the country of origin. Copies of many of these papers are kept at the Geological Society of London.

8. Preliminary sources of information for site investigations in Britain are recorded in detail by DUMBLETON and WEST (1970). A comprehensive guide to geological reference sources (maps, books) is given by WARD and WHEELER (1972) and by WOOD (1973). A detailed description of all available maps is given by WINCH (1974). The International Union of Geological Sciences publishes the IUGS Newsletter which lists new geological maps in a systematic manner. The last four references are worldwide in scope.

9. Major libraries throughout the world are listed in THE WORLD OF LEARNING, published annually.

Institutions

10. Government institutes of geology and allied sciences provide the main reference source for geological information. They publish maps and literature, store unpublished data, and may provide a specialist advisory service through their staff members.

11. Most countries now possess an equivalent to the Institute of Geological Sciences in Britain, although this may be called a Geological Survey Department, Bureau, or Department of Mines and Mineral Resources. Evidently, the work produced by such an institution will be in the language of that country and a technical translation of high quality is often essential.

12. Universities usually have a Department of Geology or Earth Sciences and may possess information not available elsewhere, comprising research work in progress and unpublished theses, dissertations and reports. Many American Universities, for example, have sent students to developing countries for one or two year periods of study to be followed by the publication of a MSc or PhD thesis. The Departments of developing countries are usually most actively concerned with the primary or secondary geological mapping of their country. British universities have research interests overseas: details of staff and their

research interest are published annually by DEPT OF EDN AND SCIENCE. The address of overseas and universities and staff are given annually in THE WORLD OF LEARNING.

13. Local Museums frequently serve as depositories for, and sources of, geological data, both published and unpublished.

14. Certain national organisations in any country accumulate specialised geological information. Examples of these in Britain are shown in Table 14B.

15. Similar organisations exist in some developing countries but for many of these areas particularly useful information may be contained in unpublished reports of oil companies, mining companies, and civil engineering firms. The oil companies have produced geological maps for many otherwise unsurveyed areas and some of these have been published. A request for geological information, particularly about near-surface formations, may be received sympathetically depending on company policy and circumstances.

16. Besides the methods of obtaining geological information mentioned above, there are data retrieval organisations and geological abstracting services which will provide information for a fee.

17. *Satellite photographs* (*see* Section 7.1, paragraph 15). A catalogue and price list of satellite photographs can be obtained from Audio-Visual Branch, National Aeronautics and Space Administration, Washington DC, USA.

REFERENCE LIST—CHAPTER 7

BLYTH F G H and DE FREITAS M H, 1974 —*A Geology for Engineers*, Edward Arnold, London (6th Edn).

CP 2001, 1957 —Site Investigation. *Code of Practice 2001*, British Standards Institution, London (under revision).

DEPT OF EDN AND SCIENCE —*Scientific Research in British Universities and Colleges*. Vol. 1. Physical Science, HMSO, London, published annually.

DUMBLETON M J and WEST G, 1970 —Preliminary Sources of Information for Site Investigation in Britain. *Ministry of Transport RRL Report LR 403*, Transport and Road Research Laboratory, Crowthorne, Berks, England.

EARLE K W, 1965 —*The Geological Map*. Methuen, London.

GEOL SOC, 1972 —The Preparation of Maps and Plans in terms of Engineering Geology. Geological Society Engineering Group Working Party report. *Q.Jl Engrg Geol. 5.*

GEOL SOC, AMERICA —Bibliographies and indices of the Geological Society of America. Published as 12 paper-bound monthly issues in a volume year.

HIMUS G W and SWEETING G S, 1968 —*The Elements of Field Geology.* University Tutorial Press, London (2nd Edn).

LAHEE F H, 1961 —*Field Geology.* McGraw Hill Book Co., London (6th Edn).

LEGGET R F, 1962 —*Geology and Engineering.* McGraw Hill Book Co., London (2nd Edn).

WARD D C and WHEELER M W, 1972 —*Geologic Reference Sources.* The Scarecrow Press, Metuchen, New Jersey, USA.

WINCH K L (Ed), 1974 —*International maps and atlases in print.* Bowker, London.

WOOD D N (Ed), 1973 —*Use of Earth Sciences Literature.* Butterworth, London.

WORLD OF LEARNING —Two volumes published annually. Europa publications, London.

CHAPTER 8

TERRAIN EVALUATION

Section 8.1. STAGES OF TERRAIN EVALUATION

1. *Introduction.* Terrain evaluation describes the processes and techniques used to assess an area of ground which is of interest to engineers. This application is not restricted to the location and design of civil engineering structures, but may be used for other purposes, to assist in the choice of suitable vehicles, either for construction or cross-country mobility, to estimate the effect of relief on vehicle operating costs, or to assess the advantages of terrain and vegetation in providing cover for military operations. The scale of the evaluation can vary from a national, or even international, assessment, down to a particular project. Whatever the scale and purpose, the process can be separated into four related stages.

2. *Project definition.* The object should be clearly understood to ensure an effective terrain evaluation. These objectives taken in conjunction with the required accuracy, which in itself is a function of the scale, will determine the features which have to be mapped and defined. In some cases one particular feature may be the most important, e.g. depth of overburden for quarries or spacing of major trees for vehicular access. Most investigations, however, are more complex than this, involving the identification of several factors.

3. *Mapping.* The extent of the features described must be shown in some way. For a small investigation it may be sufficient to define a uniform range of properties which applies to all of the ground. Usually some sort of subdivision of the area is required and a map or annotated air photograph may be the most convenient method of defining this. The type of mapping chosen will depend on the purpose and scale of the project. Where little initial effort is appropriate, as in some reconnaissance surveys, the mapping will be based on existing sources of information (*see* Chapter 7). For more detailed investigations the mapping units can be chosen to suit the scale of the survey and to express the relevant properties of the terrain that affect the particular project.

4. *Unit description.* The first stage of the project will have established the main features of the terrain and the required level of accuracy. The processes of mapping the units and defining their properties are interrelated, since a provisional mapping unit which is extremely variable may be divisible into smaller more uniform units. The decision to subdivide such a unit will be justified if the resulting increase in accuracy of the survey will pay for the extra effort required to map in greater detail. The information used to define the units may be obtained

171

using the site investigation techniques described in the following chapter; the frequency of these measurements will depend upon the variability of the terrain and the required level of accuracy. In many cases indirect methods of estimating the properties will be used, including: geophysical investigations (Chapter 6); literature searches (Chapter 7) leading to a correlation of known and unknown areas, especially by air photographic interpretation; and the use of other forms of remote sensing to measure particular properties of the ground. The use of air photographic interpretation and the classification of ground into units of landscape are two techniques which have become identified with terrain evaluation as they have proved effective over a wide range of conditions.

5. *Presentation.* For any survey to be effective there must be full communication between the surveyor and the user. For engineering surveys the normal method is to present a map and a report detailing all relevant results obtained. Where there is a large amount of information it is usual for the surveyor, who is more familiar with the material, to extract the most important information and prepare a summary report. When there is no time limit to a project, such as a national data store, then it becomes more difficult to define the appropriate method of presentation. A map is a convenient method of summarising the properties of a large area, but the cost of conventional map making is fairly high which means that it is difficult to justify frequent revisions. If data is recorded on a computer it is possible to reproduce it graphically either in its original form or, more usefully, by computing average values for limited areas. This process is most easily applied to the mapping of one specific characteristic at a time, although to obtain a complete understanding of a problem it may be necessary to produce several maps of different features; this is the basis of parametric classification described later. Because of the expense and inflexibility of map production it will normally be used to show general features from which specific detail must be interpreted from the map legend, or by reading a report. As a result of these shortcomings alternative methods of recording and presenting information have been investigated, particularly data stores. A data store can range from a collection of file boxes to an elaborate computer programme. A simple system can work well provided the amount of data is limited, or it is possible to accept considerable simplification of the information, and that it is all managed by an experienced operator. When it is required to file all information and process it in many different ways, possibly unforeseen when the system is set up, then the use of a computer will provide a more effective solution.

SECTION 8.2. TERRAIN EVALUATION TECHNIQUES

1. *Remote sensing methods.* The techniques particularly relevant to terrain evaluation are those that gather ground information by indirect methods, or present data in such a way that a reasonable interpretation may be made with a high degree of confidence. Remote sensing is the general name given to all measurements of the Earth's surface made from aircraft or satellites. Most of the techniques depend upon the recording of energy from part of the electro-magnetic

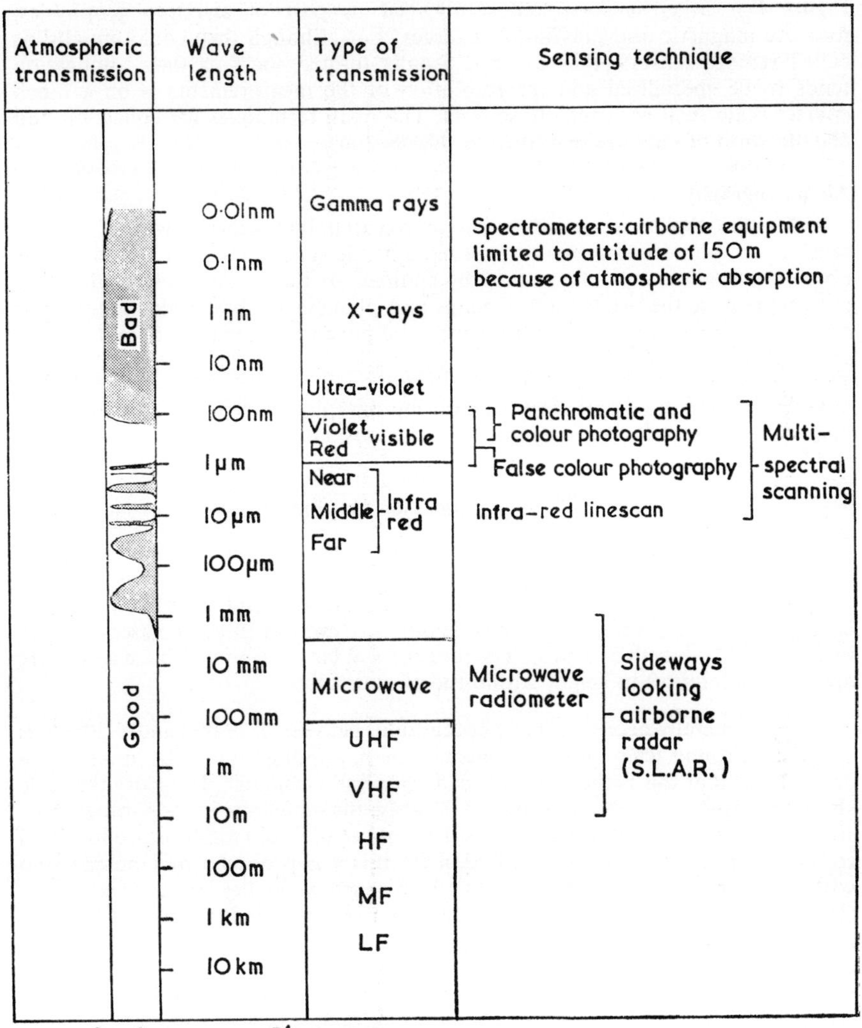

Atmospheric transmission	Wave length	Type of transmission	Sensing technique
Bad / Good	0·01 nm	Gamma rays	Spectrometers: airborne equipment limited to altitude of 150 m because of atmospheric absorption
	0·1 nm		
	1 nm	X-rays	
	10 nm		
	100 nm	Ultra-violet	Panchromatic and colour photography
	1 µm	Violet / Red visible	False colour photography — Multi-spectral scanning
	10 µm	Near / Middle / Far Infra red	Infra-red linescan
	100 µm		
	1 mm		
	10 mm	Microwave	Microwave radiometer — Sideways looking airborne radar (S.L.A.R.)
	100 mm		
	1 m	UHF	
	10 m	VHF	
	100 m	HF	
	1 km	MF	
	10 km	LF	

Note: $1\overset{\circ}{A}$ (Ångstrom) = 10^{-4} µm (micron)

Fig 78.   Remote sensing techniques related to wavelength of the electromagnetic spectrum

spectrum ranging from gamma rays, through the visible spectrum to radar (*see* Figure 78). Other sensors, well established as part of airborne geophysics, measure magnetic and gravitational forces; but although they could equally be called remote sensors, they are not usually termed such, as their application tends to be specialised and the resolution of the measurements is on a much coarser scale than with remote sensors. The main techniques for collection and classification of data are described in this section.

Air photography

2. Despite the interest and extensive research into other forms of remote sensing, air photo interpretation still remains the most important technique for obtaining engineering information by indirect methods. As most engineering projects require the production of maps and plans, overlapping air photographs are often commissioned for photogrammetric purposes, or may be obtained from existing air photo libraries.

3. *Scale of photography.* The scale of photography is determined by the flying height of the aircraft and focal length of the lens, i.e.

$$\text{Scale} = 1 : \frac{\text{flying height}}{\text{focal length of lens}}$$

The flying height is largely determined by the required accuracy of mapping; as a rough approximation the flying height should not exceed one thousand times the specified level of accuracy of height measurements. Until recently most aerial cameras used the six inch lens (focal length 154 mm) or in some cases a twelve inch lens (focal length 310 mm). The longer focal length is preferable where there are great differences in height within the photography.

4. It is generally accepted that photography between 1:20 000 and 1:30 000 is the best all round scale for photo interpretation, although it may be more than a coincidence that this is the range of scales normally available. In theory the scale should be large enough to show the smallest significant feature in the photography. In practice this may not be feasible, as the number of photographs is proportional to the square of the scale. One solution for this was proposed by Goosens who wished to map the occurrence of deep narrow gullies in alluvial areas (GOOSENS 1964). Although these could have been identified at a scale of 1:20 000 he preferred to use 1:40 000 scale photographs which were adequate for all other tasks, and from these to identify all areas likely to contain gullies. The occurrence of these was then confirmed by visual inspection by light aircraft. Another example of the effect of scale on interpretation was a photo-geological study of the Kainji dam site on the river Niger.

Detailed photographs at a scale of 1:6 000 existed and were used to map the position of outcrops; but it was found that for this area of peneplaned crystalline rocks the high quality photographs at a scale of 1:40 000 were preferable for locating the major geological features such as faults, shear zones and even geological boundaries.

5. Procedure for examining air photographs. The normal method of examining air photographs is to make a print lay-down or mosaic. This is very important for geological investigations as it should be possible to pick out large scale regional patterns which do not show up clearly on individual photographs. Similarly, the first definition of land systems is seen more clearly on a mosaic than on many separate prints. As an alternative to the print lay-down, satellite imagery or radar mosaics may be used to obtain an overall view of the area. The lower resolution of these systems, which suppresses minor ground detail, can be advantageous on a large scale, and the absence of tonal variations between the individual prints of a photo-mosaic assists in tracing features over a long distance.

6. A stereoscope is needed to obtain a three-dimensional image of the ground. When examining a pair of photographs with a stereoscope the first step is to mark the centre of each photograph, known as the principal point (*see* Figure 79) using the fiducial marks found at the corners or edge of the print. The site of the principal point (**PP**) on each photograph is found on the other photograph and marked as the transferred principal point (**TPP**). If these four points are then aligned parallel to the axis of the stereoscope, at the correct distance for the particular instrument, an image of the relief in the photographs can be seen. An important feature of this image is that the impression of the relief is exaggerated so that trees appear higher and slopes appear steeper than would be expected. This exaggeration, which normally appears to be about $\times 3$, is an important aid to the effective examination of the terrain, helping the interpreter to see minor changes in slope and very small differences in elevation or relief. In mountainous terrain this degree of exaggeration is a disadvantage as it is difficult to distinguish steep from very steep slopes; in these conditions the use of a camera with a longer focal lens reduces the exaggeration, and may be preferable.

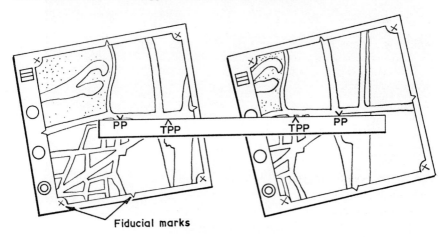

Fiducial marks

Fig 79.   Linking up a pair of air photographs

175

Fig 80.   Annotated air photograph (after DUMBLETON and WEST 1970)

NOTES

Location: Theale, Berkshire   *Scale: 1:12 000*   *Date: 1 January 1965*

This area shows a valley with river features and river deposits, bordered by an area of higher ground consisting of the Chalk with overlying sand and clay of the Reading Beds. Man-made features include roads, railway, canal, pipeline, drainage measures and domestic and industrial buildings.

The River Kennet K has associated with it a number of ox-bow lakes and abandoned channels. Below the river in the photograph the flood plain is occupied by Alluvium, and is crossed by the railway R and by numerous streams and abandoned stream channels. Most of the fields are under pasture, and the water-table must be high. One of the streams passes through a small area of peat, where drainage measures can be seen at P. A soil survey showed that here the peat is between 1 m and 3 m thick. Theale T is situated on Valley Gravel clear of the Alluvium of the flood plain of the River Kennet, and the A.4 Trunk Road A–A', which passes through Theale, also avoids the flood plain as far as possible.

In the lower part of the photograph the Chalk rises from the edge of the flood plain, which is bounded by the unsurfaced road U and the stream S. The Chalk forms a well-drained area of rolling topography, free from surface drainage, and covered by a thin layer of Reading Beds except where it is exposed along its margins bordering the road and the stream. The fields are larger, and are under arable cultivation indicating a drier or lighter soil. At D and D' there are depressions due to swallow-holes caused by solution of the chalk.

The area illustrates how topography, drainage and land-use can be recognised on air-photographs, and can be used as pointers to the geology and soil conditions. It also shows how air photographs may assist in the detection of unfavourable natural features such as areas of peat, old river channels which may contain soft silty materials, and swallow-holes which may lead to subsidence. Mapped man-made features such as roads, canals and buildings can be examined as they are on the ground, and recent unmapped developments such as building construction can be located. Less obvious man-made obstruc-tions can also be seen, such as the buried pipe-line L–L', and the site of a projected power-line Q, where a swathe has been cut through a wood.

Photograph by Fairey Surveys Ltd for Sir Alexander Gibb and Partners

177

7. To assist the recognition of topographical features air photo keys can be prepared, consisting of typical photographs annotated to show the points of interest. An example is shown at Figure 80. These keys may be classified according to the particular information that they contain, but increasing use is being made of land system analysis to provide a classification system of the features shown in air photos.

8. The use of such a key can assist in the second stage of interpretation, where predictions are made about conditions that cannot be observed directly. This process of deductive reasoning obviously improves with experience and appropriate training, but the provision of a key and the appropriate background information assists both the experienced and untrained interpreter to make more accurate deductions about the ground conditions.

Other methods of remote sensing

9. Aerial colour films aim to reproduce the colours as seen by the eye, and panchromatic films represent the full tonal range of the visible spectrum by equivalent grey tones. All other forms of remote sensing measure some specified part of the electromagnetic spectrum and, although the final product is often presented as a photographic print, the image cannot be interpreted conventionally. The techniques used can be conveniently divided into two groups; those that record the optical image directly onto photographic film, and those that generate an image which may be recorded photographically or onto magnetic tape.

10. The direct photographic techniques are limited to the visible and near infra-red part of the spectrum (Figure 78); outside this zone it becomes impractical to shield the film from fogging. One recording method is to use optical filters to limit the light to the frequency required and record on a black and white film. Cameras taking four simultaneous images are common, and systems with up to nine images have been used. By subsequent projection through filters, each corresponding to the original frequency, it is possible to recreate an image with true colours from three or more suitable images.

11. *Use of false colour.* An additional record can also be made of the infra-red band on film which, when developed, makes a visible presentation. When this infra-red image is added to the other bands, then a 'false colour' film will result. The normal 'false colour' film contains three emulsions, one of which is sensitive to the near infra-red wavelengths, and thus the colour produced on the film will depend on the amount of near infra-red energy which is reflected. Where a strong contrast in the reflectivity of infra-red is correlated with an important feature to be identified then the false colour film helps to identify the boundaries. An early use of this was to detect camouflage from the surrounding vegetation and this has led to the use of this type of film to examine variations in vegetation. Healthy vegetation reflects the infra-red more strongly than when it is diseased, and similarly vegetation near streams or other sources of groundwater can be identified. In the interpretation of the stability of slopes the presence of high

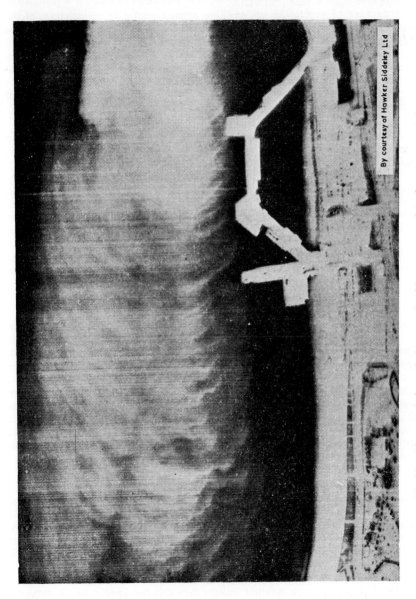

By courtesy of Hawker Siddeley Ltd

Fig 81. Linescan infra-red photograph of hot water outflow from power station

179

moisture content is a potential sign of danger, and in some cases it is possible to detect damage of vegetation on unstable ground owing to disturbance of the roots. This will show in the reflection of infra-red in the leaves long before the usual observation of trees leaning over due to soil creep. The presence of bare rock on such a slope is also easily seen, normally as a bright blue. Another useful feature of infra-red photography is that it provides a greater contrast between water and damp soil than normal photography, which leads to a quicker and more accurate survey of the extent of water, e.g. to assess the extent of flooding.

12. *Scanning detectors.* Other remote sensing devices use some sort of detector to produce an electronic signal, from which an image is built up. The two main systems are infra-red linescan equipment and side-look airborne radar (SLAR). The scanning equipment measures both emitted and reflected radiation and, by using suitable filters and detectors, it is possible to limit the measurements to certain spectral bands. Multi-spectral scanners simultaneously record bands from the visible and infra-red portions of the spectrum. Most experience has been obtained using instruments with one detector, such as the infra-red linescan detector. This equipment is normally designed to record in the middle or far infra-red sectors with wavelengths of 2–5 and 7–14 microns respectively. Beyond 25 microns transmission of infra-red may be considered impossible; at longer wavelengths atmosphere absorption is again reduced and it is possible to use this type of sensor to record at micro-wave length (*see* Figure 78).

13. The radiation measured by all detectors includes reflected solar radiation and emitted radiation. The emitted radiation is determined by the temperature of the object and its emissivity, which can vary with soil type, moisture condition, vegetation coverage and surface roughness. However, assuming a constant or known emissivity, it is then feasible to plot variations in temperature, and this is particularly useful in detecting currents in water. The water discharges from industrial works will be warm, but springs of fresh water discharging into the sea will usually be cooler. The recording of emissivity by linescan equipment must be made at night when the reflected solar radiation is absent (*see* Figure 81). The near infra-red recording, either by photography or sensor, is made in day time to measure the reflected solar radiation.

14. *SLAR sensing.* The SLAR form of sensing records the reflection of a radar pulse transmitted from the aircraft. Thus the system is independent of external energy and because the wavelengths used are not affected by cloud it is possible to obtain imagery at any time of the day or year. This ability to produce an image where conventional photographic techniques cannot be used means that SLAR provides an ideal form of reconnaissance survey. It is possible to produce an accurate mosaic which can be used for interpretation. Typical scales of radar imagery available commercially are 1:100 000 to 1:250 000 with a resolution between 10 and 30 metres. Smaller objects than this can appear on the image if they are strong reflectors, and it is possible to enlarge the original material.

These mosaics are suitable for the identification of regional and geological features and for a preliminary identification of terrain units. Lateral overlap of radar cover can be used to give a stereoscopic image, which will give a more accurate interpretation of the terrain.

Section 8.3. TERRAIN CLASSIFICATION AND MAPPING

1. *Classification methods.* The first stage in a mapping project is to decide the appropriate type of survey unit. There are two different approaches to this in terrain evaluation, the parametric and the landscape classifications. The parametric classification is based on the measurement of factors appropriate to the particular study, such as slope class, vegetation, or the extent of a particular rock type. The landscape classification is based on the main geomorphic features of the terrain, which are easily identified by air photo interpretation. In practice there is overlap between the two approaches in that parametric measurements are often used to substantiate a landscape classification and a geomorphic approach may be the only way of extending a parametric survey into an unknown area.

Parametric methods

2. Where the objectives of the survey can be clearly related to some readily identified feature of the terrain then a parametric type of classification should be used for survey. In practice the classification will normally involve more than one terrain feature and mapping units will have to be established based on subdivisions of these features. These subdivisions will reflect the intended use of the system, but inevitably involve some degree of generalisation. For example, in subdividing steepness of slope for assessment of trafficability, 45 degrees represents the maximum gradient for tracked vehicles, and the shift from second to first gear in wheeled vehicles takes place at around $26 \cdot 5$ degrees. These two parameters could then be used as criteria for mapping slope classes, although, owing to the inherent variability of ground, it is usually necessary to make generalisations involving the combination of classes.

3. The recording of quantitative data for this type of survey involves a considerable effort in the field, although the use of classes with arbitrary boundaries can speed up this process. The requirement for large amounts of data could be met by remote sensing systems which can be set to identify the presence of certain features. A more usual example is the use of semi-skilled survey assistants on field surveys who can be quickly trained to identify specific features and make simple measurements. From these measurements maps can be drawn for each terrain factor showing the boundaries of the classes. By superimposing different maps it is possible to build up a range of factor complex maps, with units determined by the intersection of factor boundaries. This tends to give a very complex map, because in theory the number of factor complex units is 2^n where 'n' is the number of factor classes recognised. A more common approach to parametric mapping is to identify areas of apparently uniform terrain, usually by means of

air photo interpretation, and then to measure the various factors for each area. In this approach the use of terrain features is only used to assist the placing of boundaries.

4. In the landscape approach to mapping the terrain units are identified and classified as such and the range of other parameters of interest to the user is defined in relation to each terrain unit. This may mean that the variability of a particular parameter of interest on a terrain unit is unacceptably high for some purposes. In such a case a parametric subdivision will be needed, although it must be realised that this will involve a very large increase in survey effort as the boundaries cannot be related to visible features. In effect this means that detailed investigations, such as are described in the following chapter, will use a parametric classification, and more general surveys can be of either type. In the absence of existing information, such as geological or soils maps, a landscape classification is likely to provide the most efficient basis for survey.

Landscape classifications

5. The increasing use of air photography, particularly at the reconnaissance stage, has led soil surveyors in different countries, working in different disciplines, to consider the merits of landscape units as a basis of mapping. At first soil surveyors used the air photographs to assist them in placing the boundaries round their normal soil units, such as the roadmaking gravels for the civil engineer, zones of trafficability for the military engineer, or soil series for the agriculturalist. In much of this early work the relation between soil type, land-form and soil behaviour was emphasised—'similar soils are developed on similar slopes under the action of weathering of similar materials' (BELCHER 1948). It was also noted that the different patterns of features seen in photographs corresponded to major changes in soil conditions. Detailed examination of these patterns showed that the differences could be caused by a variety of features, such as soil type, affecting the air photo tone, drainage intensity, vegetation, land use, etc., all features relating to soil conditions. Thus it was realised that the air photo patterns provided a breakdown of terrain into mapping units which could be defined in terms of surface expression, mainly topography and vegetation. It was also felt that this approach offered the opportunity to evolve a basic form of mapping which was not biased by the user's requirements. In this way it is possible to make an engineering evaluation of a map which may originally have been commissioned for agricultural purposes.

6. *Nomenclature.* The concept of survey by land classification was first developed and used in Australia to express the agricultural potential of ground (CHRISTIAN 1958). Subsequently these concepts were expanded and developed in the United Kingdom (WEBSTER and BECKETT 1970) for use overseas. The following system of nomenclature is generally accepted (BRINK et al 1966). The units are, in decreasing order of size:

Land Zone.

Land Division.

Land Province.

Land Region.

Land System.

Land Facet.

Land Element.

For most purposes users are concerned only with the last four units, and it is convenient to describe the last three units of the hierarchy first, as the classification of land regions and higher units is based on the land system.

Units of landscape classification

7. *Land system.* The basic pattern of terrain features, which can readily be identified on aerial photographs (*see* Figure 82) is called a land system. To convert a pattern identified on the photographs to a land system, it is necessary to define the geology, climate and range of small topographic units called land facets. The pattern persists to the limits of the geological formation upon which it is developed, or until the prevailing land-forming process gives way to another. At this point a new land system is developed, characterised by a different group of land facets. Although land systems may be provisionally mapped on print lay-downs at scales of about 1:100 000 they can only be properly defined by describing their complement of land facets. Land system maps are usually prepared at scales of 1:500 000 or 1:1 000 000, but more detailed maps may be necessary in complex terrain.

8. *Land facet.* The land facet is the basic unit of mapping. It is also defined by its geology, water regime and topography, but in a much more restricted way than the parent land system. A land facet has a simple form, on essentially homogeneous substrate, and a single water regime (both surface water and groundwater). Minor variations of these features are allowed, e.g. a 'homogeneous' parent material may consist of alternating beds of sandstone and shale, or the depth of water table may vary seasonally or according to the shape of the ground. All such variations are predictable and the important factor is that the materials developed on the facet are fairly uniform, such that a pedologist would map its soils as an association of soil series, and an engineer would accept a single design specification for a section of road built on it. Land facets vary in size, and may be mapped on air photographs at scales between 1:10 000 and 1:60 000. In humid areas the larger scales are generally necessary, while in arid areas the smaller scales are adequate.

9. *Land element.* It frequently happens that a very small feature of the landscape is of particular significance to a proposed scheme. The feature is too small to be mapped, but is nonetheless important enough to warrant a special category. In this case a land element is recognised, the smallest unit of landscape that is normally significant. For example, a hill slope may consist of two land elements, a steep upper slope and a gentle lower slope. To an engineer each slope element

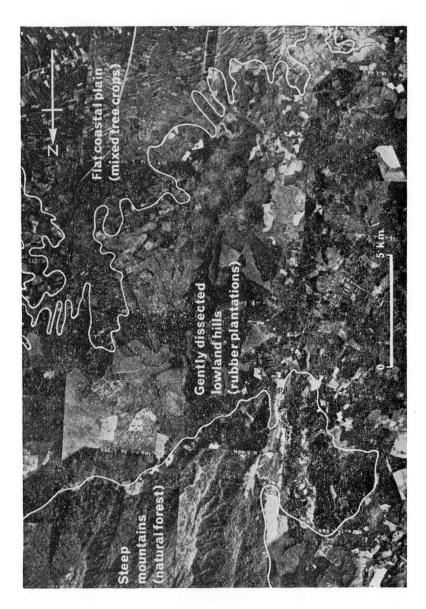

Fig 82. Air photo mosaic showing changes in pattern of terrain features (W. Malaysia)

will be treated differently when considering slope stability or amounts of cut and fill. Other examples of land elements are very small river terraces, gully slopes and small rock outcrops.

10. The land system, land facet and land element are the main units of the terrain classification, and the relationship between them is shown in Figure 83. The occurrence of one or more land elements in a particular facet is predictable, although they are not necessarily always present. Similarly, within the land system, land facets occur in a certain pattern, but some minor facets may not

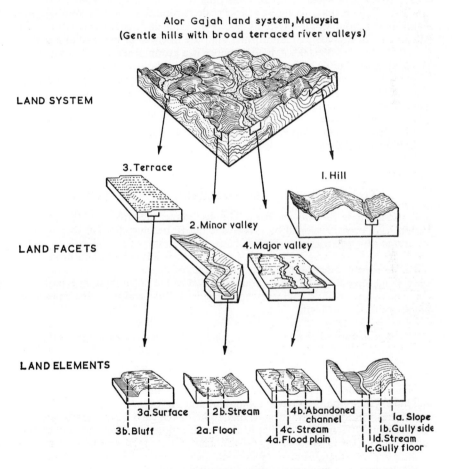

Alor Gajah land system, Malaysia
(Gentle hills with broad terraced river valleys)

LAND SYSTEM

LAND FACETS

3. Terrace

I. Hill

2. Minor valley

4. Major valley

LAND ELEMENTS

3a. Surface   2b. Stream   4b. Abandoned channel   Ia. Slope
3b. Bluff   2a. Floor   4c. Stream   Ib. Gully side
4a. Flood plain   Id. Stream
Ic. Gully floor

Fig 83.   Diagram to show the relationship between land system, land facet and land element (*see* Table 15)

185

appear in all parts of the system. In other cases it is necessary to recognise a variant land facet which has unpredictable differences from the standard facet. Such differences generally occur below the surface and have little or no surface expression and therefore the frequency or extent cannot be predicted. Its presence may be forecast from a knowledge of the area, but it can only be mapped by field investigation. Thus a land element is a different part of a land facet which is definable and predictable; a variant is a different kind of a land facet which, though definable, is not predictable.

11. *Higher land units.* Of these units only the land region is likely to be used in engineering evaluations such as large scale feasibility studies. A land region is made up of a group of land systems having the same basic geological composition and an overall similarity of landforms. They are likely to be mapped at a scale of 1:1 000 000 to 1:5 000 000. A land province is a group of land regions belonging to a major structural, geomorphic, or sometimes lithological, unit. Appropriate mapping scales are of the order of 1:5 000 000 to 1:15 000 000. The land division is a major morphotectonic structure, such as a fold mountain belt or a continental shield area. The land zone is a major climatic region of the world; an investigation has been made of the extent to which analogous land systems occur within a land zone, for example, the hot deserts (PERRIN and MITCHELL 1969).

SECTION 8.4. **THE USE OF TERRAIN EVALUATION**

1. *Applications for the engineer.* The techniques described in the previous section have been developed by specialists, but to become effective they must be integrated into engineering practice. Thus, it is necessary for engineers to understand the basis of these techniques, to realise their advantages and limitations, and to use them where appropriate. The scale of operation can range from the land system mapping of a whole country down to the positioning of a particular site. The relative role of engineer and specialist will depend upon the complexity and scope of the problem.

2. The compilation of a land system analysis of a large area is best performed by a team having a variety of experience in such disciplines as geology, geomorphology, ecology and soil sciences. If the main objective of the survey is some particular use then the relevant specialist is included in the team as this may influence decisions affecting the details of mapping. In general it has been found that land systems maps prepared for agricultural surveys are suitable for engineering surveys, although it was found that some of the early land classifications in Australia needed to be subdivided for engineering purposes. In a materials survey for a road in Malawi the existing land system map was used as a guide to the air photo identification of the land facets. These were then sampled and tested, leading to a pavement design for each individual facet. In this way it was possible to reduce the amount of field work to a minimum on a rational basis.

Land system maps which could be used in this way have been published in many overseas territories for agricultural, engineering and other land use surveys (*see* Chapter 7).

3. *Form of evaluation.* Nearly all terrain mapping is at land system level; land regions are not usually detailed enough, and land facet maps are too detailed for normal publication. To convert the land system map to a terrain evaluation it is necessary to describe the relevant engineering features of the land units. Most land system maps are accompanied by a report giving the basic information used to establish the classification of the survey area. The occurrence of land facets is normally shown on a block diagram (*see* Figure 18), cross-section, or a map; maps are more often used in flat alluvial areas where relative relief is very small. The facet descriptions will normally define the slope and soil profile with vegetation and water regime usually described where appropriate. Further information may be available according to the purpose of the original survey (*see* Table 15). An important feature of the report is that it should enable the user to identify the terrain units in the field. The land system may be identified from the map, but to distinguish the land facets and land elements a detailed description of identifying features is needed. A ground photograph may be useful, but the most effective method is to provide an annotated stereo-pair of air photographs and a list of distinctive features used in air photo interpretation. In this way the land facet descriptions, arranged according to land system, can be used as an index for air photo keys.

Engineering terrain evaluation

4. The present position of landscape mapping is that many areas have been surveyed but there is rarely much engineering information included in the descriptions. Thus the engineer has to extend the survey to make an engineering terrain evaluation. The descriptions of the land systems and land facets normally contain the necessary background information needed to make a preliminary assessment of the soil types and conditions in the area. From this it is possible to identify the facets, or elements, which are likely to need more detailed investigation, either because they are likely to be troublesome, or because they are advantageous. For example, in locating a road line certain alluvial soils may have a low bearing strength or be liable to lead to differential settlement. On sloping ground the construction of cuttings can also lead to failures if the slopes are too steep, or the area has particular features which could lead to potential instability, e.g. spring lines or evidence of past landsliding. On the other hand certain soil types associated with land facets or land elements may be useful for construction purposes, such as river terraces, or sand bars in an area of marine alluvium.

5. It is not usually necessary to map all of the facets in the area completely, but only to identify and map those relevant to the actual project. It is then essential that there is sufficient field testing to establish the relevant properties of each land unit and also the variability of these properties. Although the basis of

Table 15. ENGINEERING CHARACTERISTICS OF THE ALOR GAJAH LAND SYSTEM, MALAYSIA (*See* Figure 83)

Land facet	Form	Soils, materials and hydrology	Engineering properties and comments
(a)	(b)	(c)	(d)
1	*Hills:* **a.** Slope. Very gentle to gentle slopes (5)-10-12°, occasionally smooth but usually bumpy and irregular in detail. Overall straight or gently convex; often slightly concave in upper portion to give a more prominent hill top.	Up to 1 m of red or red-brown sandy clay over nodular or massive laterite up to 4 m thick over a silty clay mottled zone. Weathered rock may occur below about 4 m. The irregular development of laterite on the slopes tends to form the benches on the bumpy slopes, and it may outcrop on the steeper parts of the slope.	Main soil type (CH). Under normal drainage conditions for less than 150 commercial vehicles per day total pavement thickness should be 250 mm. In newly opened cut high natural moisture content may prevent compaction leading to lower bearing capacity. In alternating cut and fill, important to use the nodular laterite for the top layers of till. Mottled zone in new cuttings is susceptible to erosion. Recommended angle of slope between 40° and 50°. Nodular laterite (GC) suitable for sub-base but not for base unless stabilised with cement. Suitable for gravel roads, therefore useful in stage construction. Easily won from borrow pits.
	b. Gully. Sides 5 m deep. 20°, unstable. Bottom 6-10 m wide, flat, with a small stream (2 m wide) incised up to 2 m. 100-200 m and occasionally 750 m long.	As 1 **a.** Temporary stream.	Culvert across stream.

2	*Minor valley:* 30–100 m wide; gently concave or flat bottom with steep (20°) margins to facet 1; narrow stream a few m wide.	Variable soils, the weathering products of local rocks—mostly clays and silty clays. Poorly to well-drained, depending on materials.	Wet plastic soils in minor valleys. Fill usually imported from adjacent hill (facet 1 **a**). Important to use nodular laterite in upper layers. Slopes of embankment may need protection in new construction to prevent erosion. Culverts across streams.
3	*Terrace:* Level or very gently undulating, up to 120 m across but varies considerably along its length. Not continuous for more than about 2 km.	Variable according to local materials, usually light buff-coloured sand or clayey sand with poor profile development. Usually freely-draining, depending on composition.	Good subgrade and fill material.
4	*Main river valley:* **a.** Floodplain. Flat 100–300–(600) m wide.	As Facet 3, but wetter with impeded drainage. Usually flooded for padi.	Weak soils requiring fill embankments for road, about 1·5 m above ground level.
	b. Abandoned meanders and river channels. Sinuous, about 20 m wide; no topographic expression.	As **a.** but wetter. Probably composed of finer materials.	As **a.**
	c. River. Few 20 m wide; meandering.	Permanent flow.	Bridge crossing required.

189

land system mapping is that there is a direct link between land-form and soil type, the accuracy of this assumption should be tested in a survey for each land unit.

6. In addition to recording the properties of soils it is possible to link other engineering factors to the terrain analysis. A complete evaluation of soil type and strength enables the preparation of broad engineering designs, such as bearing pressure for the foundations of light buildings, together with recommended construction thickness of road pavements, even with suggestions for base material. Recommendations for slopes of cuttings can be made and also for the construction of embankments, e.g. maximum height of embankment to avoid overstressing an alluvial soil, or the necessity to use a retaining wall in steep terrain. Calculations of earthwork quantities in neighbouring land systems have shown significant differences and the relation of earthwork quantities to land systems should show an improvement over the typical classification of 'mountainous', 'rolling' and 'flat', used to make preliminary estimates. An example of the way data may be recorded is shown in Figure 84. Observations during construction, and when the works are finished, should also be related to the system, such as the performance of equipment on the site, and maintenance problems that arise on completion, which may lead to alternative designs in the feature. In this way a complete engineering evaluation can be built up of the land system and its units (DOWLING and BEAVEN 1969).

7. *Preliminary land system classification.* In the absence of an existing land system map, it will be necessary to make a preliminary land system classification. This may involve the use of a specialist consultant, although in many cases an adequate classification can be made by a geotechnical engineer familiar with air photography. The advantages of preparing a terrain evaluation over a wide area is that it can delineate the range of conditions likely to be encountered at an early stage of the survey before a project becomes tied to a particular site or route. An example of such an evaluation is shown at Figure 85. The later stages of the survey to gather the information about relevant land facets will then follow the same programme described for the evaluation of an existing land system map.

Use of data stores

8. Terrain evaluation provides the techniques to improve the efficiency and accuracy of preliminary surveys, so as to enable the subsequent site investigation to be correctly located and directed towards the relevant problems. Terrain evaluation also provides a rational method correlating known and unknown areas and thus transferring information and experience from one project to another. This ability to correlate between areas is the basis of proposed engineering data stores using terrain classification. From experience it is known that a substantial amount of information could be re-used on different projects if the relevant data could be extracted. Considerable thought has been given in several countries to the most suitable design for such a store (BECKETT et al 1972).

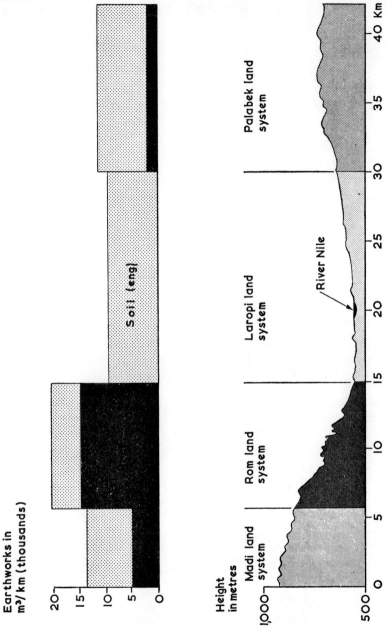

Fig 84.   Earthworks on a road in Uganda

191

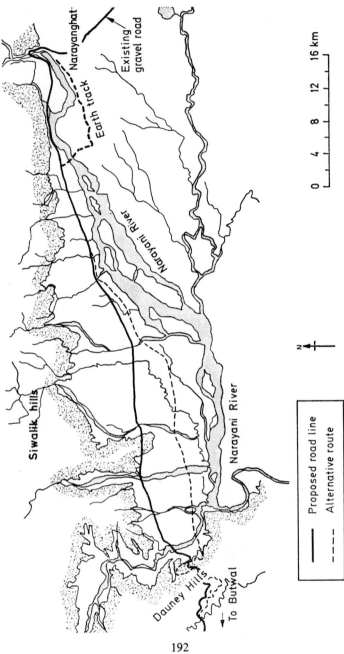

Fig 85. Road project in Nepal

NOTES ON FIGURE 85:

1. This section of the East–West Highway in Nepal provides an example of the use of terrain evaluation at an early stage of a project. As a result of studies certain sections of the route were realigned, and the search for road base materials was organised on a rational basis.

2. The section of road runs roughly east–west for 56 km, constrained on the north by mountains, and on the south by a river. In between the lower slopes of the mountains and the flood-plain of the river lies a complex arrangement of terraces with a well-defined piedmont fan system extending southwards from the foot of the mountains.

3. Air photographs of the area, at a scale of 1:12 000, were examined, first as a photo mosaic, and then stereoscopically. The main terrain types were delineated and potential sources of gravel, old stream channels, terraces and areas liable to inundation were marked on the photographs. Particular attention was paid to identifying the best crossing points for the three large rivers and several smaller rivers. The original line traversed low-lying terraces and flood-plains of fine-grained plastic soils used for wet-padi cultivation and involved the crossing of wide ill-defined water courses.

4. A more northerly route was traced out traversing elevated well drained gravel-bearing fans and crossing the principal water courses at points where they are narrowest and most stable, thus permitting bridges above flood level at a reasonable cost. Rough quantities were taken off by scaling from the air photographs and these showed the northerly route to be considerably cheaper.

193

9. To be effective a data store must be capable of accepting diverse information, such as maps, reports and test results, and of providing information on a wide variety of subjects. The large amount of data being produced and existing on files creates a problem of data handling. One approach to this problem is to ignore all previous data and build up a data store from new projects. This approach is favoured in South Africa, and information is filed according to terrain units which are broadly subdivided under geology, erosion cycle and relative relief. The alternative approach is to base the data store on a land system map, which could either be national or local. The cost of setting up such a store would soon be offset by savings in more efficient and effective surveys. In the meantime local systems based on the terrain brief and using a simple method of data storage can prove effective.

REFERENCE LIST—CHAPTER 8

ALLUM J A E, 1966 —*Photogeology and Regional Mapping.* Pergamon Press, Oxford.

BECKETT et al, 1972 —Beckett P H T, Webster R, McNeil G M and Mitchell C W. Terrain evaluation by means of a data bank. *Geog Jl* 138, 4, pp. 430–456.

BELCHER D J, 1948 —The engineering significance of land-forms. *Bull No. 13* Highw. Res. Bd. pp. 9–29. Nat. Res. Council, Washington D.C.

BRINK et al, 1966 —Brink A B A, Mabbutt J A, Webster R and Beckett P H T. Report of the Working Group on land classification and data storage. Mil. Engng. Expt. Estab. Christchurch, Hants. Report No. 940.

CHRISTIAN C E, 1958 —The concept of land units and land systems. *Proc. Ninth Pacific Sci. Cong.*

DOWLING J W F and BEAVEN P J, 1969 —Terrain evaluation for road engineers in developing countries. *Jl Inst. Highw. Engrs.* 14(6) 5–15.

GOOSENS D, 1964 —The use of aerial photography for the development of soil resources in South America. *Tenth Int. Cong, Int. Soc. Photogrammetry.* Photogrammetria Vol. 19 (1962–1964).

MITCHELL C W, 1973 —*Terrain evaluation.* Longmans, London.

PERRIN R M S and
MITCHELL C W, 1969

—An appraisal of physiographic units for predicting site conditions in arid areas. Mil. Engng. Expt. Estab., Christchurch, Hants. Report 1111, Vols I & II.

WEBSTER R and
BECKETT P H T, 1970

—Terrain classification and evaluation using air photography: a review of recent work at Oxford. *Photogrammetria*, Vol 26, pp. 51–75.

DUMBLETON M J and
WEST G, 1970

—Air photograph for road engineers in Britain. Tpt. and Rd. Res. Lab, Report, LR 369.

CHAPTER 9

SITE SELECTION AND INVESTIGATION

SECTION 9.1. GENERAL ASPECTS OF SITE SELECTION

Topography and land use

1. The form of the land surface is a function of geology and climatic conditions. In areas underlain by rocks of differing hardness and durability, the landform can be a direct measure of bedrock geology. On the other hand, in regions which have been subjected to glaciation or deep weathering, a cover of drift or residual soil may mantle, and so obscure, the underlying geology. Over-steepening of valley-sides or cliffs, or an increase in precipitation, can give rise to slope instability processes which form a part of the natural erosional cycle. Particular aspects to be taken into account in site selection are:

    a. Influence of bedrock geology on form of rockhead.

    b. Depth of penetration, and effects, of weathering.

    c. Distribution and type of overburden.

    d. Stability of valley-sides.

2. Modifications to the natural erosional cycle or ground surface by artificial means can give rise to significant engineering problems. For example, over-steepening of slopes by excavation may lead to landsliding, and urban construction can result in an increase in the rate of run-off and consequential risk of flooding. Natural patterns of vegetation are a valuable guide to sub-surface conditions and specifically to variations in groundwater conditions. In areas of uniform geology, vegetation changes can be directly related to the proximity of the water-table to the ground surface. Equally, zones of better drainage formed by coarse grained overburden or more fractured rock, can be identified.

Application of geology and hydrogeology in the prediction of ground conditions

3. During initial site assessment it is of importance to establish a preferred means of access because this may affect the choice of techniques which can be applied in site investigation and preliminary works contracts. In many cases partial access may be obtained from existing roads and tracks. Topography is a major factor which determines a preferential route although this may need to be modified in relation to the ground. In general, access routes should avoid low ground with a shallow depth to water-table which may be subject to flooding, thick organic deposits such as peat, unstable ground and areas exposed to severe weather conditions. Natural barriers can be created by rivers, deep valleys, steep

196

hillsides and major rock outcrops. It is of importance to predict both the relative ease with which rock and soil materials can be excavated, and the feasibility of using specific plant. Careful consideration should be given to the relative advantages of alternative routes. For example, it may be preferable to excavate a road by blasting rock on steep hillsides rather than by digging potentially unstable scree on a gentler slope.

4. The type and depth of the foundation selected for a specific engineering structure is determined both by the requirements of the structure and by the underlying geology. All buildings need a stable foundation which has adequate strength and minimum deformability. The heavier structures, such as power stations or multi-storey blocks, require, in preference, a rock foundation. However, if rock is not present below the site the structural load may be spread onto a raft or carried to considerable depth on piles. Shallow foundations, whether on rocks or soils, normally take the form of a pad or strip footing. The main factors, therefore, which need to be established at a site are the thickness and properties of the overburden cover, the properties of the bedrock and influence of weathering, and the depth of the water-table. The allowable bearing pressure for the possible foundation materials needs to be determined by either assessment of exposures and rock cores, or by *in situ* testing. Typical values for the maximum bearing pressure for different soils and rocks, as based on CP 2004, 1972, are presented in Table 16.

TABLE 16. **TYPICAL EXAMPLES OF ALLOWABLE BEARING PRESSURES**

	Types of soils and rocks	Maximum bearing pressure
(a)	(b)	(c)
I Rocks	Sound igneous and metamorphic rocks	10 MN/m²
	Massive limestones and sandstones	4 MN/m²
	Hard shales, soft sandstones	2 MN/m²
	Clay shales	1 MN/m²
	Hard, solid chalky limestones	0·6 MN/m²
	Thinly bedded or fractured rocks	Assessed after inspection
II Soils	Compact well-graded sands and sandy gravels	0·4–0·6 (Dry)　　0·2–0·3 MN/m² (Submerged)
	Very stiff clays	0·4–0·6 MN/m²
	Firm clays and sandy clays	0·1–0·2 MN/m²
	Loose uniform sands	0·1–0·2 (Dry)　　0·05–0·1 MN/m² (Submerged)
	Very soft clays and silts	0–0·5 MN/m²

1 MN/m² = 10 bars = 10·2 kgf/cm²

For rock foundations, weathering, fracturing and faulting can have significant influence on the selected bearing pressures. If the sequence is layered, as in the shale and sandstone alternation of the Coal Measures, variations in rock properties within the limits of a site will influence foundation selection. The dip of bedding, dominant jointing and faulting must be allowed for on sloping sites. Foundation conditions in superficial overburden materials are determined by the grading, moisture content and state of consolidation of the potential foundation. Soft cohesive soils, such as the post and late-glacial sediments in estuaries, and poorly consolidated granular sediments, of alluvial or glacial origin, may require special precautions. It is possible artificially to improve the properties of such materials *in situ* by drainage, or compaction by vibration with or without the use of sand or water. There may be a choice between founding the structure at high level in soil using a low bearing pressure and a deeper foundation on rock using a higher stress level; such alternatives are best reviewed in the light of the predicted construction costs.

A knowledge of groundwater conditions is relevant to foundation selection and the dewatering of excavations. Potentially deleterious minerals, such as gypsum or pyrites, may require the use of sulphate-resisting cements or other precautions. Hazards in foundation construction occur where the site is subject to subsidence or is potentially unstable.

5. The determination of the distribution of ground and surface water is an important aspect of most site investigations. The outcrop of the water-table, below which the ground is saturated, is typically represented by a spring line marked by seepages and marshy vegetation. The groundwater system is fed by percolation from precipitation and typically the sub-surface water flows downwards and away from hilltops. Artesian pressures may exist in areas topographically below the spring line. The perviousness of the surface determines the extent to which percolation or seepage occurs. In consequence, run-off tends to be high in areas underlain by clay and infiltration is high where exposed bedrock has open fissures or is a cavernous limestone. Artificial introduction of fluids by the process of infiltration may give rise to groundwater pollution. Pervious surface conditions will influence both the long term, average distribution of the water due to precipitation and short term flooding associated with high rainfall. Regions with steep valley-sides and underlain by impermeable rocks will consequently be associated with immediate flooding hazards, whereas the flow in rivers traversing limestone and chalk country will be largely determined by the groundwater level with maximum flow occurring some weeks after heavy precipitation. An additional factor arises when the surface rocks are soft and friable, and unprotected by stable topsoil or vegetation. In the lower part of catchment areas where the river profile is flatter the suspended sediment or bed-load tends to be deposited. Many such river beds illustrate a continual interplay between sedimentation and erosion. When groundwater flows through a granular material, excavation may of itself lead to obstruction of water flow by fine material in suspension in flooded borrow pits, clogging the pores in the ground. For instance this happens in gravel pits in plateau gravels in the Thames valley.

6. Engineering construction involves the excavation and possible re-use of soil and rock. Rock, such as gneiss, granite, limestone and sandstone, can be used for rip-rap, rock fill or building stone providing the discontinuity surfaces are widely separated and large blocks can be extracted. Similar rock types in a more closely fractured but still sound condition, can be used for concrete aggregates and roadstone; gravel deposits yield important sources of aggregate. Most sands used in construction as fine aggregates or filters occur as sedimentary deposits but it is possible to process such materials from rock. Rock and soils are used extensively as fill materials in the construction of embankments and rock bases. It is common practice in road construction to balance the cut-and-fill so that as much as practicable of the rock or soil excavated is re-used in construction. Such re-use may introduce special problems in handling, moisture control and placing. Dumps of mine waste can form valuable sources of fill. Natural cement-making materials include limestone, clays and natural mixtures of these materials. Pozzolans such as volcanic ash, fly-ash or ground brick, can be valuable additives to concrete which may improve quality and reduce cost.

Avoidance of pollution

7. The disposal of waste materials is a matter of increasing public concern, and it is important that the method of disposal adopted should not result in the creation of a potential hazard. Domestic, industrial or radioactive waste must be deposited in such a manner as to minimise pollution. The major pollution hazard arises from the redistribution of the waste products by groundwater flow. Domestic and non-toxic industrial wastes are commonly used in landfill projects, either on sites of old mineral workings or low lying ground marginal to a river or the sea. It is necessary to assess with care the potential hazards of disposal particularly if the waste deposit is situated partially below the water-table (Figure 86). If toxic wastes have to be placed in excavations they must be protected from nearby aquifers by thick impermeable clay layers. Sub-surface disposal of liquids in deep wells has been used in some countries. Waste spoil resulting from mineral working or engineering construction should be placed in stable mounds on level sites. The foundations to such spoil mounds must be composed of materials of adequate strength and the water-table should be well below the base of the mound.

Geological hazards

8. Mass movement results from the loss of stability in a hillside caused by natural or artificial over-steepening, overloading, subsidence, a rise in water-table or earthquake shocks (ECKEL 1958). The several processes of slope instability are discussed in Chapter 10.

9. Subsidence and ground collapse arise from a number of artificial and natural causes of which the following are among the more important: artificial extraction of fluids (oil, water, brine, etc) and gases, underground mining, solution of rocks, sub-surface mechanical erosion (piping), consolidation (gravity compaction)

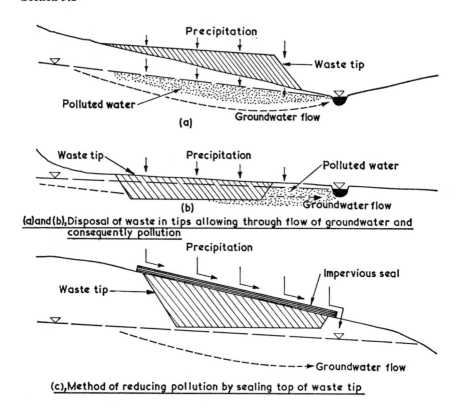

(a)

(b)

(a)and(b),Disposal of waste in tips allowing through flow of groundwater and consequently pollution

(c),Method of reducing pollution by sealing top of waste tip

Fig 86. Influence of groundwater conditions on pollution associated with waste disposal

of soft sediments, tectonic deformation (possibly associated with specific earthquakes), volcanic activity and the thawing of frozen ground (CIV ENG 1959). Some of the more spectacular effects of subsidence arise from the artificial processes but on a longer timescale geological processes can be of major influence. Pumping of groundwater may give rise to regional settlement as the water-table is successively depressed. Severe results of such pumping occur in areas underlain by sediment covering limestones which contain voids. The drop in water-table leads to a loss of support to the overburden and consequent ground collapse, which may be locally catastrophic (Figure 87). Solution of deposits of rock salt, by natural or artificial means, can lead to subsidence and this may continue for a significant period after the cause of solution has been removed. Sub-surface mineral extraction can be operated in such a manner as either to prevent subsidence or to permit controlled subsidence.

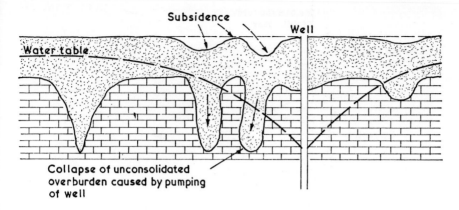

Fig 87. Artificially induced subsidence caused by pumping from limestone

Total extraction of flat-lying seams may lead to major settlement but this can be reduced if back-stowage of waste is used. Partial extraction can be carried out, so leaving pillars of stable rock to support the overlying rock and overburden. Block caving methods, as are used in many ore bodies, can also contribute to ground collapse. Considerable caution is needed in engineering construction over old shallow mine workings or those to be worked in the future, whether they are shallow or deep. The gravitational compaction of soft sediments results primarily from loading and consequential drainage of pore water. External vibrations can give rise to settlement, and certain soils, such as loess, are subject to collapse if flooded. An important factor in regional subsidence is continued deformation of the Earth's crust, possibly resulting from mountain-building or over-stressing by an icy cover during the Pleistocene. Regional subsidence at the moment accounts for most of the average rate of rise of sea level in SE England of 2–3 mm per year and this situation has resulted in the decision to construct a Thames Barrier to reduce the risk of damage from future storm surges.

Earthquakes

10. Individual earthquakes result from a fault movement within the Earth's crust which leads to the release of energy from the focus of the earthquake (Figure 88). The magnitude of an earthquake is the total energy released at the focus, whereas the intensity of an earthquake is a measure of the effects of the shock at the ground surface in the region around the epicentre (the point on the ground surface above the focus). As the energy travels out from the focus, there is attenuation and so the intensity of the earthquake reduces with distance from the focus (*see* Chapter 3). In major earthquakes, displacement on faults may extend to the surface and this can be detected by a step in the topography or offset in a stream or fence, depending upon the direction of fault movement.

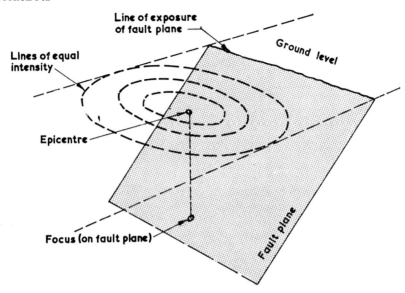

Fig 88. Location of focus and epicentre of earthquake

The prediction of earthquake activity must be based firstly upon an assessment of the seismicity of the region, which is deduced from the location and magnitude of past shocks, and the probability of the scale of the events in the future. The prediction must then be related to the ground response and for a specific structure to the structure-foundation response. Earthquakes can give rise to long period waves at sea and these are known as tsunamis. Such waves can build up over shelving shores and cause widespread damage in low-lying coastal areas. Earthquakes have also been generated by the artificial injection of water in wells at depth and the filling of large reservoirs (induced seismicity).

11. Current volcanic activity is experienced in the same general regions of the Earth's surface as the earthquakes described previously. The main hazards which arise from volcanic activity are associated with the eruption of lavas, ash clouds and the risk of catastrophic destruction of the area round a volcanic centre. In most instances, the potential hazards are clearly recognised locally but prediction is difficult or unreliable. Modern techniques can detect evidence of movements in molten magma at depth by changes in the gravity or magnetic field, or heat flow and consequential variations in ground temperature. Similarly, the techniques of earthquake prediction can be applied to volcanoes. Particular hazards occur when a hot ash cloud is ejected from the volcano and flows very quickly down the ground slope (nuées ardente), or when an ash cloud is converted to a mud flow by heavy rainfall. Nevertheless, controlled volcanic activity can be of value if natural steam from thermal springs can be harnessed to generate electricity.

SECTION 9.2. GEOLOGICAL FACTORS APPLICABLE TO
ENGINEERING WORKS

Choice of routes for communication and transportation

1. The main geological factors which need to be taken into account in route location may be summarised as follows:

   a. The occurrence of unstable ground which is, or could be, subject to landsliding.

   b. Unstable or soft foundations which are liable to collapse or excessive settlement.

   c. Control of ground and surface water particularly in areas underlain by impermeable rocks.

   d. Sources of construction material, and re-use of excavated rocks and soils.

   e. Design of structures such as embankments, cuttings, tunnels and the assessment of foundations for bridges.

The risk of foundation instability, resulting from either mass movement or settlement, is probably the major hazard encountered in road construction. Unstable slopes, or areas which could be activated by engineering works, can normally be identified from conventional site investigations coupled with careful field mapping and inspection of aerial photographs (*see* Chapter 7). The major uncertainty is that the mechanism of failure may be misinterpreted and so the instability could be analysed in an inappropriate manner providing a false impression of security. Such a situation arises when borehole samples do not give an indication of a critical slip surface. Foundation settlement may occur in soft, compressible sediments such as alluvial or estuarine silts and clays, organic deposits such as peat, and collapsing soils, such as loess and certain weathered rocks. Catastrophic collapse can occur as the result of sub-surface cavities, which may result from sub-surface solution of soluble rocks, such as limestone, or near-surface mine workings. In either case the risk of collapse can be recognised from general geological considerations but specific location of cavities can be difficult.

2. Roads are normally designed on the basis that there should be as complete balance of cut and fill as practicable. If such re-use is to be adopted in areas where there is an extensive cover of overburden, or soft rocks are to be excavated, it becomes of considerable importance to ensure that the fill material is of acceptable quality and not liable to changes in properties during bulk handling.

The moisture content of the fill determines to a large extent its workability by plant and the extent to which settlement may continue after construction. In view of the large volumes of material handled over relatively short periods, it is also essential that the mode of excavation can be predicted with some certainty, and the requirements for difficult excavation and blasting appreciated. The consequential effects on construction operations which arise from misappreciation of material properties can be considerable. Local sources of material, possibly won from the road excavations, can often provide suitable sub-base and base course materials. However, suitable materials for the wearing surface, and possibly other parts of the road structure, may have to be imported from some distance.

3. On most roads it is possible to identify particular areas where there is a major ground problem. Such difficulties may involve a large bridge, major embankment or deep cutting, a tunnel or an area of poor foundation properties. It is common practice to carry out separate investigations in such areas to ensure that proper consideration is given to these special conditions. Deep cuttings are normally designed on the basis of overall stability but some allowance is also made for the provision of drainage, surface protection, berms and rock trays to catch falling debris.

The detailed structure of rock faces cannot be predicted until after excavation and, again, it is common practice to put in extensive support to minimise the hazard of rock falls. The foundations of embankments may require special preparation and blanketing with a drainage layer of granular material. Appropriate sampling and testing of soft foundation materials will yield information on the probable settlement of the road; it may be preferable to excavate soft layers of the material prior to construction rather than cope with long-term remedial works.

Reservoirs

4. The main geological factor which influences the surface water discharge from a catchment area is the extent to which permeable rocks are present. In some circumstances there can be a major loss of surface water into the groundwater system and underground flow into an adjacent catchment area. For example, catchment areas which contain major limestone outcrops may be unreliable for reservoir construction but this is determined by mass geological structure. The exposed rock and soil materials, and the extent to which they are protected by vegetation, is a determining factor in the sediment yield of a catchment. The effective life of a reservoir, particularly in semi-arid or arid areas where soil erosion can be of major significance, is determined by the rate at which the reservoir becomes infilled with silt. Soft overburden, very weathered rocks or soft materials such as loess and volcanic ashes can be major sources of silt. If localised areas of sediment-production can be identified, then small dams can be created to store the silt, or the soil surface can be protected by re-grading and vegetation.

5. The initial selection of reservoir sites is based upon topography, coupled with a knowledge of the catchment yield in relation to the requirements of the engineering project. If the reservoir is to be constructed on the main river channel there is a minimum distance from the headwaters that a dam can be created to provide the necessary storage and yield. On the other hand, a dam can be constructed in a side valley, where the topographical and geological conditions may be more favourable, and water pumped at periods of high flow into the reservoir from the river. An ideal reservoir site consists of a wide, flat-bottomed valley floor which

becomes constricted into a narrow gorge, within which the dam can be constructed with minimum quantities of materials. In practice, most reservoir sites deviate from this ideal but the basic principle forms a valuable guide-line in the initial search. Once a series of alternative reservoir sites has been identified, geological examination of the dam and reservoir site is essential.

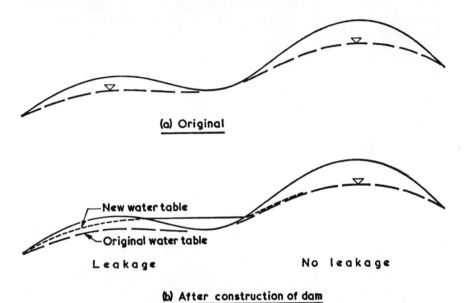

(a) Original

New water table

Original water table

Leakage   No leakage

(b) After construction of dam

Fig 89. Method of leakage from reservoirs

6. A prime requirement for a reservoir site is watertightness and this is determined by the groundwater conditions and local geology. It is common practice to associate satisfactory reservoir conditions with impermeable rocks such as shales, intrusive igneous rocks or metamorphic rocks. However, whether a reservoir is watertight or not is determined by the groundwater pressures below and around the reservoir flanks. If the groundwater pressure in the flanks of the reservoir is in excess of the top water level, then leakage cannot take place irrespective of the permeability of the reservoir flanks. Where the rock mass at depth is free-draining, or the reservoir site is perched above adjacent topography, there will be a down-ward component of groundwater flow (KNILL 1971). If the water-table, and so inevitably the groundwater pressure is less than the top water level, then leakage must take place (Figure 89), although the quantity of such leakage will be determined by the permeability of the rocks which form the reservoir basin. Leakage of reservoir water will take place at the dam site because of the steep

hydraulic gradient, but engineering measures are normally adopted to reduce this leakage to a minimum. In addition to the influence of groundwater on reservoir feasibility, the rock structure is important. For example, a sequence of bedded rocks will, if dipping away from the reservoir, tend to encourage leakage. Similarly geological defects, such as the presence of permeable or cavernous rocks (including limestones, evaporites and volcanic rocks), fault zones, permeable overburden and buried channels can contribute to leakage, although this may be localised and simpler to control and treat. If the water-table is deep, or the underlying rocks are very permeable, a natural blanket of impermeable materials can provide satisfactory reservoir conditions. The main risk in such circumstances is that the reservoir pressure will be such that the blanket fails and major leakage takes place.

If the reservoir level is liable to fluctuation, there is also the risk that a lowering of water within the reservoir will give rise to a drawdown failure in the blanket, leading to rupturing and leakage.

7. The choice of dam type is related to the foundation geology and the availability of construction materials. It is most important that site investigations and geological studies should be related to the type of dam to be constructed. For example, the Malpasset dam in France, an arch structure which failed in 1959, was sited on poor quality schists which would have been a more appropriate foundation for a gravity dam; indeed the early investigations at Malpasset were related to the construction of a gravity dam. Concrete dams of arch or buttress type and the large simple gravity dams require a rock foundation so the depth of overburden or weathered rock should be at a minimum. Embankment dams constructed from rock or soil materials may be built on deformable foundations such as deep overburden or soft rocks, or can be built in areas where the availability of suitable concrete-making materials is restricted. Once the dam type has been selected, appropriate foundation levels are chosen; special *in situ* or laboratory testing may be required to ascertain the stability of the structure. Seepage of water below the dam, and around its flanks, needs to be reduced and controlled by cut-off and drainage works. The cut-off may take the form of an excavated trench, backfilled with clay or concrete, a grouted cut-off formed by injecting grout mixtures from boreholes or an impermeable blanket laid on the reservoir bed. The detailed design and location of such cut-offs requires careful investigation and evaluation particularly in permeable rocks, or deep overburden. It is current practice to provide drainage, by means of boreholes and adits, downstream of cut-offs in order to reduce uplift pressures on the base of the dam and to reduce the risk of deleterious flow of groundwater through the foundations or base of the dam. Materials for dam construction are commonly won from the immediate vicinity of the reservoir. A concrete structure will require a source of coarse and fine aggregates which could be provided from either a quarry in rock or suitable alluvial deposits. The design of an embankment is determined by the properties of the locally available fill; however, unless it is hard and sound, the overall profile of the dam is determined by the properties of the foundation material. An embankment requires an impervious core, filter layers and shell materials, which are commonly free-draining. All dams have associated engineering works which

are primarily concerned with the control of water flowing out of the reservoir. Such works may require the provision of tunnels or channels to carry water; particular attention is needed in those areas where discharge of water takes place in view of the risks of induced erosion. Very often the largest structural works are associated with the spillways required to pass floods.

River works

8. The geology and topography of a catchment area can have a significant influence on river flow. If the catchment contains a significant cover of permeable overburden, or is underlain by aquifers, there will be a tendency for percolation during periods of rainfall, thus reducing the river flow. However, the river discharge may be maintained by groundwater feeding the river system. Thus there is, except for evaporation, a redistribution of run-off during the year rather than a loss of water. The shape and form of the catchment will also influence river regime. An individual storm which follows the line of river flow will result in a higher peak than one which crosses the catchment. Further factors of importance include the drainage density (length of river per unit area), the ratio of overland to channel flow, the slope of the ground surface and the gradient of individual streams and rivers.

9. Most river valleys have had a complex geological history during the past few thousand years. Many valleys were excavated to depths well below the present sea level during the Pleistocene and, with the rise in the sea associated with ice melting, the valleys became flooded and subsequently back-filled by alluvial deposits. The lowermost parts of such buried valleys are often composed of coarse materials laid down as channel deposits. The overlying sediments are generally finer grained and, particularly in flood plains associated with a meandering river, the pattern of alluvial types may be complex. This overall situation can give rise to difficult foundation conditions in valley floors where excavations or foundations have to be constructed in the heavily water-bearing or soft alluvial sediments. On valley sides, above the main level of the river, there may be river terrace deposits which accumulated in the valley flow when the river was at a higher topographical level at some period in the past. Such terrace deposits can form important sources of sands and gravels and provide well drained working areas.

10. One of the major problems which arises in river valleys is the influence which artificial changes in the valley will have on the river regime. It is normal practice to increase the control on a river as it flows through a built-up or developed area in order to minimise changes in its course. However, the natural pattern of river flow in a flood plain involves an interplay between erosion and deposition, coupled with the more catastrophic effects of a flood when the river banks become over-topped. Artificial changes to a course of the river tend to reduce the areas subjected to erosion and flood flow may be artificially contained. Streams may also be regraded to reduce their transporting capacity. Sediment banks can be built up within the river bed and these may need to be dredged at regular

intervals. Scour can be increased downstream if the river water is cleaner or the velocity increased. Engineering structures, particularly bridges can be subject to erosion of their supports. Localised scour of this type is often generated during floods.

Coastal works

11. Waves, and to a lesser degree currents, control the coastal process of erosion and accretion. Short period waves are formed by winds blowing over the sea surface, the steepest waves being caused by local winds and the longer swell by winds blowing a considerable distance off shore. As the waves approach the coast, shoaling causes wave refraction and the wave crests wheel to become more nearly parallel to the shoreline. Where the coastal morphology is uniform, a relatively straight beach will tend to form. Where the coastal structure is more variable, headlands may form against which wave energy will tend to be concentrated with partial protection to the beaches around the intermediate bays and the stable coastline will be crenellated. Offshore a platform (in rock) or a bar (in sand or gravel) will tend to form at about the point of breaking of the dominant storm waves. Bars vary in size and in position with the wave attack, and to a lesser degree with tidal movement. On a length of coast, long-shore drift of beach material will tend to be in the direction of the dominant component of wave momentum along the shore. Where the waves arrive sensibly parallel to a sandy shore, currents may also influence the direction and magnitude of littoral drift. If the drift is interrupted by manmade barriers or by natural features, the beach material will accumulate and as the angle of coastline changes with this accretion so will the rate of drift be reduced.

Away from the foreshore, as the depth of water increases so does the influence of currents, as opposed to the influence of waves, control sediment movement. Another possibility which must not be overlooked in selecting a coastal site is that appreciable onshore and offshore movement of material may occur under the action of waves and currents.

12. Sea action causes not only movement of material but also preferential movement of one size in relation to another and thus to size sorting. Generally, therefore, as one proceeds from the top of a beach into deep water a reduction in material size occurs. On exposed shores a stable beach will not be composed of material finer than a medium sand (0·2 mm), since the finer fraction is carried out to sea and ultimately deposited in deeper water, the size being related to motion near the bed. Beach material is derived partly from erosion by the sea of shore and submarine banks and partly from rivers. Off the mouth of a river, as the flow rate decreases, sediment tends to drop out and to form a bar. In the absence of other factors the bar will be parallel to the general shoreline; but if there is also long-shore drift, the bar may become angled towards the shoreline on the updrift side and may even be partially attached to the shoreline. The orientation of banks offshore is sometimes explained by the underlying geology and sometimes by a tendency to be aligned with the prevailing currents.

13. Relatively rapid changes in sea level will leave evidence of cliff line and wave platform. A drop in sea level may lead to a bar being formed on the platform and to a lagoon behind. This is the probable explanation for Chesil Beach, Dorset.

14. Where beach material is inadequate, waves will directly attack the cliffs, with the risk of falls and landslides depending upon the nature of the cliffs. Where longshore drift controls the line of beach it may be improved by the construction of groynes, but this will reduce the movement of beach material down-drift (Figure 90). There may be a very delicate balance between erosion and accretion; beach behaviour may be very sensitive to slight changes affecting exposure. Sea walls constructed to protect the coastline may increase the rate of erosion of the foreshore and will affect beach supplies if these were previously derived from the unprotected shore. Unless the position can be held with confidence, flexible revetments may be preferred. Breakwaters or groynes are no solution if beach material is being lost offshore.

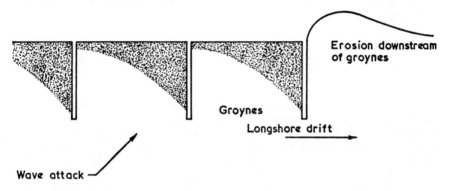

Fig 90. Longshore drifting

15. Harbours are generally constructed from the shore into deep water and their construction must take account of the changing circumstances to be encountered, including changes in scour or deposition caused by the construction. Offshore, consideration must be directed to variations in seabed and how these will be affected by any structure. Pipelines may be buried by dredging or by jetting and they should be set below the lowest seabed level likely to be experienced. In areas of fairly strong currents, sand dunes may occur on the seabed leading to appreciable fluctuation in seabed level.

Tunnels

16. Since the cost of a tunnel is so influenced by the nature of the ground, site selection will first entail a desk study of alternative routes on the basis of the known geology followed by stages of site investigation as the route (or routes) is

defined in greater detail and as the particular geological risks and hazards become better defined. The stages of a site investigation should conform to the following general pattern:

a. Study local geology and consider the relevance of local geological history and tectonic movements. Assess the validity of the evidence upon which geological maps are based and in particular the data from which geological sections are constructed.

b. Consider the principal areas of uncertainty in the geological structure relevant to all practical alternative tunnel routes within the imposed conditions (some alternative routes may be dismissed as economically impractical before starting upon site investigation).

c. Plan the site investigation to complement reliable known information on structure and on hydrogeology, also to assess suitability for possible systems of tunnelling by sampling, testing and recording. Do not overlook the benefits of a few large diameter (up to 1 metre) boreholes for direct examination, *in situ* testing and for subsequent inspection.

d. Design the testing programme to determine values of parameters of direct application to tunnel design and construction. Consider simple tests or geophysical logging methods to use in classifying and zoning the ground type, in order to be able to define the variability of the ground and the extent of application of the results of the more elaborate tests.

It is not possible to provide rules on the frequency of spacing of boreholes. At one extreme there may be sedimentary rocks so uniform in quality over a wide area that it is only necessary to be able, by identification of specific marker beds, to establish and confirm a continuity of sequence by means of a few boreholes, together with control of lithological variation. At the other extreme there may be igneous intrusions and metamorphosed rocks of such complexity as to necessitate a method of tunnelling tolerant of a wide range of possible circumstances, however well the ground may be investigated. Geophysical studies should be considered, to complement and extend the information obtained from direct geological evidence.

17. It may be necessary to acquire considerable information about rock strength and quality in order to design the scheme of tunnel construction. The cost of abandoning a highly mechanised scheme of tunnelling may be considerable and each such scheme can only tolerate a certain range of variability of the ground. If possible, the directions of dominant joint systems should be established as the amount of overbreak, and therefore the cost of excavation and the thickness of concrete in the linings will be affected by the direction of tunnelling relative to the direction of the joints.

18. Special hazards to be guarded against are squeezing rock, swelling rock and the possibility of encountering water in excessive volumes and pressures, high temperatures at depth, or inflammable or noxious gases. For tunnels beneath mountain ranges the costs of vertical boreholes can be very great and inference

may have to be drawn to a large degree from surface features; here it is necessary to combine the best geological advice with appropriate experience of encountering unsuspected difficulties. The capabilities of inclined and near horizontal drilling have advanced to the stage at which such techniques may well be considered for exploration of the ground along the proposed line of tunnel up to, say, 2 kilometres in length.

19. In areas of seismic activity it is usually found that maximum seismic motion reduces with depth below surface and that the danger to a tunnel arises only where it crosses a reactivated fault. Where such a crossing is inevitable the design of the tunnel (using a flexible or double skin of lining) must take account of the likely effect.

Section 9.3. SITE INVESTIGATION METHODS

Introduction

1. *Aim.* The aim of geological site investigation is to determine the distribution of the various types of soils and rocks in the area of the proposed works, and the physical properties of such soils and rocks in so far as they are relevant to those works. It is not normally sufficient to identify the strata by their stratigraphical terminology alone, but the soil or rock should be described.

2. *Sequence.* The first step, the desk study, is the examination of available geological maps, reports and memoirs as listed in Chapter 7. This will be followed by a preliminary field investigation to determine the main details of the geology of possible alternative sites. When the site has been selected, a fuller investigation follows in one or more phases to determine the detailed design. During construction further investigation may be required to discover the extent of local phenomena exposed during excavation or to elaborate previously-known phenomena where the design has been deliberately left tentative, e.g. the extent of a grout curtain or series of drainage wells at a dam site.

3. *Planning.* Each stage of the investigation must be planned primarily to determine the details required for that stage. In the preliminary stage the general distribution of rock and soil types, their main characteristics, and important phenomena such as large faults, old landslides or buried channels must be determined. Close control by a field geologist should enable the necessary information to be obtained at minimum cost. The detailed investigation which follows will include all necessary sampling and *in situ* and laboratory testing. A regular pattern of boreholes is frequently adopted, at such a spacing that interpolation between them can be based on sound geological judgement, and where doubt exists further examination must be made. The investigation should be carried out under the control of an engineer or engineering geologist fully conversant with the details of the proposed scheme.

4. *Presentation of results.* A full report, with all necessary drawings, tables and plans, and properly presented, is an essential part of every site investigation. This should include: a geological plan and sections, showing all exposures and the location of all boreholes, trial pits and adits, adequate logs of all boreholes, pits or trenches; the location and results of all sampling and testing with a note on the methods employed; and an appraisal of the results written in descriptive form in clear engineering terms. Photographs of the site, taken from ground or air, with marked-up overlays, are frequently of great use (GEOL SOC 1972).

Methods of investigation

5. *Trial pits and trenches.* In soft ground such as alluvium or very weathered rock, trial pits or trenches may be excavated down to water-table. Pits and trenches are frequently the cheapest way of examining soft natural deposits or made ground above the water-table and are especially valuable where the ground is very variable.

6. *Trial headings.* Where the terrain is steep, trial headings may be driven into a hillside, using normal tunnelling techniques. Such headings permit a detailed examination of the ground *in situ* which is more valuable than borehole evidence since the local orientation of strata and fractures can be measured accurately. The headings can also be used to obtain large samples for *in situ* testing and as drill sites. They are particularly useful for example in the abutments of proposed arch dams where rock deformation is important, and at tunnel portals where a decision on cut-and-cover or driven tunnel is to be made. Extended trial headings may be driven on the site of large-scale underground schemes such as road tunnels or underground power stations, and may then also serve as access tunnels.

7. *Boreholes.* Most site investigation work below the surface is carried out by boreholes. The depth to which boreholes extend must depend upon the requirements of the site. For major dams and underground power stations depths of 300 metres in bedrock may be needed. Even superficial deposits may have to be proved to great depths; during preliminary drilling for dam sites in the Peace River Valley in British Columbia alluvial deposits were found to extend to a depth of 150 metres and at Tarbela Dam in Pakistan bedrock was proved below 230 metres of heavy boulder alluvium in the Indus River bed. While most site investigation boreholes are vertical, the advantages of inclined or even horizontal holes may be considerable, particularly in rock. Where near-vertical bedding planes, joint systems or faults exist only inclined holes will give a complete picture of the geological pattern at depth. Inclined holes may be required to obtain oriented samples or to facilitate oriented *in situ* testing. Drilling methods are described in paragraphs 17 to 26.

8. *Shafts.* Exploratory shafts, which are in principle large trial pits, are used in jointed and weathered rock, where the engineer needs maximum information on ground conditions, and where core or sample recovery from boreholes may be too little to be of use.

Shafts enable the rock to be visually inspected and samples obtained for testing. In some cases pendulums or slope deflection equipment may be installed to monitor rock movements over a period. Such shafts may be in the form of boreholes of about 1 metre diameter or $1\frac{1}{2}$ to 2 metres square in plan, with depths up to 30 metres or more. Wooden platforms and ladders are provided.

9. *Geophysical exploration.* Geophysical methods described in Chapter 6 may be used in site investigation. Such methods will reveal anomalies in certain properties of the soil or rock, but the interpretation of those anomalies in the absence of other geological information is unsatisfactory. All relevant anomalies should be tested by drilling or other visual examination.

10. The great advantage of geophysical methods is the speed with which traverses can be undertaken so that visual evidence from a few boreholes can be extended rapidly over a large site and subsequent drilling thus restricted to areas where anomalous readings are obtained. A few boreholes will enable the geophysicist to interpret his data more accurately and even to comment on characteristics of the strata such as the best means of excavation. Rippability is broadly related to seismic velocity (*see* Figure 91); but measurements made on specimens in the laboratory will relate normally to sound rock, whereas the material on site may be jointed or weathered.

Survey

11. Geological information, borehole locations etc, are best plotted in the field on a base plan prepared from a suitable topographic map or from aerial photographs. Depending on the required accuracy, results of the site investigation can then be added, using pacing or taping and a hand compass or by mounting the plan on a plane table and using resection or a tacheometric alidade. Suitable scales for most site work are 1:500 or 1:1000. Any geological information gleaned during the desk study should be added to the base plan before field work begins.

12. If no suitable base plan can be prepared in advance normal surveying procedures must be used. Plane-tabling is laborious but effective since the details are plotted in the field. If triangulation by theodolite is used, a round of tacheometric observations can be done at each survey set-up. The geologist should participate in the survey and ensure that results are plotted before the field party leaves the site.

13. Where reliable published maps or stereo pairs of vertical air photographs are available, contours can be plotted on the base plan, otherwise they must be prepared from spot heights obtained in the field from tacheometric observations or other means. Contours should preferably be drawn in the field where interpolation between spot heights will be more accurate. Photogrammatic machines are necessary for all but the simplest photographic surveys, providing data input for computer plotting.

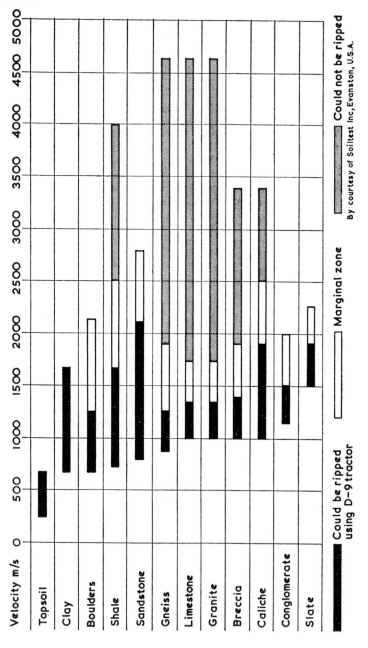

Fig 91. Rippability chart

14. On steep slopes geological features can be located in three dimensions by taking stereo pairs with a photo-theodolite. The accuracy depends on the length of base-line available. The technique is particularly suitable for surveys of joints in cliff faces for stability analysis.

15. *Methods for lakes and seabeds.* Investigations over open water are fundamentally the same as on land, with appropriate modifications to drilling and sampling techniques. The surface of the seabed can be sampled by grab, drop sampler or piston sampler, depending upon the nature of the bottom and available equipment (TOMLINSON 1954). Drilling is undertaken from dumb barges, rafts or pontoons in sheltered water and larger drilling vessels in more exposed conditions, the vessel being held on station by anchor or by automatically controlled propulsion units. The former positioning system will usually be preferred for shallow water, long swell or strong currents. Where floating craft are used in tidal waters, drilling is through a telescopic casing, the upper smaller diameter being suspended from the vessel and the larger diameter anchored in the seabed. Self-contained drilling rigs are available for obtaining cores from relatively shallow boreholes, the rig standing on the seabed, operated from a parent vessel. For the most exposed and deepest water and for the deepest boreholes, oil drilling techniques are used.

16. Geophysical surveys are particularly useful over water, for complementing boreholes and for extending and interpolating their information. The Sparker and Boomer forms of seismic survey, developed for use over open water, are described in Chapter 6. Echo sounding and Sidescan sonar will provide information on seabed topography.

Drilling methods
17. *Probing and hand-auger.* The simplest way to make a hole is to drive a metal probing rod by hand or using a hammer; this may serve to determine the depth of soil over rock or the depth of soft deposits such as peat or alluvial mudflats, the probe being calibrated by comparison with probe holes driven adjacent to test pits. Probing at close centres fills in the details between pits. Hand-augers, with extension rods, enable holes 30–50 mm diameter to be drilled with sample recovery. Hand-driven or powered post-hole augers drill larger holes, 150–200 mm diameter, to depths of 6–10 metres, but they cannot be used if boulders or large cobbles occur.

18. *Shell-and-auger.* This is a percussive rig equipped with a friction winch to lift and drop the drill string. In sands and gravels a 'shell', consisting of an open-ended cylinder with a cutting-edge and flap-valve, is used to break-up and recover the soil. In cohesive soils a 'clay-cutter' is used, similar to the shell but without the flap valve. An auger is not now normally used with this rig. Chisels serve to break-up boulders, and casing can be used to keep the hole open. This type of rig is one of the cheapest, simplest and most frequently used drills for site investigation in soil and weathered rock.

19. *Continuous flight auger.* This is a rotary drill, using a continuous auger with extension pieces of similar type so that there is no need to withdraw the rods when drilling. The rig can be mounted on a lorry or self-propelled tracks for mobility in suitable terrain, and will drill holes at any angle. Although a continuous disturbed sample is recovered the sample location is not accurate and in variable soils only an approximate geological profile is obtained. More accurate information can be obtained by lowering a sampling device through the hollow stem of the auger; drilling is interrupted but the rods need not be withdrawn.

20. *Rotary drilling.* The most common method of drilling investigation boreholes in solid rock is by rotary core drilling. A ring-shaped bit studded with industrial diamonds or, in soft rocks, tungsten carbide inserts is rotated at high speed to cut an annulus in the rock, the necessary pressure being applied by the weight of the drilling rods or by screw or hydraulic feed. The core thus isolated is collected in a core barrel immediately behind the bit while the cuttings are flushed to surface where they may be collected or discarded.

The flushing medium is usually water, but where erosion of the core is to be avoided, air-flush may be adopted. However air is not as effective as water for cooling and lubricating the bit. The annulus may also be cut by using a plain bit and feeding chilled shot down the hole as a cutting medium; this method is only used for large-diameter holes (over 150 mm) where the cuttings are collected in an open container or 'calyx' above the core-barrel; hence the alternative names calyx or shot-drilling. (*See* Table 17.)

21. In unstable ground casing is normally used to keep the hole open and standard dimensions have been established for a nesting series of casings and corresponding bits. Alternatively, mud can be used as a flushing medium which, properly constituted, will prevent the hole collapsing. In deep holes for oil and natural gas exploration the mud also serves as a counter-weight to prevent loss of oil or gas encountered under pressure.

22. A double-tube core barrel is normally used, the inner tube being mounted on bearings so that it does not revolve with the drill string. This improves core-recovery by comparison with the single tube core barrel which is seldom satisfactory for site investigation. For maximum core recovery in soft or broken ground a triple-tube barrel may be used. The triple-tube core barrel incorporates a detachable liner so that the integrity of a sample is preserved during transit (sealed) to the laboratory. It is also claimed to give better core recovery in soft friable rocks, boulder clay, and some other types of overburden. Normally the drill string must be removed and the core extracted each time the hole advances by the length of the core barrel (usually 3 metres); but for deep holes wireline equipment may be used. Here a detachable inner barrel may be exchanged by means of a wire running within the rods, which are larger in diameter than the usual ones, without withdrawing the drill string. The heavier string needs a more powerful drill, and

a larger hole must be drilled for the same core size, but for holes over about 150 metres depth the method is more economical, providing it produces samples acceptable for the purpose.

23. For faster exploration a rock roller bit may be used. Three toothed wheels rotate and crush the rock which is transported to the surface as chippings in the drilling medium, usually mud. This method may be used where very hard rock, such as chert, is encountered, and in overburden, where boulders prevent the more usual shell-and-auger equipment being used down to bedrock.

24. *Jackhammer or wagondrill.* These are percussive drills, hand-held or wagon-mounted respectively. Cuttings are removed by air or water flush, being collected more easily by the latter method. In fractured ground the flushing medium may be lost and no sample is recovered. The method is useful for shallow holes in rock, particularly to locate discontinuities such as open or clay-filled fissures, zones of soft rock, or old mine workings, in otherwise sound rock, or to make holes in which to install instruments or carry out *in situ* tests.

25. *Wash-boring.* A non-rotary method of driving a pipe through overburden is by surging it up and down while using ample flushing water. Samples may be collected unless the return water is lost, but they are not accurately located. The method is useful for checking the depths of boulder-free loose deposits above bedrock in river and marine investigations.

26. *Becker drill.* This is a powerful lorry or trailer mounted machine, which uses a diesel hammer to drive a non-rotating double-walled drive pipe in deposits containing large boulders, down to depths of about 60 metres.

The smaller rigs can drill angled holes up to about 30 degrees off vertical. Air or water flush is used to remove cuttings which can be collected for sampling and more accurate samples can be taken through the inner pipe. A diamond drill can also be used through the drive pipe to continue the hole into bedrock or to drill and blast large boulders.

Sampling

27. Samples consist of quantities of soil or rock removed from the ground for examination and testing in the laboratory. Small samples may be taken by hand from ground surface and in pits, or man-sized shafts or tunnels, or extracted from within the groundmass through boreholes. Large samples may be obtained from borrow pits or quarry trial blasts. Water samples are also taken as required.

28. Samples are known as 'disturbed' or 'undisturbed', but the latter have changed stress and pore-water pressure conditions and may suffer from end or side disturbance during extraction, especially in cohesive materials. All methods of sampling lead to a greater or lesser degree of disturbance and new methods are frequently devised in an effort to minimise this.

29. *Disturbed samples.* These are normally the by-product of drilling operations, e.g. cuttings from percussive or rotary drills, wash-borings or materials extracted in a shell or auger. The accuracy of the location of such samples depends on the method of drilling. Their content may also vary; if the samples have been carried by water flush, fines may be removed in suspension, or if the flush is inadequate, coarsér material may fall back and be ground into smaller particles by the bit. Grab samples from the surface or pits are also considered as 'disturbed' but are more likely to be a complete sample of the foundation.

30. Disturbed samples serve for identification, but are of little use for testing. They may be used for grading analysis of granular materials, with due regard for the probable loss of fines, and the method of taking the sample should be stated in the report. Samples are usually stored in airtight tins or jars, or in bags, and suitably identified.

31. *Undisturbed samples.* These are taken by special devices and, in soils, are not normally continuous. The usual device in UK for soil sampling is the standard open drive sampler of 4 or $1\frac{1}{2}$ inch (100 or 40 mm approx) internal diameter. These consist of open-ended metal cylinders 18 inches (450 mm) long, with a separate cutting-edge screwed to one end and an extension piece screwed to the other. The sampler is driven into the ground for about 600 mm, using a sliding hammer or, for better control, a hydraulic ram. Following withdrawal the drive shoe and extension are removed and replaced by screwed caps after the sample ends have been trimmed and coated with paraffin wax. The part of the sample in the extension piece represents disturbed soil at the bottom of the drill hole and is discarded. In the laboratory the sample is carefully extruded by a ram and is then available for the various soil mechanics tests (BS 1377: 1967). The sample tube is available for re-use.

32. Piston samplers are more elaborate forms of drive sampler, with either fixed or floating pistons which serve to close the lower end of the sampler until it is in position at the bottom of the hole and thus prevent contamination of the sample. The tube is always driven by pneumatic or hydraulic action. In soft soils a fixed-piston sampler can be pushed down without a previous hole being bored, and a sample then taken at the desired depth.

33. A cutter-liner system for soft sediment cores has been developed whereby a short cutter is pushed into the sediment and the core so obtained is surrounded by a plastic sheath, unrolling continuously within the slightly larger core barrel, the inner side of which is greased. Side disturbance is thereby reduced (SLY 1966).

34. Rocks can be sampled by using core barrels in rotary boreholes, as described in paragraph 20. The more usual standard sizes for site investigation are listed in Table 17. Dimensions are given in millimetres and in the case of HX and X-ray these are approximate.

TABLE 17. STANDARD CORE DRILL DIMENSIONS

Designation	Hole diameter (mm)	Core diameter (mm)
(a)	(b)	(c)
HX	100	75
NX	76·2	53·9
NX Wireline	76·2	43·6
BX	60·3	41·2
BX Wireline	60·3	33·3
AX	49·2	28·5
·EX	38·1	22·2
X-ray	30	17·5

The cores are extracted from the core barrel and stored in the correct order in wooden or metal coreboxes with the top and bottom depth of each run clearly marked. Selected portions of core may be waxed to preserve the moisture content. For rock cores subject to deterioration colour photographs should be taken of wetted clean fresh cores.

35. With care core-drilling provides a continuous sample, but 100 per cent core recovery in fractured or weathered rock is seldom obtained in the smaller sizes and at least NX size should be used in such ground. The integral sampling method (ROCHA 1970 and Figure 92) gives a complete oriented sample with all fractures and broken material correctly positioned. A 25 mm hole is drilled ahead of the main NX hole for 1 to 3 metres, an oriented metal or plastic rod is grouted into this, and the sample is then cored with normal NX equipment.

36. There are other methods of obtaining oriented samples which are simpler, but not so effective. In all cases they require transference of the orientation from the surface to the *in situ* core.

37. In stiff clays, sands or friable rock a modified form of double-tube core-barrel, called the Dennison sampler, can be used to obtain samples in holes of 150 mm or larger diameter. A thin sheet-metal liner, 500 mm long, inside the inner barrel serves to contain the sample during handling and transit to the laboratory. The degree of sample disturbance is high.

38. Undisturbed samples can be taken in trial pits or headings by carefully removing the soil around and above the sample, which is then encased in a box before final separation from the ground, inversion, and capping.

39. Accurate sampling is the basis of all site investigation and it is better to have too many samples than too few. Samples require the greatest care in logging, handling and preservation. The taking of samples at fixed intervals or pre-determined depths is seldom satisfactory except in homogeneous deposits and

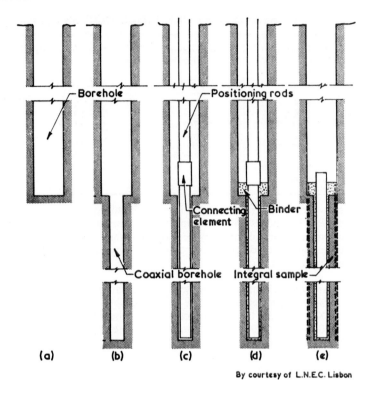

By courtesy of L.N.E.C. Lisbon

Fig 92. Integral sampling method

even there unexpected variations can occur. Regular sampling may however be required for specific tests. An engineer or geologist, well briefed in the objects of the investigation and experienced in site investigation techniques, should always be on the site and the precise position of sample localities should be left to his judgement. Colour photography can be used to record the appearance of samples which have not been preserved; a colour key should be incorporated in each photograph.

In situ testing

40. Various tests can be carried out in a borehole, either as part of the sampling procedure or independently, at specified locations as the borehole progresses or after the hole is complete. Tests may also be carried out on the ground surface or in open excavations.

41. The recorded observations of the drilling crew form an important part of the investigation. These include penetration rate, obstacles encountered, water level observations, and any other observations, particularly where no continuous sample has been taken.

42. *The Standard Penetration Test* (SPT) (BS 1377 1967 Test 18) is one of the most common comparative tests used during the drilling operation. A standard split-spoon sampler, 2 inches OD, is driven into the ground at the bottom of the hole for a distance of 18 inches and the number of blows of a standard weight, falling through a specified height, needed to drive the final 12 inches is recorded (the metric equivalents are not yet in use). This is the N-value. Table 18 (TERZAGHI and PECK 1967) shows the relationship between the N-value and the relative density of sand. The material recovered in the spoon serves as a disturbed sample. The device can be used in soft rock but boulders render it inoperative.

TABLE 18. **RELATIONSHIP BETWEEN N-VALUE AND RELATIVE DENSITY OF SAND**

N–Value	Relative Density
(a)	(b)
Below 4	Very loose
4 to 10	Loose
10 to 30	Medium
30 to 50	Dense
Over 50	Very dense

43. *The Dutch Cone Penetrometer* is also widely used. The cone is attached to rods protected by an outer sleeve. The cone is thrust down through a standard distance (75 mm) and the thrust recorded to obtain the end resistance. The sleeve is then pushed down through the same distance to obtain the friction effect. Readings are generally taken every 200 mm to obtain a continuous profile which can be used for piling and foundation calculations. The loading is static as opposed to dynamic loading of the SPT, and is thus more reliable in fine grained soils.

44. *The Vane Test* (BS 1377: 1967 Test 17) measures the shear strength of clays. A four-bladed vane, 50 mm diameter and 100 mm long, is pushed ahead of the borehole and then rotated at a constant rate of 10° per minute. The torque required is measured and the shear strength calculated, with rotation recorded against torque where peak and residual strengths are required.

45. A thermometer or remote-reading thermocouple may be used where underground fires are expected, where the freezing process is being used to control water inflow (*see* Chapter 10, Section 2), or to ascertain the extent of permafrost (*see* Chapter 2, Section 7).

46. *In situ properties of rock.* A comparatively recent development is the measurement of various rock properties *in situ.* The design of large structures in rock, such as underground chambers or large tunnels, is facilitated by a knowledge of the natural state of stress in the ground and changes in that state as the construction proceeds, and also by determination of the stress-strain relationship and the yield characteristics. These are usually expressed by the coefficient of deformability, which is a figure combining the coefficient of elasticity (recoverable strain) and the permanent deformation due to plastic flow, closure of fissures under load, and other non-recoverable strain.

47. Pressuremeters or dilatometers can be used to determine deformability provided that conditions are isotropic or orthotropic. A cylindrical device fitting closely within the borehole is expanded hydraulically and the deformation related to pressure. The test can detect zones of high compressibility in rock such as chalk or marl, which cannot easily be identified by other means. Poisson's Ratio can be obtained from tests on core samples or, often, approximately estimated.

48. Photoelastic glass plugs or strain gauge rosettes can be glued to the bottom of boreholes. The change in strain or stress when the natural stresses are relieved by over-coring or otherwise can be measured directly. The 'door-stopper' plug has sufficient strain-gauges incorporated to obtain the complete natural state of stress from one operation. Alternatively, the devices can be left in position in otherwise inaccessible locations and read at regular intervals through a telescope to monitor ground movement.

49. Geophysical logs of boreholes are records of changes in ground physical characteristics measured along the borehole, using the processes described in Chapter 6 and interpreted in the same way. Changes in porosity or water-content, density or degree of fracturing, can be derived by the use of combined emission and receiving instruments lowered down the borehole. Where a hole is uncased and at least 60 mm in diameter, film or television cameras can be used to study the ground, or in shallow holes a borehole periscope will suffice. Borehole calipers indicate the variation in diameter in one or more planes but usually lack orientation or a reference axis. Electronic devices giving a continuous record of inclination can be combined with the logging device.

50. *Groundwater.* Water level observations should be made during sinking, usually last thing at night and first thing in the morning together with rate of any noted inflows of water. The permeability of the strata may be measured by pumping-in or pumping-out tests, by measuring either the rate of recovery of water level after artificial lowering, the shape of the cone of depression as measured in nearby observation boreholes, or the pumping rate required to maintain a steady water level above or below the natural water-table. Water may be pumped under pressure into selected stages of a borehole, isolated by packers, to determine the local permeability (EARTH MANUAL 1963).

51. The local movement of underground water can be measured, and zones of high permeability located, by inserted flowmeters into boreholes. Local movement of groundwater into boreholes can also be predicted by temperature logs and large scale movements can be detected by injecting fluorescein dye, radioactive isotopes, or salt solution into boreholes and measuring the intensity of flow against time at points downstream. When using fluorescein, the unaided eye is not sufficiently sensitive for measurement; samples should be taken and compared under an ultraviolet lamp ('black light') with standard solutions.

52. *Bearing capacity.* Where it is required to determine bearing capacity of the ground at surface or in a pit, a square or circular plate is loaded by jacking against a platform weighted with Kentledge (iron blocks or similar heavy material) or against the resistance of cable anchors or tension piles installed in the ground. For large scale tests tanks filled with water can be used. It should be remembered that plate loading tests only provide information about the ground within a depth equal to about one-and-a-half times the diameter of the plate.

53. Shear tests are performed by excavating around a block of ground, encasing the upper and lower parts separately in concrete, and then jacking the upper part laterally until movement occurs, with the normal load applied by Kentledge or ground anchors. When testing natural discontinuities great care is needed to ensure that shearing will occur along the desired plane.

Sample and core logging

54. The site investigation report should include a log of each borehole. This should show the reference number, location, and orientation of the hole, the diameter, and the type of drill and bit used, rates of drilling, quantity and pressure of drilling water or mud, quantity of backfill when important to know approximate effective borehole diameter, together with details of any casing inserted. The log should include a diagrammatic indication of the strata encountered, using the symbols recommended in the Code of Practice (CP 2001: 1957), with all significant changes of strata indicated, preferably to scale, and accompanied by an adequate description. The location, inclination and (if known) orientation of bedding planes should be stated or sketched, also any palaeontological evidence which may be useful for interpretation (for instance locating discontinuities such as faults or landslip surfaces). The locations of samples and *in situ* tests should be marked in their correct positions, together with water-level observations. The log should be signed by the geologist or soils engineer responsible.

55. If geophysical observations have been made they should be plotted on the borehole log, together with any derived data such as density, modulus of elasticity, or permeability, alongside a diagram of the strata.

56. If only disturbed samples are obtained, the log must be based on these and on the driller's recorded observations. This must be made clear on the log so that its degree of reliability may be indicated.

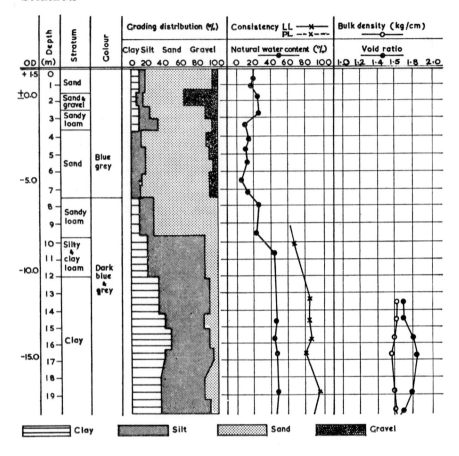

Fig 93. Soil log (after MORIMOTO and MISE 1963)

57. The results of tests on undisturbed samples are usually listed separately, grouped under each type of test, and grading curves are again usually kept separate. This practice enables a quick appreciation of the range of values to be made, but the connection between the sample location and variations in value is not immediately apparent. A soil log similar to Figure 93 gives a much clearer picture but cannot be prepared unless sufficient samples have been taken. Curves showing SPT or Dutch cone test results can, with advantage, also be added to such a log.

58. *Core logging in rock.* Since a more-or-less continuous sample is recovered, the log of a cored borehole in rock requires less interpolation than a soils log.

224

Few laboratory tests are usually necessary and the log is mainly a description of the strata with a diagrammatic column denoting the rock type and a reference to the 'core recovery' in each length drilled. Where known, major gaps in each length of core should be indicated. Such a log has only limited engineering use unless the terms used in the description are properly defined which is by no means always the case. In the past the core recovery has been expressed as a percentage of the length drilled, with no indication of the state of the recovered material, while terms such as 'closely spaced', 'weak', or 'highly weathered' have been loosely applied.

Recommended procedures are described by a working party of the Geological Society (GEOL SOC 1970).

59. Core recovery data should distinguish between solid core, broken pieces obviously from core, and irregular fragments. D.U. Deere (DEERE et al 1967) has suggested that the total length of core recovered in solid pieces exceeding 100 mm in length should also be measured and expressed as a percentage of the total length of each run; this is called the 'rock quality designation' (R.Q.D.), and has certain engineering applications not fully defined. It is probable that the figure of 100 mm quoted may have to be adjusted for core sizes below BX before the R.Q.D. can be properly related to other data.

60. *Description of rock samples.* Any description of rock core must include the degree of weathering and fracturing. Standardised systems of nomenclature have been recommended by the Geological Society (GEOL SOC 1970) and also by Franklin and others (FRANKLIN et al 1971). Similar terminology has been used to describe bedding plane spacing. Table 19 is suggested for standard usage but whether this or another system is used the terms should be defined in the site investigation report until acceptable standards are laid down.

TABLE 19. RECOMMENDED TERMINOLOGY FOR SPACING OF DISCONTINUITIES

Spacing (mm)	Terminology	
	Planar structures (e.g. bedding, laminations, foliations, flow bands)	Discontinuities (e.g. joints, faults)
(a)	(b)	(c)
> 2 m	Very thick	Very widely spaced
2 m–600	Thick	Widely spaced
600–200	Medium	Moderately widely spaced
200–60	Thin	Narrowly spaced
60–20	Very thin	Very narrowly spaced
20–6	Thickly laminated	⎱ Extremely narrowly
< 6	Thinly laminated	⎰ spaced

225

61. **a.** The strength of a rock may be determined with a degree of accuracy governed by the purposes of the investigation. It may be sufficient merely to judge it by tapping with a hammer, or more elaborate tests may be required. The point-load strength may be measured in the field (FRANKLIN et al 1971) or laboratory tests on carefully prepared samples may determine the compressive strength (uniaxial or triaxial loading), the tensile strength (Brazilian test) or the shear strength (shear box test).

b. Breakdown due to slaking can be measured by a test also described by Franklin (FRANKLIN et al 1971). Samples obtained from cores can also be used for elaborate laboratory tests to determine creep (long-term irrecoverable strain) and the stress-strain relationships for comparison with *in situ* methods described in paragraph 46. Sonic velocity tests may be made for comparison with borehole or surface geophysical surveys described in paragraph 49 and Chapter 6.

c. Tests made on small intact samples of sound rock in the laboratory will be unaffected by the fracturing and weathering which affect the rock mass in the field and due allowance for this must be made when interpreting the results. The ratio between laboratory and field results has been used in dam design (SERAFIM 1964) and for estimating grout consumption (KNILL 1970). The ratio normally tends towards unity with increasing depth from the surface.

d. Strength and weathering classifications of rocks with appropriate sampling methods are described in Chapter 4.

62. All cores should be examined and identified by a geologist. Although a hand lens and simple field tests will often suffice, it may be necessary to prepare thin sections for identification. Thin sections may also be used for studying weathering properties of rocks. Pedantic use of petrological nomenclature is inappropriate on core logs. On the other hand local terms such as 'fakey blaes' should not be used, although it may be necessary to know their significance (for correct terms *see* Chapter 4). Fossil evidence is seldom needed for engineering purposes but may assist correlation in faulted areas.

63. *Description of soil samples.* Soil descriptions should be based on CP 2001 and Chapter 5 of this book. Visual description should be supplemented by grading curves and index properties where possible. Local terms such as 'Thames ballast' should not be used unless accurately described in the accompanying report.

REFERENCE LIST—CHAPTER 9

BS 1377: 1967 —Methods of test for soils for civil engineering purposes, *British Standard 1377*, British Standards Inst., London.

CIV ENG: 1959 —Report on mining subsidence. *Inst. Civ. Engrs.*, London.

CP 2001: 1957 —Site Investigations. *Code of Practice 2001*, British Standards Inst., London (under revision).

CP 2004: 1972 —Foundations. *Code of Practice 2004*, British Standards Inst., London.

DEERE D U et al, 1967 —Design of surface and near-surface construction in rock in Failure and Breakage of Rock. C Fairhurst (Ed) *American Inst. Mining Engrs.*, New York, 237–302.

EARTH MANUAL, 1974 —*Earth Manual*. United States Bureau of Reclamation, Washington D.C.

ECKEL E B, 1958 —Landslides and engineering practice. Special Report 29, *US Highway Research Board*.

FRANKLIN J A et al, 1971 —Logging the mechanical character of rock. *Trans. Instn. Min. Metall.*, A.80, A.1–9

GEOL SOC, 1970 —The logging of rock cores for engineering purposes. Geol. Soc. Working Party Rpt. *Q. Jl. Engng, Geol.*, 3, 1–24.

GEOL SOC, 1972 —The preparation of maps and plans in terms of engineering geology. Geol. Soc. Working Party Rpt, *Q. Jl. Engng. Geol.*, 5, 295–382.

KNILL J L, 1970 —The application of seismic methods to the prediction of grout take in rocks. *British Geotechnical Soc. Jl.*, 93–100.

KNILL J L, 1971 —Assessment of reservoir feasibility. *Quart J. Eng. Geol.*, 4, 355–365.

MORIMOTO T and MISE T, 1963 —Grouting into soil under vacuum conditions. *Proc. Conf. Soil Mech. Found. Engng*, Budapest, 421–8.

ROCHA M, 1970 —A new method for the determination of deformability in rock masses. Vol. I Congr. Intern. Soc. Rock Mech., Belgrade.

SERAFIM J L, 1964 —*Rock mechanics considerations in the design of concrete dams in state of stress in the Earth's crust.* W. R. Judd (Ed), Elsevier, N.Y. 611–650.

SLY P G, 1966 —A new cutter-liner system for soft sediment cores. *Engng. Geol.* 1(4) 343–344.

TERZAGHI K, and PECK R B, 1967   —*Soil Mechanics in Engineering Practice.* John Wiley, New York (1948) 2nd Edition.

TOMLINSON M J, 1954   —Site exploration for maritime and river works. *Proc. Inst. Civ. Engrs. 3*, No. 2, Pt. 2. 225–272.

ENGINEERING IN ROCK AND SOIL

SECTION 10.1. GEOTECHNICAL ASPECTS

1. Many man-made structures are so large that the ground which supports them must be considered as part of the structure; and the safety and performance of the structure are often determined more by the ground conditions than by the design of its constructed parts. In many projects the natural ground features, or those features as modified by excavation, are parts of the works from which satisfactory performance is demanded. The natural site materials, soil and rock, may be excavated and subsequently used in the work.

2. The engineers' geotechnical studies must therefore include *in situ* soils and rocks as well as the geotechnical processes—defined in engineering geology as processes used to modify the properties of soils and incoherent rocks to make them suitable for engineering operations.

3. These studies are embraced by the two disciplines—soil mechanics and rock mechanics. The former, and older, science has developed as a branch of civil engineering, whereas the more recent rock mechanics is usually considered as an extension to, or a particular application of engineering geology. But there are no precise dividing lines between engineering, soil mechanics, rock mechanics, engineering geology and geology.

4. The history of soil mechanics may be traced through many contributions including particularly Coulomb in the eighteenth century and Rankine in the nineteenth. Emergence as a normal part of the civil engineer's mental and physical tool kit is largely attributable to the work of Terzaghi, particularly to his 1939 lecture 'Soil Mechanics—a new chapter in Engineering Science' (TERZAGHI 1939). From the early needs to estimate ground loads on walls, safe bearing loads on foundations and the stability of slopes by simple methods, have emerged complex elasto-plastic conceptual models and the principle of 'critical state' (SCHOFIELD and WROTH 1968). Too much of soil mechanics may be criticised as having developed without regard to practical problems, particularly to their geological aspects. It is generally good engineering to use geomechanical theories no more complex than may be justified by confidence in the reliability of the available physical data about the ground.

5. The practical necessity for extending these studies to include the study of rock masses has been given impetus by the increase of civil engineering works, and the urgency prompted by spectacular failures such as that of Malpasset Dam

(France 1959) and the Vajont rock slide (Italy 1963). The fundamental difference between soil mechanics and rock mechanics is that of scale. Soil particle sizes are such that the soil mass for the civil engineer may be treated as a continuum; in fact, one of the fundamental criticisms of soil mechanics is that it has tended to pay inadequate regard to the small-scale fabric involving inter-molecular forces. By contrast, most rock mechanics problems are related to the properties of discontinuities which play a dominant role in the behaviour of rock masses. It is almost true to say that soil mechanics is the study of the strength of soils while rock mechanics is the study of the weaknesses of rock.

6. The principles of soil and rock mechanics are dealt with in many textbooks but, as with all scientific studies, advances continue to be made and the engineer is advised to peruse specialist journals and papers to take advantage of new research and practical developments, especially in regard to site investigations which form the basis for engineering concept and design.

Section 10.2. GROUNDWATER CONTROL

1. Where a scheme of construction entails excavation below the level of the groundwater table, it may be necessary to control the inflow of groundwater. The works to be constructed may of themselves raise the water table. Control of groundwater may, at one extreme, be necessary only for a part of the construction phase or, at the other, be a necessary feature of the permanent works. Concern may be less with inflows than with groundwater pressures, for example, against the base or walls of a structure or affecting the stability of a slope or cutting.

2. Where an excavation is to be in clay or in rock of low mass permeability, groundwater inflow is unlikely to be serious and will usually be small in relation to surface water. The possibility of artesian pressures in underlying or adjacent strata should not be overlooked. These could give rise to inflows by way of fissures or boreholes; an alternative possibility is that uplift, piping or boiling may affect the overlying layers. Where there is a possibility of such occurrences, site investigation should be adequate to establish a stable depth of ground in relation to subjacent water pressures; this may be to a depth below excavation comparable to the breadth of the excavation.

3. Where excavation is in soil or weathered rock, the most important factors concern the permeability of the ground and the relative groundwater level. In unweathered rock, the joints and, especially, fault zones will usually govern the water movement although cavernous (karstic) rocks or coarse sandstones may be highly permeable even when they appear to be intact. Special problems may arise on account of heterogeneities in the ground, or wide variations between the permeabilities of adjacent layers. Such factors may cause local water flows or pressures appreciably in excess of those calculated by flow nets based on assumptions of isotropy. Steep groundwater pressure gradients, which may occur adjacent to shafts or tunnels, in weakly cemented porous rock may cause instability or disintegration of the ground.

4. In loose uniform soils the coefficient of permeability k (the flow in cubic metres through an area of one square metre for a hydraulic gradient of unity) can be estimated approximately from Hazen's Law:

$$k = 10^{-2}(D_{10})^2 \text{ m/s}$$

where D_{10} is the grain size (in mm) on the grading curve corresponding to 10 per cent retention.

The permeability of rock is usually expressed in metres/second but is occasionally expressed in Lugeons, representing the flow (in litres/minute) into a metre length of NX sized borehole at a pressure of 10 atmospheres (1MN/m^2) above the groundwater. In a homogeneous medium, one Lugeon represents a permeability of about 10^{-7} m/s.

5. Control may be by exclusion or by groundwater lowering. It may be adequate to lengthen or modify the seepage path to retain a safe reduction of the hydraulic gradient but caution is then especially necessary to foresee possible local concentrations. Piping may occur if the hydraulic gradient is sufficient to lift particles of the local material which may then be displaced by the water flow. Piping will occur earlier and faster in fine grained cohesionless soils and is most likely in material of single size rather than graded. A special risk of piping should be foreseen where water flows in a permeable soil immediately beneath a stratum or obstruction of low permeability.

Control by cut-off

6. There is a wide choice in the type of cut-off for an excavation, each type being suitable for a certain range of geological situations, as described below:

a. *Steel sheet piling.* Applicable to all soils, apart from very stiff clays, cemented soils or where gravel is present in sizes greater than, about, 75 mm. The heaviest sheet piling may be driven for short distances through soft rock. Piles may be pitched over water as well as land. Sheet piling is not totally impermeable although, in fine soils, leakage through the clutches may decrease with time. Where piling becomes unclutched on account of boulders or other cause of local hard driving, additional local ground treatments may be necessary.

b. *Sector piling.* Reinforced concrete bored sector piles are established by forming a first set of piles with somewhat less than a diameter's separation, followed by a second set of intermediate piles in the same line, providing an overlap. A cut-off may be constructed by such means in virtually all types of ground provided the fresh concrete is not liable to be affected by chemical attack or strong groundwater flows. The piles are normally 500–1000 mm diameter, but smaller (75 mm) piles made with grout have been used in pervious seams in otherwise sound rock.

c. *Diaphragm walling.* Diaphragm walls are constructed in two stages. First, for each panel of the wall, a trench is kept topped up with a bentonite mud

231

as it is excavated; second, concrete is placed by tremie to displace the mud, reinforcement, if required, being inserted in the form of cages in the bentonite. The purpose of the bentonite is to stabilise the sides of the trench, which it achieves by a combination of three processes; the pressure exerted by the mud, the shear strength of the bentonite gel and the caking of the mud against the side of the trench (in permeable ground) (ELSON 1968). The surface of the bentonite must be above that of the groundwater and shallow concrete guide walls are required to stabilise the top 1–1½ metres of the ground. Diaphragm walling is appropriate to a wide range of soils and may be incorporated into the permanent works as side walls or as a cut-off beneath dams. Plastic concretes, incorporating bitumen or bentonite, may be used where some relative ground movement is expected. The trench width is usually between 600 and 750 mm and the maximum depth may be 100 metres or more, although the most rapid methods of excavation are confined to lesser depths. For temporary cut-offs in granular soils, a soil-cement mixture may be used for the walling, or the originally excavated material may be mixed with sufficient finer soil to form suitable backfill.

d. *Grouted membrane.* A zone of low permeability may be formed by grouting granular soils. Cement or clay-cement grouts may be used for gravels, with low viscosity chemical grouts for finer soils (ISCHY and GLOSSOP 1962). In variable soils, several grouts may be used successively, since the cost of a grout mixture increases very markedly with the fineness of the ground that it will penetrate (PERROTT 1965). The permeability of a membrane may be an order of magnitude more than that of a soil-cement diaphragm wall (say 10^{-6} m/s against 10^{-7} m/s or even 10^{-8} m/s). Sleeved grout pipes (tubes-à-manchette) or similar techniques permit selection of the particular length of grout hole from which the ground is to be injected. A grouted membrane may be used in jointed or decomposed rock, sometimes as an extension to a piled or diaphragm cut-off. Grouting is also widely used in tunnelling, either to reduce inflow into an excavated tunnel or to allow tunnelling to be carried out through a previously grouted zone of ground. The object may be to provide a hood, an annulus or a block of treated ground. Grouting in rock depends upon the condition of the joints, particularly as to their width, continuity, frequency and filling. Cement grouts may be used for fissures greater than $0\cdot1$ mm, with sand-cement mixtures selected for wider fissuring (LOUIS and WITTKE 1971). Generally cement grouts cannot be injected into rock with a permeability less than 10^{-7} m/s (1 Lugeon).

e. *Ground freezing.* There are two basic types of freezing process. The traditional method entails the circulation of cold calcium chloride brine through piped circuits in cased boreholes. This process has usually been used for shaft sinking (COLLINS and DEACON 1972) and requires a period of several months freezing time for each installation. The limitation on depth is one of accuracy of drilling the boreholes and acceptable pressure of the circuit. The requisite thickness of frozen wall may be calculated for any specific circumstances of external pressure, by relating the strength of frozen ground to its temperature and the soil type and structure. The lowest brine temperature is usually about $-20°C$ but may be as low as $-35°C$.

A more rapid freezing process circulates liquid nitrogen at temperatures down to $-150°C$. This process is particularly effective for sinking a shaft through a water-bearing stratum of limited thickness, or for driving a tunnel through a fault zone filled with fine material, since it provides a positive if costly barrier in a situation where variable ground may cause grouting to be unreliable. Where groundwater flow is excessive, however, pre-grouting may be needed to make freezing feasible.

7. It will thus be seen that there is a wide choice of means of constructing a cut-off and that two or more expedients may be associated. Apart from the question of economics in construction, the type of barrier should be chosen in relation to the risk of ground movements impairing its function.

Control by groundwater lowering

8. Before embarking on any system of groundwater lowering, the consequences to the ground and hence to other works or structures nearby should be considered. Removal of water from the ground leads to the formation of a draw-down curve which may be readily calculated for known or assumed permeabilities. In alluvium permeability parallel to the bedding is normally higher than across the bedding. Since the actual permeability may vary from that predicted, it is prudent to check that the proposed dewatering system will perform adequately over the possible range of conditions.

The draw-down curve about a single well is known as a cone of depression. There are methods for assessing the effect, for an array of wells, of superimposing two or more such cones (*see* Chapter 12).

The effectiveness of a large diameter well may be increased by a series of radial filtered near-horizontal boreholes from near its base. If no water is to be allowed to enter the excavation the water-table must be lowered throughout to below the level of the excavation.

9. Filters may be introduced to prevent removal of fine particles at the point of pumping or natural discharge. A simple filter may take the form of a wire screen or a fabric mesh. A particulate filter medium may be provided to separate fine material from coarse material, in which case the grading of the filter material must lie between defined limits for the filter to be effective. The grading will depend on the character of the filter material (EARTH MANUAL 1974). Where a filter is liable to undergo relative movement or settlement it should be well graded and of generous thickness.

10. As a general rule surface water should be excluded from a groundwater drainage system to avoid blockage by fine material. For this reason it is normally more satisfactory to collect groundwater directly through a filtered drain than to allow it to emerge as a spring at the surface.

233

11. *Drains.* The stabilisation of clay slopes may be assisted by construction of counterfort drains, aligned with the slope, and possibly by cut-off drains along the top and bottom of the slope, where these latter do not in themselves lead to reduced stability of the slope. The drains may be filled with filter material arranged in vertical layers, or lined with porous blocks. Such drains do not normally intercept large quantities of water apart from the surface run-off, but they mainly serve to lower the water-table and hence increase effective stresses across surfaces of incipient slipping. If drains are required to be closely spaced, a herring-bone pattern is usually selected.

12. Where a slope contains aquifers or where it is likely to be affected by surface washouts in weathering, the filter medium is best provided as a blanket conferring protection and surface loading.

13. When designing a scheme for an excavation, thought should be given to the levels and positions of inflows from known aquifers so that the water may be collected as nearly as possible at source and led or pumped away. For example, a garland drain may intercept flow from an aquifer situated some distance above the base of an excavation.

14. Frequently, particularly in roadworks, excavations are made along hill-sides through superficial weathered ground which provides a drainage path for intermittent flow from precipitation. A cut-off drain should always be considered for such a situation during construction as well as for the permanent work.

15. *Wells.* Direct drainage of permeable soils may be achieved by a system of vacuum well points (60–75 mm diameter) or by means of submersible pumps operating in cased borehole wells, of 300 mm or more diameter (*see* also Chapter 12—hydrogeology). Whereas water is drawn into a well point from near its tip, which must be within 5–6 metres in level of the pump, a borehole may drain the ground from one or more aquifers extending over a considerable depth. Well points are normally inserted by jetting, only practicable in fine-grained soils. Generally, deep wells are preferred when permeability at upper levels is not excessive and particularly where drainage of a permeable underlying layer may have a dramatic effect on the pore pressures and hence the stability of overlying ground, of which the permeability is two or more orders of magnitude lower. For example, during construction of the Clyde Tunnel, on account of an undetected erosion channel, it became evident that a sheeted cofferdam excavation was inadequately sealed in what was expected to be 1–3 metres of stiff glacial till. Attempts to drain the silt by external deep filter wells were unavailing until one such well penetrated fissured Carboniferous Sandstone beneath, when immediate relief was achieved. Whether or not a well is protected by a fabric or gauze filter, the need for external filter bands should be considered in relation to the factors mentioned in paragraph 9.

16. In heterogeneous or cross-bedded material the object should always be to extend the zone of influence of dewatering by drawing water from the coarsest

and most permeable strata. Consideration of the geological structure and history should help to indicate the degree of continuity of such layers. Evidently greater continuity can be expected in bedded material than in colluvium. In saturated soils, stability may be improved by drawing water away from the face of the slope, a matter principally of lowering pore pressures and hence raising effective stresses.

17. On sloping ground, an exposed aquiclude or layer of low permeability may prevent percolation of water perched above. In suitable circumstances, such as when highly permeable pockets exist, drainage may be by near-horizontal boreholes, inserted by drilling or thrusting, in which pre-formed lengths of filter are placed around a perforated pipe before withdrawal of the casing. Where ponding water is a major threat to ground stability, drainage may be by large diameter holes or by headings.

18. *Electro-osmosis.* Drainage of ground of fairly low permeability may be accelerated by the application of an electrical potential along the intended drainage path. Electro-osmosis relies upon the pressure gradient imposed in saturated ground by the passage of an electric current. The effectiveness of the process depends upon the permeability, and the specific resistance of the ground, the flow per unit area (in metres) being related to the coefficient of osmotic permeability, k_e and the potential gradient (TERZAGHI and PECK 1967). For most soils, k_e lies in the range $0 \cdot 4$ to $0 \cdot 6 \times 10^{-6}$ m/s. The process is found to be most effective in stabilising silt-sized materials where the permeability k is of the same general order as k_e. Electro-osmosis is specifically a short-term expedient, since secondary electro-chemical effects tend to reduce conductivity local to the anode and permeability local to the cathode. It can be used for stabilising ground by control of the direction of seepage gradients.

A process which utilises the phenomenon of electro-osmosis is that of electro-chemical consolidation, whereby chemicals, usually silicates, are diffused into the ground by electro-osmotic action. In fine grained soils ground treatments may be undertaken by this means relatively rapidly without having to resort to excessive injection pressures.

19. *Use of caissons.* One expedient for overcoming groundwater problems during construction is by the use of caissons or monoliths. During the sinking of caissons there is a choice as to whether or not to use compressed air. In practice the choice will be largely determined by the geology, affecting not only the feasibility of excavating under-water without collapse of adjacent ground, but also the control of even settlement. An important factor in sinking a caisson concerns the frictional effects of the ground against the wall. These can be minimised for a caisson sunk on land by using a technique which ensures a continuous thin bentonite mud diaphragm (*see* paragraph 6) above the level of the cutting edge. The use of compressed air in relation to the ground for tunnelling, where it remains a most important expedient, is discussed in greater detail in Section 10.7, paragraph 21.

SECTION 10.3. STABILITY OF NATURAL SLOPES

1. Superficial erosion by wind and water generally tends to produce, via local intermittent instability, a progressively more stable state of the Earth's surface, ultimately represented by planation producing peneplains. Reversals of this trend are provided by tectonic processes, glacial valley cutting, by rejuvenating rivers and, all too frequently, directly or indirectly, by the activities of man.

2. Knowledge of local geological history is of paramount importance in determining areas of incipient instability and areas which can tolerate only limited interference by man before becoming unstable. Failure may occur on account of natural processes, including earthquake shock, or by artificial factors, affecting the balance of marginal stability. For example, the frequency of major rock fall activity in Norway has been closely related to the periods of intense freeze-and-thaw in the autumn and spring.

3. Examples of marginal stability may be seen in the valleys of South Wales, the hillsides of which have been affected by numerous and widespread landslides since the last period of glaciation. At the present day these same hillsides serve to support spoil heaps from the coal mines. In addition, subsidence resulting from mining may alter the groundwater drainage pattern. Several of these factors for instability contributed to the catastrophe in 1966 at Aberfan in Wales (ABERFAN 1968). HIGGINBOTTOM and FOOKES (1970) discuss aspects of periglacial instability and define associated terminology.

Features of low stability slopes

4. Features of low stability may be recognised by cambering, gulls, valley bulging (these usually of periglacial origin—*see* Figure 94) also surface ridging (across the slope) or, more spectacularly, by evident landslides or cliff falls. Where failures have occurred, the slipped masses of material have a hummocky, irregular surface which is a general characteristic of slope failures in soil materials. In relatively fresh slips it may be possible to identify the fragments of the original ground surface, displaced from the top of the slope, sometimes back-tilted within the slipped mass. The internal arrangement of landslides may be complex (HUTCHINSON 1968) and can involve several slip surfaces within one slipped mass (multiple) or several smaller slides succeeding one another down a slope (successive), resembling a cascade (Figure 95).

There is no sudden demarcation between natural landslides and those provoked by man. On the contrary, there is a continuous gradation in the extent of man's contribution to failure; each may entail a similar failure mechanism and call for similar methods of analysis. The position of the water-table, and specifically the water pressure generated on the slip surface, can significantly influence slope behaviour. In many regions past climatic conditions have been associated with higher rainfall, and by implication, more active mass movements than at present. The stability of excavations is discussed in Sections 10.4 and 10.5.

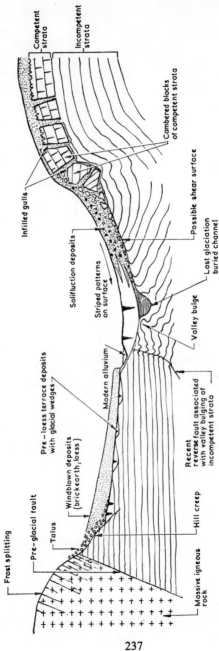

Fig 94. Section of valley showing important periglacial features and deposits
(after HIGGINBOTTOM and FOOKES 1970)

237

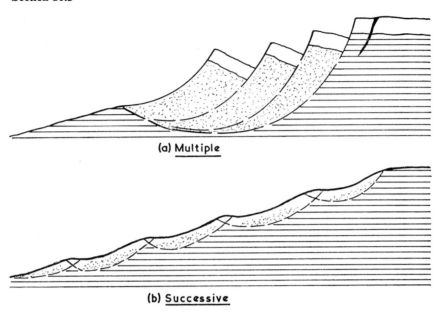

(a) Multiple

(b) Successive

Fig 95.   Types of rotational failure

5. Instability of a ground slope may be shallow or deep, the designation being expressed in terms of the ratio of the maximum depth of the surface of failure below ground level to the length of the landslip (measured in the direction of movement).

Types of failure

6. Several terms are used to define the type of failure, the commonest of which are described below:

a. *Creep* is a slow gravity movement ($<0{\cdot}01$–1 metre/year), affecting soil or weathered rock, parallel to the surface slope. This motion may be assisted by diurnal temperature variations or by alternate freezing and thawing, leading to solifluction; the rate of creep is often sensitive to seasonal rainfall, the motion assisted by cyclical swelling and shrinkage. There is usually no well defined surface of failure, the motion gradually reducing with depth. Symptoms of continuing creep are surface cracks, irregularities, distinctive variations in surface gradient, surface extensions (or contractions), mis-alignment of tracks or hedges and in angular movement of trees or structures. Soil creep that has occurred in the past will tend to destroy the original fabric of the ground, which may have a marked effect upon the drainage characteristics (ROWE 1972). Trial trenches reveal such phenomena and will indicate variations in the dip of strata, especially of thinly bedded shales, caused by creep. The strength of clays affected by creep may be reduced by such movement towards their fully softened or residual values (EARLY and SKEMPTON 1972).

b. *Mudflows* can be generated when a slipped, hummocky mass formed by an old landslide is artificially re-mobilised, often by falls. Many mudflow deposits rest on well developed shear surfaces and, in part at least, are slab-like transitional failures. Flows can be developed following rock falls or slides if air, water or snow provide a suitable medium in which the debris can travel. The term 'mudflow' has been used in the past to describe two different types of soil movement, here distinguished as mudslides and mudspates (HUTCHINSON and BHANDARI 1971).

(1) *Mudslides* represent relatively slow mass movement of saturated or partly saturated fine material, associated with the occurrence of shearing failure along the boundary slip surfaces. Mudslides commonly occur on slopes of 5 to 10 degrees or greater. (The rate of motion is usually variable or intermittent, affected by seasonal short term variations of water-table. The movement will usually not exceed 10 m/day and, typically, the rate of motion of the mudslide is no more than 5–25 metres/year).

(2) *Mudspates* and *lahars* (volcanic mudflows) represent a relatively rapid mass transport of material predominantly in suspension with, in consequence, a high water/soil ratio (MUIR WOOD 1971). Mudspates may occur on gradients as flat as one degree. By the nature of the physical difference between the two types of mudflow the velocity of a mudspate must be considerable, greater than the terminal velocity of the soil particles in water, (>1 m/second for coarse sand with higher minima for coarser material). The motion is usually infrequent and sudden, e.g. following heavy rainfall or collapse of a steep earth slope.

c. *Rotational and translational slides.* The manifestation of instability of the ground may take many different forms, dependent principally upon the geological structure and history (SKEMPTON and HUTCHINSON 1969). The typical conditions are set out in a simplified manner in Table 20.

d. *Flowslides* occur when a sudden disturbance of a material causes a tendency for closer particle packing. The resulting increase in pore pressure leads to a consequent reduction in effective pressure between particles, with full or partial liquefaction. While the mechanism is not yet completely understood, it appears that the material need not be fully saturated and that the pore pressures may be the sum of those due to pore water and to pore air. Evidently, however, a loosely packed material with a relatively high water content on a steep slope provides the appropriate ingredients for a flowslide. Sudden movement or shock, which may be caused by local landslip, explosion or seismic tremor, is required to trigger off the motion. Flowslides may affect steep slopes of granular material on the seabed (or flat slopes of material with a metastable structure) usually off the mouth of a large river.

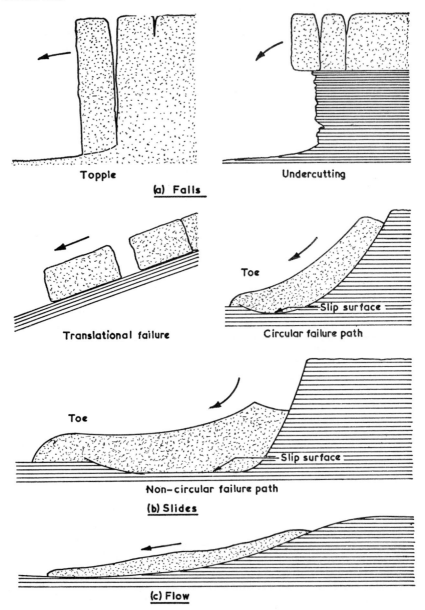

Topple   Undercutting

(a) Falls

Translational failure

Toe   Slip surface

Circular failure path

Toe   Slip surface

Non-circular failure path

(b) Slides

(c) Flow

Fig 96. Types of mass movement

TABLE 20. **CRITERIA FOR TYPES OF MASS MOVEMENT**

Serial No.	Description	Typical nature of ground	Typical conditions	References to figures
(a)	(b)	(c)	(d)	(e)
1	Falls	Rock	Geometry largely controlled by jointing pattern	96 (a)
		Scree slopes	Surface movement related to weather	
		Dry cohesive **ground**	Where subjected to basal erosion	
2	Rotational slides	Soft homogenous clays	Circular slides	96 (b) and 95
		Weathered clays	Planar shallow slides	
		Silty or over-consolidated clays	Rotational (non-circular)	
3	Compound slides	Non-cohesive soils over-lying soft stratum	Planar back to slip surface and planar base (simple, multiple or successive)	95
		Clays or silts overlying firm ground	Rotational back to slip surface, and remainder planar	
4	Translational slides	Where weak layer occurs near and parallel to surface of slope or band of soft clay	Failure in base of weathered rock or in gouge-filled rock joint	96 (b)

The motion of this variety of flowslide may be explained initially as a liquefaction and subsequently as a mass transport or turbidity flow similar to the mudspate. Such a flowslide may be subsequently recognised by the manner in which the material is redeposited, grading in size from the largest particles at the base. It is possible that volcanic dust may become unstable due to fluidisation by escaping gas and the apparent slides on the rims of lunar craters have been explained in this way.

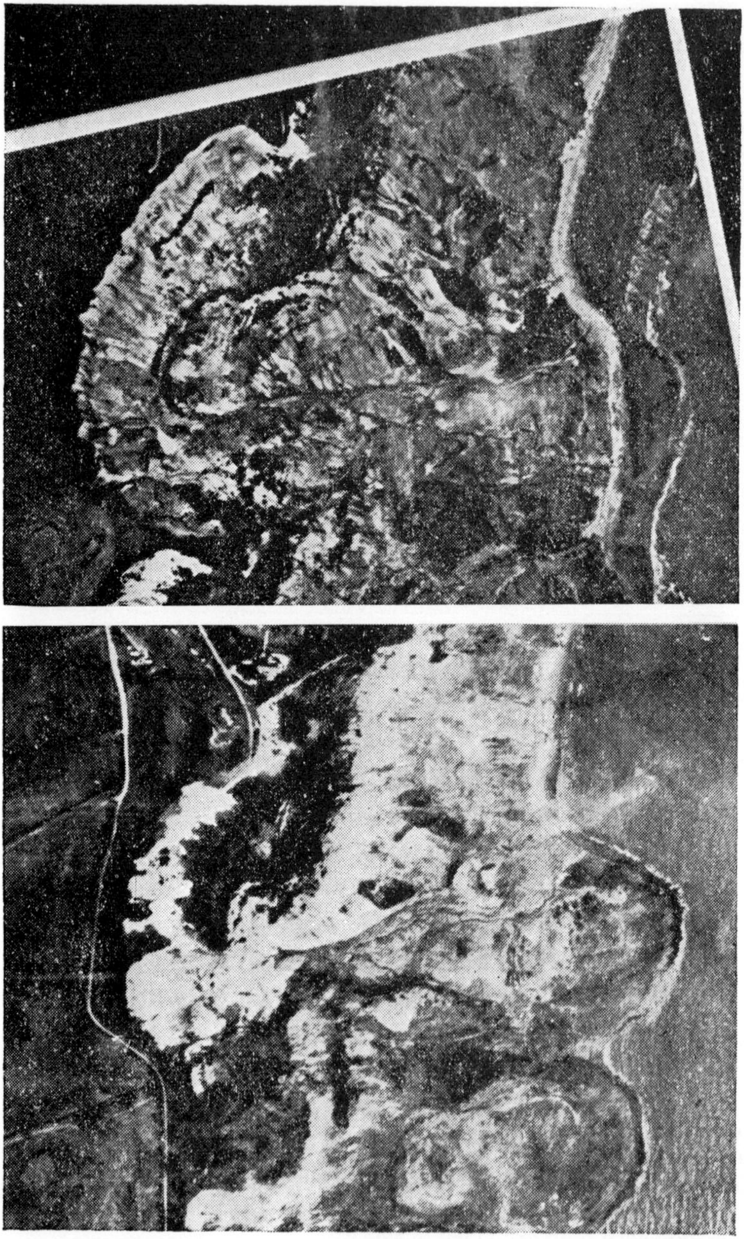

Photo: R.A.F.

(a) Situation in 1958

Photo: Fairey Surveys

By courtesy of D. Brunsdon

(b) Situation in 1969

Fig 97.　Coastal landslide. Black Ven, Charmouth, Dorset

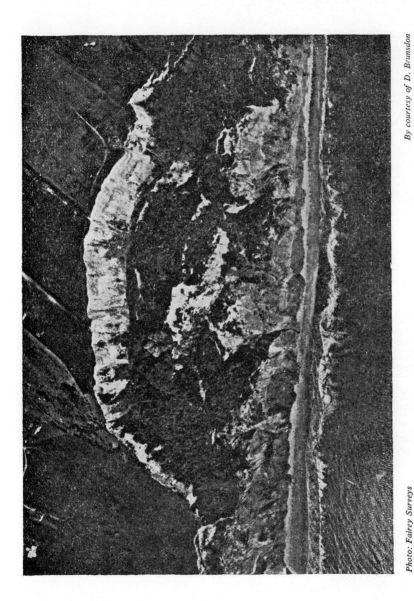

By courtesy of D. Brunsdon

Photo: Fairey Surveys

Fig 98.   Compound rotational slip. Fairy Dell, Stonebarrow Hill, Dorset

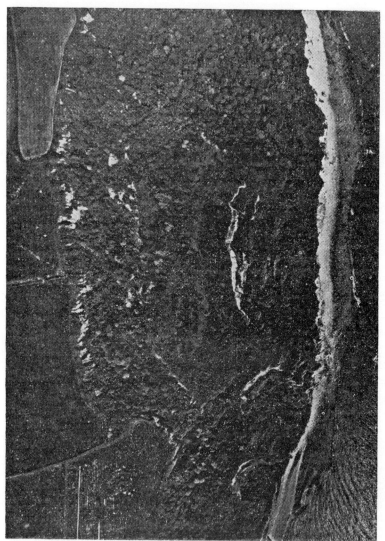

Photo: Meridian Airphotos

By courtesy of D. Brunsdon

Fig 99.   Old compound slip with small new slips. Landslide nature reserve, Bindon, Devon

7. All forms of slide may be provoked by changes in loading, by chemical or physical action leading to loss of strength by weathering, solution or erosion, or by water pressures. It is often stated that water causes lubrication of slip surfaces, but this is a secondary mechanism. The principal action of water is in the reduction of effective stresses across the slip surface, leading to reduced shear strength.

The computation of stability is beyond the scope of this book. It is important, however, to stress that whatever method is used, based on slip circles, vertical slices or wedge analysis, the accuracy of the answer is not greater than the reliability of the data or the correctness of the representation of the physical problem. For landslides in colluvial materials, the variation of the strength of the soils may be considerable and predictions of stability should recognise the importance of examination by a trained observer of the principal relevant features.

8. Some typical landslides are illustrated in Figures 97–99.

Stabilisation

9. Stabilisation of slopes entails the modification of one or more of the factors leading to instability: in consequence, remedial works will usually be concerned with adjustment to the loading, possibly by transference of load from the top of the slope to the toe, or with drainage. In order to be able to assess adequately the degree of benefit of remedial works, it may be desirable to monitor ground movement, also water levels and flows. Simple methods of surveying of indicator pegs or more elaborate use of strip gauges or photogrammetry serve to record surface movement, while the depth of the slip surface and sub-surface movement may be examined by the use of boreholes, using slip indicators (pipes or plastic tubes plumbed from the surface which become constricted at the level of failure) and inclinometers or extensometers respectively. When designing works in the vicinity of incipiently unstable slopes the effects of the works on loading or groundwater drainage (above and below ground) should be assessed. Fossil landslides should be treated with respect; marginal reductions in safety factors may be enough to remobilise the landslide with risks of multiple or successive failures.

Section 10.4. STABILITY OF ROCK IN OPEN EXCAVATIONS

Factors influencing slope stability

1. Rock type has an important influence on the behaviour of rock slopes not only because of the direct effect of the rock material but also because jointing controls its mass properties. Possibly the most important factor to be taken into account is the durability of the rock material with regard to weathering. Although no precise guide lines can be given, any rock with an unconfined strength of about 30 MN/m² or less must be viewed as potentially subject to weathering. Simulation of the conditions to which the rock is exposed may provide a useful guide to its

245

durability. Small specimens may be subject to freeze-and-thaw tests, or experiments in which a mineral salt is permitted to crystallise inside the pore spaces of a rock. Alternatively, there are various forms of slaking test which involve observing the change in condition of the rock as it is subjected to wetting-and-drying cycles, or to attrition in water. However, the most appropriate method of assessing the influence of weathering on a rock is to observe the results of weathering in natural exposures. In such cases the age of the rock face and, thus the rate of weathering, should be determined.

2. The stability of most rock faces is controlled by the natural discontinuities which define the structure of the rock mass. Such discontinuities include those formed at the same time as the rock, such as bedding in sedimentary rocks, mineral banding and some tension jointing in igneous rocks; foliation, schistosity and cleavage in the metamorphic rocks; also joints and faults, which are formed after the rock has lithified. From an engineering viewpoint, it is the strength and orientation of these discontinuities which will determine the behaviour of the rock. Particular consideration has to be given to those discontinuities on which displacement has taken place, so leading to the creation of a weak plane within the rock mass. Where movement has taken place along a number of discontinuities represented by jointed surfaces, the resulting shear planes will form low strength surfaces. Shear planes are generally very smooth, often polished, and grooved by slickensides. If the displacement has taken place within a layered sequence of rocks, a pre-existing weak layer may be re-used and consequently its strength further reduced. Such a situation arises in folded sandstone-shale sequences. In the investigation of rock stability it is important to collect information on the characteristics of the various sets of discontinuities. Such information includes:

a. Orientation: Angle of dip and dip direction.

b. Frequency: Average separation between joints or number of discontinuities present in unit distance.

c. Number of joint planes.

d. Extent: Length or area of either individual discontinuities or groups of discontinuities.

e. Degree of opening: Width of the opening of the discontinuity.

f. Infilling material.

g. Surface shape of discontinuity: Surface roughness which may be determined on several scales.

There may be practical difficulties in obtaining all this information; alternatively there may be limitations on the accuracy of the data collected and their presentation in statistical terms. It is normal practice to rely upon the interpretation of natural and artificial exposures, together with information obtained from rock cores.

3. The orientation, frequency, extent and continuity of the joints when combined together determine the size range of the natural fracture-defined rock

blocks and, in addition, their shape. One can take as an example a rock mass with three essentially rectilinear sets of joints. If the joint separation for each set is the same, then the joint blocks will be cubes; however, if the joint separation for one or both of the vertical sets is smaller than for the horizontal set, then the joint blocks will be columnar.

Toppling is characteristic of a rock mass structure which contains steeply inclined discontinuities trending approximately parallel to the rock face. Either columns or slabs topple forward with tension cracks developing behind the moving rock mass. Once the cracks open slightly, surface water can discharge directly into the rock mass and heavy rainfall can give rise to a rapid increase in groundwater pressure.

4. Translational failure takes place on one plane or a set of planes within the rock mass. A simple slab slide may occur with movement down-dip along the bedding plane. Wedge failure occurs when a rock slab fails by sliding along two joints in intersecting planes; clearly a three-dimensional stability analysis is required.

5. Groundwater has an important influence on the stability of rock slopes. Most groundwater flow in consolidated rocks is restricted to discontinuities, and so the properties of discontinuities may be critical to the volume and direction of flow. The orientation of discontinuities or dominant layering in the rock structure will significantly influence the flow pattern of the groundwater. The presence of a dominant set of joints can result either in good drainage or in retardation of groundwater flow, if the slope of the ground is at right angles or parallel to the joints respectively. Gouge-filled joints and faults, and low permeability dykes, can form important barriers to flow. From the viewpoint of stability, it is the groundwater pressure which is the dominating factor. Piezometers are often installed in boreholes to measure the pressure distribution and so determine the existing flow pattern.

6. Under similar conditions, steeper and higher slopes will tend to be less stable. In general terms it is possible, in relatively uniform rock conditions, allowing for variations in structure and groundwater behaviour, to delimit fields of instability and stability on graphs, plotting slope/height relationships for natural and artificial slopes. The examination of present-day slopes, and the back-analysis of failed slopes by calculation, can provide a valuable adjunct to field studies. Detailed variations in slope geometry can influence stability. Thus a ledge can act as a trap for surface water on the face and, unless some form of drainage is provided, can result in seepage and so contribute to groundwater flow immediately behind the face. Natural undercutting of the face will similarly encourage failure. Because of the effect of arching, a concave face viewed in plan will be more stable, but a convex slope in contrast will encourage failure, because less side support is provided to the rock mass. The shape of a proposed excavation needs to be taken into account when considering the possible mechanisms of rock failure.

7. All rock at depth is in a state of stress which is progressively relieved as the load of overburden materials is removed. Rock exposed at the surface is therefore in a destressed condition and this can result in joint widening and an increase in permeability. In some rock types, such as granite joints formed by destressing tend to parallel the topography and are referred to as sheet joints. Stress relief fracturing may be a specific cause of rock instability. Another factor of importance is that a concentration of stress can lead to swelling of the rock mass and consequent fracture opening; such a situation can develop at the toe of a steep slope inside a gorge or at the base of a cliff.

8. External vibrations may influence near-surface rock stability, particularly if the rock is already in a marginally stable situation. Blasting can lead to both artificial fracturing of a rock mass (which may contribute to instability) and to vibrations which may give rise to falls.

Stabilisation methods

9. It is important to distinguish between the stabilisation of a natural slope and the design of a slope which is still to be excavated, and which could require support in order that the total volume of excavation might be reduced. Equally, the long-term function of the slope is of relevance.

For a quarry, open-pit or construction site, a slope may only be required to be in existence for a limited number of years and so the life of the slope, in practical terms, is relatively short. On the other hand, in areas open to public access, or excavated slopes above permanent structures and roads, it is necessary to consider long-term stabilisation of the slope. Every problem of slope stabilisation must be considered on its merits. For a continuing engineering operation, plant is available and it is possibly simpler to rely on cleaning-up below failing slopes, rather than the installation of expensive support works.

10. Both the scale of the actual or potential instability, and the mode of failure involved (*see* Figure 100) must be considered in devising a stabilisation system. For very large slopes, involving large volumes of material and possibly considerable total heights, the methods of stabilisation involved are limited to modification of the slope angle and drainage. Changes to the form of the slope can be effected by moving part of the weight from the upper part of the unstable mass or by placing material at the base of the slope to form a toe weight. Drainage, provided by drilled drainage boreholes or adits, can be an efficient method of improving stability.

Where either method is adopted, it will be of importance to establish the factors which have contributed to movement in order to ensure that all precautions are taken to avoid the same situation being repeated. For example slope failures can be generated by undercutting at the toe of a cliff by marine or river erosion; in such a situation, part of the stabilisation should involve protective works to reduce or prevent erosion.

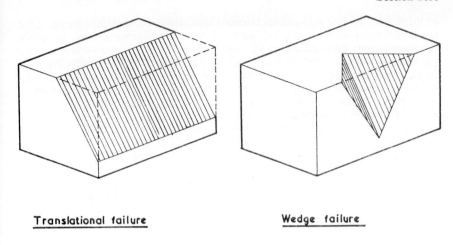

Translational failure   Wedge failure

Fig 100.   Types of slope failure in rock

11. In designing stabilisation measures for rock slopes, economies may often be made by differentiating between the behaviour of different layers or sections of the rock. Thus, for near horizontally bedded sedimentary rocks, it may be possible to achieve a steeper overall face slope by alternating berms (or low angle faces) in the weaker areas with steeper (possibly vertical) faces where strong rock, appropriately free from jointing, is exposed.

12. *Removal of rock*. The simplest method of improving stability is the removal of rock from the slope, possibly combined with the addition of toe weight. On most steep rock faces, this process can only involve the removal of loose material in a cleaning-down operation. If the rock movements are more deep-seated, bulk excavation of the rock by benching can be effective; blasting may be used providing there is no risk of creating a hazard at the base of the slope. Removal of rock as a means of stabilisation is most effective on relatively small slopes, up to about 10 metres height, or on very large slopes, where plant can be marshalled more effectively and there are engineering or economic incentives to ensuring stability. Where rock faces are to be cut artificially it is important to ensure that the appropriate slope angle is selected in order either to minimise instability or keep it within acceptable economic limits in terms of the required maintenance costs.

13. *Rock support*. Two of the commonest forms of rock slope stabilisation are the provision of surface or sub-surface support. Surface support normally takes the form of masonry or concrete retaining walls which underpin over-hanging or unstable masses of rock. Sub-surface support can be provided by mechanical anchorage including rock bolts, pins and dowels, and cables. Temporary support may be needed during remedial works and this commonly consists

of bracing made with scaffolding. Hawsers fixed around a rock face are used for both temporary and permanent support, usually requiring mortar or other protection from the weather. Rock bolts are possibly the commonest type of support used in the stabilisation of rock slopes. A rock bolt transmits the load from a face plate to an end anchorage. The length of anchor bolt may vary between 1 metre and 10 or more metres, being set into a drilled hole. Rock bolts are generally installed normal to the dominant discontinuity system in the rock, with the anchor in stable rock. The anchor may be of a mechanical nature, involving the expansion of a sleeve or of a split rod against the rock by rotation or end thrust of the bolt. Alternatively the end of the bolt may be bonded into the rock by cement grout or by a resin. The system of anchorage and the bolt loading should be related to the strength and condition of the rock. The resin is usually contained in a capsule with a hardener in a second capsule so that rotation of the bolt leads to mixing and consequent setting, the rate of the process depending upon the ambient temperature. The rock bolt may be from a single length of steel bar or from shorter lengths connected by screwed coupler.

14. Rock dowels are designed on a different basis from rock bolts, being bonded along their full length and unstressed; they are most suitable for weak laminated rock since they do not, like rock bolts, rely for their function primarily upon increasing shear strength along joints. Resin-bonded bolts may have a quick set agent for the end anchorage and a slower bonding resin for the length of shank of the bolt. Rock dowels may be in wood or steel, wooden dowels being preferred for instance where the dowelled ground has subsequently to be excavated mechanically. Indented bar is often preferred for a grouted steel anchorage to provide a good key along the full length. A typical rock bolt of 25 mm diameter may be loaded to about 0·1 MN and higher loads can be provided by cable anchorages. Cables can be installed to greater depth than rock bolts and can be used with cemented anchorages to take loads of 1 MN or more. Cables are most appropriate where stabilisation of a large mass of rock is involved, whereas rock bolts are used for stabilisation of rock faces where perhaps the outer 10 metres is unstable. Rock bolts and cables are installed in groups, say, at 2 metres centres or more. Rock support systems must be designed carefully with regard to the adjacent rock structure (PRICE and KNILL 1967). Once a basic technique has been established, detailed bolt location should be based on site inspection.

15. *Grouting.* Grouting may be used in association with rock bolting as a means of stabilisation. The main function of the grouting is to infill fractures which may have become open during weathering and destressing of the rock face. Usually cement grouting is used at relatively low pressures in view of the potential hazard of developing pressures behind the rock face. One major problem which results from grouting is that the natural pattern of groundwater flow may be altered by the reduction in permeability of the rock mass, leading to high internal pressures. If grouting is used in rock slope stabilisation it is recommended that drainage should also be provided by raking holes drilled into and through the grouted zones (Figure 101).

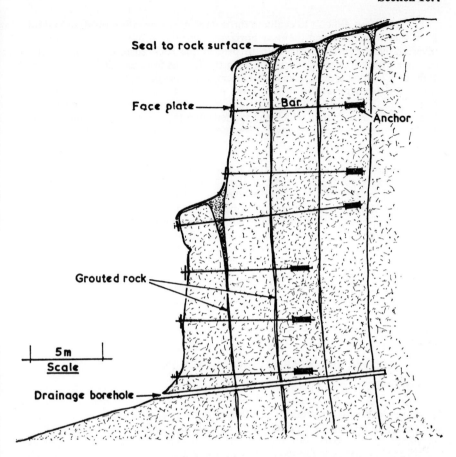

Fig 101.   Treatment of rock face by rock bolting, grouting and drainage

16. *Surface protection.* If the rock is subject to weathering, it may be necessary to provide some form of surface protection. Concrete mortar or resin spray can be used on rock surfaces which are subject to disintegration but it may be necessary to provide steel netting as a reinforcement. One of the risks of providing a protective surface is that conditions below the surface may be such that the protection layer could loosen and spall off. It is possible that drainage behind the applied layer can reduce this risk. If there are flat areas on the rock face, these need to be protected in order to reduce the risk of inflow of surface water either into the rock mass or down open tension cracks. Protective drainage works above the slope assist in directing surface water away from the rock face.

17. It is often difficult to evaluate the influence of vegetation on slope stability. Both the slope itself and the near horizontal ground above the slope need to be investigated. Vegetation can help to protect the rock surface or, alternatively, can cause rock fractures, so leading to instability, as for example in the limestone cliffs of Cheddar Gorge, England. Observations on the rock face demonstrated significant wedging action by the roots of small trees and bushes, encouraging instability. However, at the same time, a dense cover of ivy provided a net-like support to loose rock which had spalled from the face but was at the same time held in place by vegetative cover. Removal of the ivy led to several small rock falls.

18. The detailed shape of an excavated rock slope needs careful consideration. It is necessary to decide, firstly, if the slope is intended to be stable without support. If so, then it is necessary to determine from the rock structure and discontinuity strength, the optimum slope at the required height. The slope of existing natural or excavated surfaces may also be a guide. Supplementary works, such as drainage, may be regarded as an additional safety factor at this stage. Depending on the total height of the slope, it may be divided into a series of faces, separated by berms. The berms and the toe of the slope must be provided with appropriate drainage measures, protection and rock traps to collect falling debris. In strong rock, special excavation techniques, such as pre-splitting, may be used to ensure that there is minimum damage to the rock during blasting or bulk excavation. In weaker rock, which may be subject to weathering or surface erosion, protection should be provided.

Section 10.5. **STABILITY OF SOILS IN EXCAVATIONS**

Cohesive soils

1. *General.* To the engineer, cohesive soils are uncemented fine-grained naturally occurring materials. The effect of grain size is two-fold: as the particle size decreases below about 1 micron, so the surface chemistry plays an increasingly dominant role in controlling the strength of the soil; also, the rate of drainage within the soil depends predominantly on the grain size.

2. Advances in soil mechanics have shed much light on previously unpredictable aspects of the behaviour of clays. In particular, for all but short term stability problems, analysis should be based upon effective stress parameters for cohesion (c′) and friction (φ') rather than total stresses (total pressure = effective pressure + pore pressure). Special problems arise with over-consolidated clays, on account of softening under reduced horizontal or vertical loading; also, where fissures are present in over-consolidated clay, the mass strength of the ground will be less than that indicated by small intact samples. Furthermore, the effective stress parameters c' and φ' may be reduced by 50 per cent or more towards residual shear strength values of c'_r and φ'_r by previous movement along a slip surface. Disturbance by an excavation in such clays can reactivate a large scale slide on a relatively gentle slope. There is also the class of sensitive soft marine clays known

as collapsing clays (*see* Section 5.2). Before undertaking any sizeable excavation the possibility of special difficulties of such a nature should be adequately investigated.

3. Simplified theory indicates that a vertical face in clay should, in the absence of tension cracks, be capable of standing to a height given by

$$H_{crit} = \frac{4c}{\gamma g} \text{ metres}$$

where c = undrained shear strength (kN/m²)
γ = natural wet density of clay (Mg/m³)
g = 9·81 m/s²

This equation requires tension to develop in the clay and, in the absence of this capability, it may readily be shown that the critical height is reduced by one-third to $H'_{crit} = \dfrac{8c}{3\gamma g}$

4. This relationship is only valid for shallow short-term excavation. Two factors will tend to impair stability with the passage of time:

a. The reduction of lateral pressure in the ground, inducing swelling and softening of the clay.

b. Surface softening and weathering.

Both factors will cause rapid deterioration of stiff fissured clays and for such clays there is also the risk of water entering unsealed fissures, building up pressures which will tend to displace slices of the ground into the excavation. For stiff fissured clays H'_{crit} has only a transient validity.

5. On the other hand, when considering the stability of a natural or man-made clay slope, the possibility of a vertical crack existing to the depth of H'_{crit} should be taken into account since this will reduce the stability.

6. Clays may be approximately classified in terms of undrained shear strength as in Table 21.

TABLE 21. **SIMPLIFIED CLASSIFICATION OF CLAY STRENGTH**

Consistency	Undrained shear strength (kN/m²)	Approx wet density (Mg/m³)
(a)	(b)	(c)
Very soft	0–20	1·85
Soft	20–40	1·90
Firm	40–80	1·95
Stiff	80–160	2·00
Very stiff	160–320	2·10
Hard	over 320	2·20

7. When a face of clay is exposed, short term stability may be dependent on the tendency for the clay to expand, on account of the relief of lateral constraint, since this induces low pore pressures within the clay and hence higher effective stresses. In time, as suction causes migration of pore water, effective stresses are reduced and failure may occur.

8. *Unfissured clays.* Most recent alluvial and marine clays are both unfissured and normally consolidated. Most tills (or 'boulder clays') have been over-consolidated by ice pressure but remain essentially unfissured. When they contain more than say 5 per cent of clay-grade material they may be treated for engineering purposes as over-consolidated unfissured clays. Fissured tills exist but appear to be rare.

9. Most alluvial clays would fall into the very soft to firm categories of Table 21 and most glacial tills would tend to lie in the stiff to hard range.

10. *Fissured clays.* Over-consolidated fissured clays, and normally consolidated clays which have become fissured by desiccation, cause special problems in excavations. Such clays are moreover widely distributed, since they include practically all pre-Quaternary clays. Fissured clays can also form in the superficial weathered zone from the swelling and softening of some shales and mudstones which are much stronger at depth.

11. Shrinkage cracks formed during a dry period may penetrate the soil to depths of say 1 metre in a temperate climate and 3 metres in a tropical climate, and have an adverse effect on stability.

12. *Silts.* Silts when dry may stand vertically for long periods, but below the water-table they become one of the most troublesome materials known in civil engineering. Groundwater lowering by pumping is usually difficult and when this is so, chemical injection or freezing are virtually the only methods of controlling unsupported faces in silt below the water-table. Silts tend to be sensitive to vibration and this may induce them to flow from a face where they might otherwise have remained in place.

Cohesionless soils

13. Dry uncemented sands and gravels will not stand vertically but will fall back to a natural angle of repose which is an inherent property of the material and is independent of height. For a truly homogeneous material, the angle of repose should equal the angle of internal friction, but the latter depends upon the shape and packing of the constituent particles. The general order of variation can be seen from Table 22.

14. Naturally occurring sands and gravels above the water-table are often damp and surface tension tends to bind the particles together and give them an apparent slight cohesion. In these conditions they may stand vertically for a short time to

TABLE 22. **TYPICAL ANGLES OF REPOSE FOR DRY SANDS**
(after TERZAGHI and PECK 1967)

	Loosely packed	Densely packed
Round grains, single size	28°	33°
Angular grains, well graded	34°	46°

Densely packed well graded angular gravels may have an angle of repose as high as 50 degrees.

heights of a few metres before collapsing to the angle of repose. If excavated to a slope of about 1 in 1, damp sands and gravels will often stand for several weeks, provided water is not allowed to flow down the face. At the angle of repose and freely drained they should be permanently stable when undisturbed and may be grassed or otherwise protected against surface erosion.

15. Most pre-Quaternary sands and gravels have at least a trace of mineral cement, introduced by percolating groundwater. This may give sufficient cohesion to enable them to stand vertically for considerable heights, but individual cases vary so widely that no general rules can be given and each must be considered on its own merits.

16. Open excavations in cohesionless sands and gravels below the water-table are very unstable. Water issuing from the toe of the face will cause undermining and progressive collapse until a stable slope of about 20 to 15 degrees or less results.

Effects of groundwater in excavations

17. A water-table above the floor of an excavation always increases the disturbing forces acting on its sides and tends to cause a loss of stability. There is a special risk that such a condition may pass unrecognised in clays because their low permeability prevents the ingress of water in noticeable quantities, although positive water pressure may exist in the pores and fissures. Suspect locations are not necessarily flat or low-lying and it is often very difficult to determine the real hydrostatic condition in clays without the installation of piezometers for a sufficient period to allow reliable readings to be made. The evidence of surface water, or of nearby wells and boreholes, may be of value.

18. The most disturbing effects of groundwater in excavations result from the unforeseen encounter with saturated granular soils under a hydrostatic pressure, contained in otherwise impermeable ground. For example, if an excavation in clay is taken too close to a deeper water-bearing layer, the clay below the floor may be too thin to resist the uplift pressures developed. It will then heave and 'blow', perhaps carrying with it any supporting structures bearing on the floor. The underlying granular material may then 'boil' up into the excavation

255

and more or less immediate collapse and flooding will result. This is quite a common possibility in alluvial clays, which tend to be weak and to overlie sands and gravels deposited during an earlier and more energetic episode in the history of the river. To aggravate the situation, a high water-table may be present in alluvial deposits. Glacial tills may contain unpredictable bodies of water-bearing sand and gravel at almost any level. Probably only the relatively high strength of most tills has averted disaster in many cases.

19. The hydrostatic pressure in a mass of saturated granular soil may be sufficient to cause fluidisation if released in an excavation. The soil will then flow bodily, sometimes along well defined internal channels which give this effect the name of piping. Often some of the overlying soil is carried away along these channels and surface collapse results. Piping may also occur in a less sudden and severe form when the exit velocity of water from the face is sufficient to carry away soil particles in suspension, but this effect is not common in materials coarser than fine sand.

20. Water-bearing layers or lenses of granular material are a feature of many alluvial and glacial clays and they tend to flow, in the form of 'sand runs' for example, from the faces of excavations which intersect the water-table. These cause undermining of the overlying cohesive soil, which tends to collapse in a series of small rotational slips or slumps. This may result in so much breakage and disturbance of the clay that rapid softening ensues from contact with the escaping water and the slips develop into a mudflow.

Excavations close to existing structures

21. Some surface settlement should always be expected round an excavation, even when the sides are supported; damage to nearby structures may result. In addition horizontal expansion may cause damage, particularly for excavations in over-consolidated clays.

22. Pumping from excavations draws down the water-table and increases the effective overburden pressure in the surrounding soil. This may cause further consolidation and settlement of adjacent structures. The risks are greatest in organic soils, especially peats, and in normally consolidated clays.

Behaviour of some special soils in excavations

23. Old landslips and mudflows. Soils which have been involved in former slope failures are likely to inherit a permanently impaired stability and movement may be restarted by small changes in gradient or groundwater level. The original movements were often associated with the climatic conditions of late glacial or immediately post-glacial times, and soils affected by these processes may be separately shown on geological maps as slipped or unstable ground, or as solifluction or 'head' deposits. Sidelong cuttings in clays are a particular source of trouble. Old landslips and mudflows can often be recognised in the field or on

aerial photographs by the irregular ridged or hummocky ground which remains. The movements frequently occurred on low-angle slip surfaces, which part readily during excavation and have a characteristic polished or grooved appearance. If such slickensides are encountered, they must be regarded as a warning of possible instability, which at worst could cause the renewal of movement on an entire hillside.

24. *Chalk.* Most of the European chalk is a weak limestone of Cretaceous age consisting of fine sand, silt and clay-sized micro-fossil particles feebly cemented at their points of contact. The greater part of the intergranular spaces remains open, so that chalk has a much higher porosity than is usual for other rocks of comparable geological history. The white chalk contains virtually no clay and is usually highly jointed, therefore allowing free movement of groundwater. However, in North West Europe the clay content increases towards the base of the chalk, to about 20 per cent in the 'grey chalk' and up to 40 per cent in the underlying 'chalk marl'. This leads to greater density, reduced open jointing and hence to reduced permeability. In other countries, particularly in the Middle East, 'chalk' may be used as a synonym for any fine-grained limestone, not necessarily of Cretaceous age, nor of the same origin.

25. Chalk is very susceptible to frost damage, although it will usually stand indefinitely with almost no weathering at near vertical angles if protected by a capping of vegetation. However, inclined joints or bedding planes may give rise to sliding failures. It is also necessary to take account of the variability of the chalk, leading at the coarse-grained extreme to massive rock sufficiently durable for use as building stone (clunch) and at the fine-grained extreme to weak marly seams. Chalk which was near the surface in glacial times is often found to be closely jointed by frost-shattering or remoulded by shearing, such effects penetrating to a depth of 30 metres or more.

26. Because of its high porosity, chalk as dug may have a moisture content as high as the liquid limit of remoulded chalk. Consequently, solid chalk may turn to an uncontrollable slurry when worked over in wet weather. Compacted chalk fill has to be treated as a silt/clay liable to frost heave. For embankments, chalk has been found to show better stability, improved drainage and reduced settlement if placed without compaction. The engineering behaviour of chalk is discussed in some detail in CIV ENG 1965.

27. *Loess* (*see* Section 53, paragraph 8). This is a wind deposited silt which forms the surface layer over large areas of North America and Eastern Europe, extending through the USSR and Iran to China. Loess is poorly stratified and of almost uniform grain size. The particles are weakly cemented and the material when undisturbed will stand vertically to appreciable heights. The grains are loosely packed, however, and in the presence of excessive percolating water, the structure collapses with considerable settlement and instability of unsupported faces. Some of the material mapped as 'brickearth' in South East England is true loess and much of the remainder is redeposited loess.

257

28. *Peat.* Many peat deposits are fairly tough and fibrous and may stand well in excavations. They are very compressible however and are usually associated with high water-tables. Pumping from excavations in peat may therefore cause rapid and large settlements.

SECTION 10.6. **TYPES OF FOUNDATION IN RELATION TO THE GROUND**

(*To be read in conjunction with Section 9.1*)

1. For any given configuration of loading, the type of foundation should be selected in relation to the nature of the ground. The criterion of a satisfactory foundation is that it should have an adequate factor of safety against failure of the ground in shear, compression (e.g. in chalk) or in tension (in sound rock) and that total and differential settlement should be within acceptable limits. Some particular types of soil are described below.

2. *Granular soils.* Granular soils do not generally introduce serious problems of settlement under normal loading unless the loading leads to damage to the particles; this may be an important consideration for sands containing a high fraction of shell fragments, for coral sands, and for collapsing soils of loosely packed but lightly cemented sands. Damp sands, however, often exist in a 'bulked' state with a high ratio of voids (air or water filled). Such sands are subject to considerable settlement when loaded. Examples are the great depths of sand forming the plains of Belgium, and the flat areas near the coast in Cumbria. The classical solutions to founding on these sands are to use piles or to preload the ground surface before permanent construction takes place. While loading during construction is usually applied too slowly to cause appreciable rise in pore pressures in sand, high live loading on sands of loose structure may require a limitation on the initial rate of application to avoid possibility of shear failure (PECK 1969) or, alternatively, compaction by means of vibrating piles with or without a local flow of water. Dynamic compaction may also be used.

3. *Cohesive soils.* For cohesive soils, immediate settlement is a quasi-elastic phenomenon while subsequent consolidation is usually considered in two phases, of primary and secondary consolidation. When a load is applied, the relatively incompressible water in the pores of the soil takes a proportion of the load; primary settlement occurs as the water is expelled and the pore water pressure drops to its equilibrium level. Secondary consolidation is believed to be due to slippage between grains with associated shearing of the highly viscous layers which have adsorbed water. Rapid loading of soft cohesive soils may give rise to shear failure especially at structure boundaries. In soft clays, sand-wicks and sand-drains are useful measures to enable a dissipation of pore pressures leading to more rapid increase in strength.

4. *Loess.* Loess is described in Section 5.3, paragraph 8. It may occur (in Russia and Romania for example) to depths of 20 metres or more, and is a dangerous foundation material for a structure containing water, such as an irrigation canal. Measures taken to combat these difficulties are; pre-consolidation by loading, pre-inundation with water, and the support of structures on piles.

5. *Gypsum.* Gypsum is hydrated calcium sulphate (Ca SO_4 $2H_2O$). Its solubility in pure water is about 1:500, but the amount dissolved can be much greater if the water contains some salt.

The effect on engineering works can be significant—sometimes spectacular—for three reasons:

a. The solution of the gypsum from under structure foundations will lead to settlement.

b. The shear strength of the soil is reduced (by a factor of 4, for example if about 10 per cent is dissolved) leading to collapse of embankments or slopes.

c. Particles of soil are removed with the water, particularly as it travels along preferred drainage paths. Even a dense clay containing gypsum particles can be affected in this way.

6. Another adverse effect of gypsum is that it reacts chemically with the free lime in cement to destroy the cementing properties of concrete.

Sulphate resisting cement must be used if there is any risk of contaminated water moving against the concrete.

7. Gypsum occurs in many types:

a. Thick beds which are likely to include anhydrite (Ca SO_4).

b. Thinner bedded gypsum in more or less compact layers.

c. Crystals in surface layers, sometimes crusts, or re-crystallised from evaporated groundwater.

d. Gypsum powder distributed in the soil mass, or in buried layers possibly associated with terrace deposits. Gypsum powder is readily dissolved by water.

8. The nature of the deposits is determined by the geological history of the ground, and by the cycles of groundwater levels and evaporation to which the soil has been subjected. There are very large areas of the worlds surface containing gypsum deposits—in Spain, the Middle East, USSR and USA for example.

9. *Gypsum-rich soil.* Since gypsum is water soluble, seepage from a water-retaining structure can lead to significant settlement (canals in Spain for example). It is vital that the geologist and engineer should recognise these special soils in the field.

10. Fluctuations in water-table may give rise to ground movement on account of variations in effective loading and these may cause unacceptable total or differential settlements. Particular caution is required with organic soils such as peat and with montmorillonitic clays, whose natural water content may be several times the mass of the soil particles. The high settlements of Mexico City are due to the compressible montmorillinitic volcanic soils combined with a lowering of groundwater for water supply. Where a building is supported on piles, consolidation of the ground above the toe level may give rise to negative (i.e. downward) friction on the piles.

The piles should be designed to be able to accept this additional load, or they should be sleeved or externally treated to limit this effect.

11. An increasing use is being made of ground anchors, to support walls and bulkheads and to resist buoyant flotation of foundations. Evidently, high local loads are transmitted to the ground at the point of anchorage, which normally entails the concreting or grouting of tendons, and the effects of joints in rock and of compaction or consolidation of soft ground have to be considered against the requisite loading. For temporary anchors, time-dependent release may be acceptable if accompanied by periodical restressing of the anchors.

SECTION 10.7. UNDERGROUND CONSTRUCTION

General considerations

1. Underground construction comprises the sinking of shafts, vertical or inclined, the driving of tunnels, and excavations for mines and large chambers.

2. Mining and tunnelling have much in common. The main differences spring from the extent of controlled collapse accepted by the former and from the generally longer time scale of stability required by the latter. Since mining operations entail high local concentration of investment, the relative economics of ground treatment operations may be very different. For example, water problems in a mine are usually controlled by pumping from considerable depth, a practice which would be unacceptably costly for a long tunnel.

3. Geology is undoubtedly the most important single factor determining the nature, form and cost of a tunnel. The very need for the tunnel may be caused by a topographical barrier to surface communication of geological origin. The tunnel route, its geometry and structural design should each be largely dependent upon geological considerations. Problems in construction which will directly affect cost are again direct consequences of the geological structure of the ground. The most important aspect of tunnelling concerns the degree of freedom in siting a tunnel in favourable ground. The different types of tunnel have different degrees of freedom; for instance, the route of the most economic tunnel for water or gas under pressure is almost entirely controlled by geological factors. Where such

tunnels are required to connect between established points, the geological structure is paramount in determining the tunnel line. A road or rail tunnel on the other hand has to conform to very much stricter tolerances in route and level.

4. The size of tunnel affects the rate of progress; in traditional drill and blast rock tunnelling the cross-sectional area is a factor in determining the length of 'pull' of each round and tunnels smaller than 2 m × 2 m are too cramped for economy. For tunnels which may be driven at high speed by conventional or machine techniques the optimum economic cross-section may vary between 7 m^2 and 15 m^2, the optimum size generally increasing with the degree of mechanisation. Tunnelling costs vary widely in relation to the nature of the ground (MUIR WOOD 1972).

5. The needs for temporary support and for permanent lining are related to the ground and to the use of the tunnel. Requirements for tunnel supports, where the strength of the rock is high in relation to the vertical ground pressure, are related to the behaviour of the rock. Squeezing rock is rock which, under the action of the stress field around the tunnel, requires heavy supports to prevent excessive plastic deformation, including bottom heave, into the tunnel. It will impose very high loads on supports and may have to be penetrated by a series of small sized drifts.

6. For a long tunnel, an early decision to be made concerns the number of working faces with the consequent need for intermediate working shafts and drifts. Here again the economic length of tunnel depends greatly upon the method of construction and hence upon the ground. The saving in cost from reduced time needed for construction has to be weighed against increased costs in equipping and manning more faces together with the additional access and shaft sinking costs.

Rock tunnels

7. The method of carrying out the works, the design and costs are related principally to:

Rock strength and hardness.

Types and direction of jointing.

Variability in the face and along the tunnel.

Faulting and diagenesis, including possibility of reactivation of faults associated with landslides.

Initial state of loading of the ground.

Equivalent elastic moduli of the ground.

Creep (i.e. time-dependent strains).

Effects of exposure and release of load.

Extent and distribution of water.

261

8. By the time a tunnel route has been selected, the site investigation (*see* Chapter 9) will be able to provide a certain amount of information on the factors set out above and the engineer should then obtain geological advice on the degree of reliability of the available data as being representative of conditions along the tunnel. He should next consider the need for experimental adits or lengths of pilot shaft or tunnel to enable ground behaviour *in situ* to be compared against borehole records and sample tests. The risks of striking water-bearing or gas-bearing fissures will usually determine the need and extent of forward drilling during excavation, whereby the rock is proved ahead of and around the face.

9. Where groundwater is encountered by forward probing it is necessary to have some criterion whereby the need for ground treatment is determined. This must evidently be related to the nature of the rock. Small flows in soft rock or in rock with gouge-filled joints may suffice to cause major instability. If a major flow of water is encountered it may be necessary to carry out grouting from a pattern of boreholes drilled either through a temporary concrete bulkhead in the tunnel or from the surface. In the extreme, if the effects of water flow have been underestimated, the tunnel might have to be abandoned for a new route. More than three years delay was caused to the Awali water tunnel for the control of the River Litani in Lebanon due to the encounter with steeply dipping and lightly cemented sandstones highly charged with water at pressures up to $7 \cdot 5$ MN/m². Completion necessitated major ground treatments ahead of a local diversion of the tunnel line.

10. It is of great importance that detailed records are made of the rock revealed as the tunnel advances. Such records should comprise sketches of the face indicating zoning of rock, faults and major fissures, indications of movement and deterioration of the periphery, water entry and description of supports (Figure 102).

Where measurements of movement, observation holes or other special measures are adopted these should be related to the tunnel record sheets. For a recent rock tunnel in mudstones and siltstones, for example, the rock has been graded into six types and these gradings are used for all purposes of zoning. In addition, a petroscope, i.e. a small graduated periscope, is used to determine opening of joints intersected by NX size holes drilled from the tunnel.

11. The orientation of joints in rocks is best recorded by means of a stereo-graphic plot whereby a hemisphere is projected onto a plane. The orientation of a plane is represented by a dot corresponding to a normal to the plane. Thus the distribution of the dots indicates preferential orientations. The plotting convention is that the projection point is at the lower end of the diameter normal to the plane of the joint (Figure 103).

12. The estimation of existing stresses in the rock is often necessary particularly before excavation of large openings. The stress pattern in strong rock is very

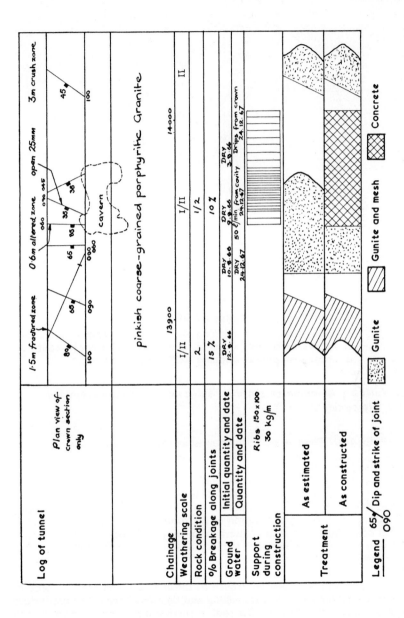

Fig 102. Tunnel geological log (after NEWBERY 1971)

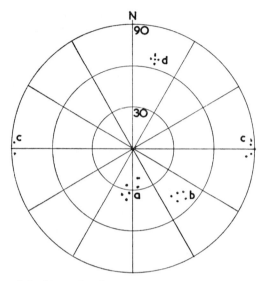

Lambert–Schmidt projection on lower hemisphere

Key

a. poles of planes dipping 30° northwards
b. poles of planes dipping 45° northwesterly
c. poles of planes dipping vertically strike north-south
d. poles of planes dipping 70° south-southwest

Note
A Wulff equal-angle or stereographic net is similar except for the dip circles

Fig 103. Point, or polar, diagram (stereographic plot of joints)

different from the simple gravity stresses and is determined by the past geological history of the site and the nature and distribution of the rocks. In many cases, the horizontal stresses can be measured in boreholes, usually by an over-coring method (Chapter 9), but interpretation of the results requires skill in relation to geological conditions. The stress pattern of incompetent rock may more simply be related to gravity stress, provided the incompetent rock exists from the surface to the tunnel level.

13. Overbreak will be largely controlled by the angle and frequency of bedding planes and joints, the method of tunnelling and the tunnel profile. In flat-bedded strata, rock will often break to a flat roof in small tunnels, but in larger tunnels

roof falls may tend to occur sometimes forming high cavities, with the rock sides developing a series of corbels, depending on rock type, fissuring, stress pattern and support system.

14. Machine-driven tunnels are less likely to require temporary support than drill-and-blast tunnels in the same rock, but support will be required where there are open fissures approximately along the line of the tunnel or where, usually across the bedding plane, variations in the stress-strain characteristics of the rock lead to local overstressing and slabbing of the roof, particularly in the haunches of the tunnel. Many practical factors enter into the choice of the type of temporary support. Traditionally, support in tunnels was provided by timber framing, later largely superseded by steel arches, which depend upon adequate blocking and packing of the rock to limit bending stresses in the arch. Struts and ties are also provided to avoid instability of straight lengths of arches considered as struts. Where the rock is reasonably sound and where a natural arch may be formed across the crown of the tunnel, rock-bolting may be used in place of steel arches. Rock bolts function in a variety of ways, the influence of each depending upon the nature of the rock. Table 23 sets out the principal contributory factors and the types of rock most favourable to each.

TABLE 23. **PRINCIPAL STRUCTURAL FUNCTIONS OF ROCK BOLTS**

Serial No.	Function	Optimal conditions
(a)	(b)	(c)
1	To enable the rock to perform more effectively in shear and in tension	Rock laminated transverse to line of bolts
2	To create an arch of rock in compression around the tunnel	Fissure planes oblique to the line of bolts
3	To support loose rock	A relatively thin band of loose rock with sound tight rock beyond

15. Pneumatically applied concrete (shotcrete) or mortar (gunite) may be used on its own or in combination with rock bolts. Shotcrete acts in combination with the rock to which it adheres to form an arch around the tunnel. Although it can to some extent fill exposed open fissures in order to provide continuity of arching, the body of rock must be fairly tight for the shotcrete to provide reliable support. Without such a condition the apparent benefits of shotcrete may be illusory and may even tend to conceal hidden and possibly hazardous defects. Another object of shotcrete may be to protect the face of the rock against weathering, against progressive sloughing of the face, or, for example in mudstone, protection against drying of the face which may lead to cracking and gradual failure by spalling. Shotcrete may be used to prevent the wash-out of a softer filling from joints in the rock.

265

16. A major objective should be to provide support as early as possible, where time and distance from the face may lead to increasing instability in the roof. Netting attached to rock bolts will help to contain small rock fragments and to avoid progressive collapse.

17. Tunnelling machines may be used in most types of sedimentary rock and in some igneous and metamorphic rocks whose unconfined compressive strength does not exceed about 150 MN/m². Particular factors affecting the economics of tunnelling by machine include the extent of jointing of the rock, its brittleness and the problems of water, which affect the handling of the spoil by the machine. Jointed rock may jam the rotating cutter head and highly variable rock may put excessive loads upon the main bearings. Most machines cut a circular profile, although some machines of lower power for a given size of tunnel can form other tunnel shapes. Machines have to be shielded if used in swelling ground; excessive swelling may eliminate the possibility of the use of a machine. Generally each type of machine can tolerate only a limited degree of ground variation. Certain new machines are designed to accept different types of cutting tool for different conditions and are in consequence more tolerant. High costs are inevitably incurred if a machine has to be withdrawn as unsuitable for a tunnel.

Soft ground tunnelling

18. Soft ground tunnels are defined as those which require immediate full support to the ground. Traditional methods have widely survived in lining soft ground tunnels; in the United States use is made of temporary supports, liner plates and *in situ* concrete, whereas in Europe cast-iron (and, since the 1930's, reinforced concrete) segmental linings have been far more widely adopted. Where the ground is reasonably predictable and relatively invariable there may be considerable savings in adopting an appropriate form of one of the newer systems of lining (DONOVAN 1969).

The lining of a tunnel in soft ground should be considered as a structural shell acting compositely with the ground; but the actual process of the behaviour is far too complex to analyse fully, depending as it does on the stages of the construction process, the standards of workmanship and the distribution of stresses, both in three dimensions and in time. The theoretical analysis can be simplified . by selecting the factors of greatest significance and ignoring the remainder.

19. The main problem in soft ground tunnelling concerns the stability of the exposed ground in the face and elsewhere. The main factor affecting stability will be the strength of the ground, the depth below surface, the grain size, the head of water and the size of tunnel. Nearly all major soft ground tunnels in Britain are now shield driven (*see* paragraph 22).

20. Where the overload ratio of ground pressure to the shear strength of the ground exceeds a factor of about 5 the face will require special measures for support. Such a criterion has to be applied to the weakest part of the soil structure, be it for instance the softest layer of inter-bedded material or the network of

fissures or slickensides ('backs') in a stiff clay. Where compressed air is used then the effective overload ratio is reduced.

21. The behaviour of the ground in the face of a tunnel is affected by the permeability, since the release of load will initially set up negative pore pressures which will gradually be restored by the inflow of groundwater. Ground of low permeability may be found to be stable in the face, so long as the rate of tunnel advance is sufficiently high. If the tunnel is stopped then the same ground may require close boarding to contain it. Where small falls in the face occur, the ground is said to be 'ravelling'. Where the ground has a fairly high permeability, i.e. coarse silt or sand, and is beneath the water-table then it will probably be running ground. Expedients for tunnelling through such ground may entail a boxed timber face, a closed face with ports through which the ground is extruded or alternatively (or additionally) compressed air may be used (*see* paragraph 23).

22. Support to the ground may be largely facilitated by a Greathead shield (Figure 104) which is used principally for reasons of safety but also for economy.

23. Where tunnelling occurs in running ground there is a choice between the following expedients which may or may not be economically practicable in any specific set of circumstances:

a. Drainage by pumping or other means may be used to lower the water-table. Such an expedient is only practicable where dewatering will not cause unacceptable ground movements. Particular care is required where dewatering of organic soils may cause high rates of settlement.

b. Ground treatment by grouting with clays, cements, chemicals or resins. The grouting medium depends on the nature of the ground; the cost of such treatment increases with reducing grain size (PERROTT 1965) (*see* Section 10.2, paragraph 6**d**).

c. Freezing has been widely used for shaft sinking through water-bearing ground (*see* Section 10.2, paragraph 6**e**) using brine as the coolant. Quick freezing with liquid nitrogen has also been used for advancing tunnels and shafts through short lengths of particularly difficult water-bearing ground. Freezing is usually found to be more expensive than its alternatives for long lengths of tunnel but it may be used in all types of ground except where there is rapid flow of groundwater, where there is a local source of heat or where the groundwater contains enough salts in solution to lower its freezing point unduly.

d. Compressed air is a most valuable expedient in tunnelling. The purpose of compressed air is to prevent flow of water into a tunnel and to stabilise the ground exposed in the tunnel face and elsewhere. For soft clay, direct support is the main object. For sand, the air displaces water from the voids in the soil adjacent to the exposed face, thus stabilising the ground by capillary attraction set up between moist grains. For intermediate types of soil,

267

Photo by courtesy of Edmund Nuttall Ltd.

Fig 104.   Greathead type shield (Tyne tunnel)

there will be a contribution from both effects. Care has to be taken in ensuring strict medical examination of all who work in compressed air, and the observation of appropriate standards of working periods and decompression to avoid risks of bone necrosis.

e. For certain types of ground the face of the tunnel may be stabilised by using a mechanical shield with transverse bulkhead, the space in front of which is kept filled with mud to balance the ground pressures. Such a shield has been used in Mexico City for tunnelling in montmorillonitic volcanic silty clays. The same principle has been used for tunnelling in gravels, with bentonite mud injected into the face and recirculated with the spoil, and may also be applied to tunnelling in sand.

24. An important feature in tunnelling concerns the control of ground movements and consequential surface settlements. At the time of designing a tunnel, the engineer should be able to predict the approximate shape and rate of development of the basin of depression above the tunnel in relation to the tunnel size, depth and the nature of ground, with allowance for unforeseen variations. It is also particularly important to consider the effects of the tunnel on water in the ground, due for example to drainage leading to variations in water-table and consequent consolidation which may be particularly severe in peat and other organic deposits. An alternative factor may be the washing out of fines in the ground. In a most extreme case, shaft sinking for South African gold mines beneath cavernous (karstic) dolomite entailed considerable lowering of the water-table and this has led to collapse and partial filling of large caverns in the rock.

Subsidence

25. Mining operations leave voids which will tend to close. Where these occur near the surface, changes in ground level (subsidence) will result. In the simplest case, that of complete extraction of a horizontal bed such as a coal seam, the entire surface will subside through a distance equal to about 90 per cent of the seam thickness, over an area rather larger than the area of extraction (NCB 1975). The area is governed by an angle of draw usually determined from previous experience in the particular ground (Figure 105). Where a seam is mined by the longwall method, successive slices are removed from a face which advances at right angles to the direction of slicing.

26. The subsidence proceeds as a wave advancing ahead of the face. First a tension wave moves through the ground, any surface structures are strained, cracks appear in buildings and fractures develop in more competent rocks, although weaker beds may accommodate the strains without fracture. A temporary increase in permeability may occur. This is followed by a compression wave which tends to close up the fractures and cracks, so that a suitably designed structure may remain structurally sound, perhaps needing minor repairs and redecoration; but at a lower level. The tension and compression waves move with the face, although complete subsidence may take months or even years.

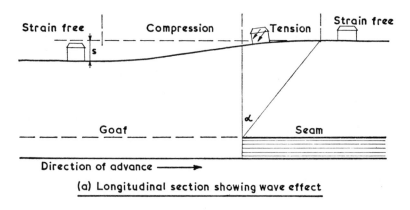

(a) Longitudinal section showing wave effect

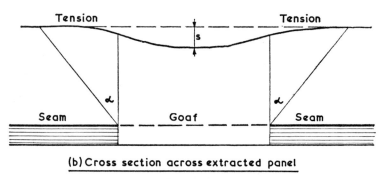

(b) Cross section across extracted panel

α = Angle of draw (35°–38°)   s = Maximum subsidence

Fig 105.   Subsidence effect

27. The rate of advance of the face will be controlled by the width and frequency of the slices. Earlier mining methods whereby pillars of coal were left in position to support the roof resulted in perhaps 60 per cent extraction only, unless the miners later extracted the pillars, a method not unknown although not always indicated on the mining maps.

28. Bed separation during subsidence may reduce the change in ground level so that deep mines may close up without affecting the surface. Other factors which affect the subsidence pattern are the number of seams worked, their dip, and any important faulting. Ground pillars may be left around shafts or main haulages, or beneath rivers, railways, or towns where subsidence cannot be tolerated, and some mined-out areas may be left permanently supported or filled

270

with waste material. Such blocks of ground, including the ground ahead of a stationary face, will be subject to tensile strains but not the subsequent compression, and this must be considered in any later construction. The 1963 vehicular tunnel under the Tyne was driven through rock where maximum strains of 0·8 per cent were expected as a result of mining operations prior to 1850 which left an unworked block beneath the river.

29. Old coal or similar workings, often unmapped, may still be held open by pillars, or may be wholly or partially collapsed following failure of the pillars over the years. They may be flooded, and dewatering may accelerate failure. Techniques are available for backfilling these to ensure stability during subsequent construction and after. The overlying disturbed ground may be consolidated by grouting.

30. Salt is frequently extracted by pumping in fresh water, allowing the salt to dissolve, and pumping out the resulting brine. The subsequent effects of subsidence, including numerous shallow lagoons ('flashes') are seen in Cheshire.

31. The removal of oil from its host rocks can lead to volume changes and subsidence, such as is believed to have contributed to the 1963 failure of Baldwin Hills Reservoir in Los Angeles, where local subsidence of almost 2 metres in 50 years is known to have occurred. Injection of water to maintain fluid pressure may inhibit compaction, but by reducing effective stresses at faults may initiate further subsidence.

At Long Beach, California, extraction of oil has caused nearly 9 metres of subsidence.

32. The lowering of the water level in an aquifer by over-extraction for water supply can also result in subsidence. In the London area, under-drainage of the London Clay caused by over-pumping from the underlying chalk aquifer had caused about 225 mm subsidence by 1931, increasing to an estimated 400 mm by AD 2000.

REFERENCE LIST—CHAPTER 10

ABERFAN, 1968 —Report of the tribunal appointed to enquire into the disaster at Aberfan, October 21, 1966. HMSO, London.

CIV ENG, 1965 —*Symposium on Chalk in earthworks and foundations.* Inst. Civ. Engrs., London.

COLLINS S P and DEACON W G, 1972 —Shaft sinking by ground freezing, Ely Ouse–Essex Scheme. *Proc. Inst. Civ. Engrs.*, Supp. VII, Paper 7506S, 129–156.

DONOVAN H J, 1969 — *Modern tunnelling methods.* Inc. Inst. Pub. Hlth, LXVIII Pt 2 (Apr), 103–139.

EARLY K R and SKEMPTON A W, 1972 — Investigations of the landslide at Walton's Wood, Staffs. *Q. Jl. Eng. Geol.* 5, 19–42.

EARTH MANUAL, 1974 — *Earth Manual.* United States Bureau of Reclamation, Washington, D.C.

ELSON W K, 1968 — An experimental investigation of the stability of slurry trenches. *Geotechnique* 18, 37–49.

HIGGINBOTTOM I E and FOOKES P G, 1970 — Engineering aspects of periglacial features in Britain, *Quart. Jl. Eng. Geol.* Vol. 3, No. 2, pp. 85–117.

HUTCHINSON J N, 1968 — *Mass movement. Encyclopaedia of Geomorphology.* Fairbridge R W (Ed). Reinhold Book Corporation, New York.

HUTCHINSON J N and BHANDARI R K, 1971 — Undrained loading, a fundamental mechanism of mudflows and other mass movements. *Geotechnique 21*, 353–358.

ISCHY E and GLOSSOP R, 1962 — An introduction to alluvial grouting. *Proc. Inst. Civ. Engrs.*, 21, (March) 463–465.

LOUIS E and WITTKE W, 1972 — Etude éxperimentale des ecoulements dans un massif rocheux fissuré, Tochien Project, Formose. *Geotechnique 21*, 29–42.

MUIR WOOD A M, 1971 — Engineering aspects of coastal land-slides. *Proc. Inst. Civ. Engrs.*, 50 (Nov), 257–276.

MUIR WOOD A M, 1972 — Tunnels for roads and motorways. *Q. Jl. Engng. Geol.* 5, 111–126.

NCB, 1975 — *Subsidence Engineers' Handbook.* National Coal Board, London.

NEWBERY J, 1971 — Engineering geology in the investigation and construction of the Batang Padang Hydro-electric scheme, Malaysia, *Q. Jl. Eng. Geol.*, Vol. 3, pp. 151–181.

PECK R B, 1969 — Advantages and limitations of the observational method in applied soil mechanics. *Geotechnique 19*, 171–187.

PERROTT W E, 1965 — British practice for grouting granular soils. J. Soil Mech. Found. Div., A.S.C.E. 91, S.M.6 Paper 4542 (Nov).

PRICE D G and KNILL J L, 1967 — The engineering geology of Edinburgh Castle Rock. *Geotechnique 17*, 411–432.

ROWE P W, 1972 — The relevance of soil fabric to site investigation practice, *Geotechnique 22*, 195–300.

SCHOFIELD A N and WROTH C P, 1968 — *Critical state soil mechanics.* McGraw-Hill, London.

SKEMPTON A W and HUTCHINSON J N, 1969 — Stability of natural slopes and embankment foundations. Proc. 7th Intern. Conf. Soil Mech. Found. Engng, Mexico, *State of the Art*, Volume 291–340.

TERZAGHI K, 1939 — Soil mechanics—a new chapter in engineering science. *Inst. Civ. Engrs. Jl.* Vol. 12, pp. 106–141. (Reprinted in A *century of soil mechanics* 1844–1946. Inst. Civ. Eng, London.)

TERZAGHI K and PECK R B, 1967 — *Soil mechanics in engineering practice.* 2nd Edn. Wiley, New York.

GENERAL REFERENCES ON ROCK MECHANICS

COATES D F, 1970 — Rock mechanics principles, Mines Branch monograph 874 (revised edition). Dept of Energy, Mines and Resources, Canada.

FARMER I W, 1968 — *Engineering properties of rock.* Spon, London.

JAEGER C, 1972 — *Rock mechanics and engineering.* Cambridge University Press.

JAEGER J C and COOK N G W, 1969 — *The fundamentals of rock mechanics.* Methuen, London.

McGREGORY K, 1967 — *Drilling of rock.* C R Books, Maclaren, Croydon, Surrey.

ORBERT L and DUVALL W I, 1967 — *Rock mechanics and the design of structures in rock.* Wiley, New York.

STAGG K G and ZIENKIEWICZ O C (Ed), 1968 — *Rock mechanics in engineering practice.* Wiley, New York.

TALOBRE J A, 1967 — *La mecanique des roches et ses applications.* Dunod, Paris.

Journals — *The Quarterly Journal of Engineering Geology.* Geol. Soc. London, Scottish Academic Press.

— *Engineering Geology.* Elsevier Scientific Publishing Co., Amsterdam.

GENERAL REFERENCES ON LOESS

BALLY, ANTONESCU and PERLEA, 1965 —Loess as foundation soil for irrigation systems. *6th Int. Conf.*, *SMFE*, Montreal.

BELES and STANCULESO, 1961 —Settlements of structures founded in highly porous soils. *5th Conf. Paris*, Vol. I, p. 587.

CLEVENGER W A, 1956 —Experiences with loess as a foundation material. *Trans. ASCE*, Vol. 123, 1958.

PECK, HANSON and THORNBURN, 1957 —*Foundation engineering.* John Wiley and Sons Inc. Chapman and Hall Ltd.

CONSTRUCTION MATERIALS

SECTION 11.1. GENERAL REQUIREMENTS FOR ROCKS AND AGGREGATES

Introduction

1. This chapter covers the general physical and mechanical properties of natural materials required for different engineering tasks, in sufficient detail for preliminary guidance in selecting suitable sources. For the design of major engineering works, detailed field and laboratory investigations would normally be necessary. Since the behaviour of construction materials during handling and placing is seldom precisely the same from job to job, field trials of excavation and compaction techniques may also be required. British standard specifications and testing techniques are quoted where appropriate.

Stone for rockfill breakwaters, rip-rap, masonry and embankments

2. *Armour stone for rockfill breakwaters.* Armour stone for these structures has to meet fairly exacting requirements. The properties which must be known for design purposes are the relative density* and the heaviest block available in quantity for armouring stone. In general, if the relative density falls below about 2·6, the durability becomes suspect and increasingly large armouring blocks, less likely to be available from the weaker rocks, are needed. Requirements vary with design conditions and standard specifications are inappropriate; useful general guidance has been given by the US Corps of Engineers (ENGINEER MANUAL 1963). The rock should be dense, fine-textured, durable and tough. Highly porous rocks such as sandstones and vesicular lavas, and those containing clay minerals, such as mudstones, are not normally suitable. Closely jointed or bedded rocks will not provide the large armouring blocks, which typically range from 2 to 20 tonnes each. The required combination of mechanical properties with block size is best in rocks of the granite and gabbro 'Trade Groups' and some rocks of the basalt, porphyry, limestone and quartzite groups. British Standard 812 (BS 812 1975) divides all rock into 11 Trade Groups, chiefly for the purposes of the general construction industry. The Trade Groups are:—Artificial, Basalt, Flint, Gabbro, Granite, Gritstone, Hornfels, Limestone, Porphyry, Quartzite and Schist. However, in any detailed work BS 812 urges more precise geological determination; for example the Gritstone Trade Group includes Agglomerate, Arkose, Breccia, Conglomerate, Greywacke, Grit, Sandstone and Tuff, all with

---

*NOTE: Relative density is used in this book in the general sense and not in the special sense relating to soil mechanics, *see* Glossary.

very different geological attributes, but often broadly similar gross engineering properties. In all reports and contracts it should be clearly stated whether a Trade Group name or more precise rock name is being used.

3. The appropriate tests are relative density, absorption and aggregate abrasion value, as described in BS 812. Resistance to repeated wetting and drying and to the crystallisation of salts in the pore spaces is very important, and can be assessed by the aggregate soundness test (ASTM C 88–73) with sodium or magnesium sulphate. A development of the slake-durability test may in future prove more convenient (FRANKLIN and CHANDRA 1972), but in its present form the test is mainly intended for comparing the durabilities of the weaker rock types. Freezing tests are advisable in high altitudes or latitudes.

4. Rock fill for breakwaters should be sought close to the construction site in view of the cost of transporting the large quantities and sizes of material. Suitable rock types occur in such varying situations that the best guide to their location will be a local geological map. If the latter is not available the local area must be reconnoitred, paying attention to prominent physical features, such as coastal escarpments, that may indicate outcrops of durable rocks. However, hard crystalline rocks quite often have a subdued surface relief and occur as a platform on which more prominent but younger and weaker rocks are built up. The large quantities often needed may rule out sources such as minor intrusive igneous bodies or steeply dipping stratified rocks, which may become quickly exhausted or unworkable, although their physical properties may be good. It is often difficult to predict accurately the largest block size likely to result from quarrying, or the proportion of armouring to hearting stone available, without full scale blasting trials.

Rip-rap hearting stone and rubble masonry

5. The requirements for revetment stone, hearting stone for breakwaters and rubble masonry are similar, but less exacting than those for breakwater armouring stone. The design of rip-rap protection is related to wave action (CIRIA RP 88). Wave action on revetments in sheltered water is comparatively limited so that individual stones can be small (typically 10–100 kg) and less dense. A minimum relative density of 2·5 features in some American specifications for dam rip-rap. All weathered rock should be excluded, together with mudstones and shales, which may soften on wetting, frost-susceptible rocks, and soluble rocks such as gypsum. The flattened shape of many stratified sedimentary rocks is often an advantage in constructing hand-placed stone revetments or rubble walls.

6. If the suitability of the material is doubtful from visual inspection, the soundness test (ASTM C 88–73) or the slake-durability test may be used, perhaps supplemented in marginal cases by one or more of the tests of physical properties specified in BS 812. The factors underlying the weathering of natural stone have been reviewed at length (BRE 1972).

7. *Embankments.* Most rocks are acceptable for embankment construction in dry situations. Closely jointed and comparatively weak rocks will be quicker and cheaper to excavate and place to a design profile. Materials containing coal, lignite or oil-shale may be liable to spontaneous combustion and should be used with care. With a high water-table, and where surface drainage is likely to percolate rapidly through the embankment, the rock fill should be selected according to the criteria described above for rip-rap.

Crushed stone aggregates for concrete

8. *General.* Specifications for crushed rock aggregates form part of BS 882, 1201 1973. The material must satisfy certain requirements in respect of strength, water absorption, relative density, particle shape, dimensional stability and freedom from reactive minerals or organic matter. Tests for the first four of the above properties are specified in BS 812: 1975. Those of special relevance to concrete aggregates are described in CP 110 1972 and in ME VOL XIV PT I: 1974. The basic tests common to aggregates for all purposes are description, simple petrological classification in terms of one of the eleven trade groups defined in BS 812, grading and silt content, relative density and water absorption. Further tests (such as one of the strength tests) are normally made only in cases of doubt or when some specified limit applies. They will depend on the use intended for the crushed stone, as described below.

9. *Strength.* Strength is usually the most important quality. In practice, provided the aggregate meets commonly accepted requirements in terms of relative density and water absorption, it is unlikely to be deficient in strength. These requirements are not defined in standard specifications, but for general guidance relative densities below 2·6 or absorption above 3 per cent may be regarded as beyond normal working limits and investigation by further testing may be needed. The usual strength tests are either the 10 per cent fines value or the aggregate impact value, both determined in accordance with BS 812.

10. The required strength is likely to be obtained from virtually all rocks, unless weathered, of the following trade groups: basalt, flint, gabbro, granite, hornfels, porphyry and quartzite. These are all chiefly igneous or metamorphic rocks. Whether sedimentary rocks of the gritsone or limestone trade groups are suitable depends on how far their porosities have been reduced by overburden pressures. In North West Europe few sedimentary rocks younger than the Carboniferous are suitable, but elsewhere the position is different, especially in the younger fold mountain areas. Rocks of the schist trade group may be adequately strong but their particle shape when crushed is likely to cause difficulties in producing good quality dense concrete.

11. Stone for concrete aggregate must be stable in the presence of water and cement. Rocks containing clay minerals, especially mudstones, but also some gritstones and members of the basalt group, are known to shrink as the concrete dries, thus destroying or weakening the bond between mortar and aggregate, perhaps with cracking.

12. Rust stains may develop from the oxidation of iron sulphides, common in igneous and metamorphic rocks, and from the weathering of some other iron-rich minerals. They are disfiguring but usually structurally harmless. Iron sulphides can be detected by immersing the suspect aggregate in limewater, since the iron then forms a green precipitate of ferrous hydroxide.

13. Flint and quartzite aggregates have poor fire-resistance since they become converted to a high temperature form of silica, accompanied by volume change and disintegration.

14. *Alkali-aggregate reaction.* A few aggregates react with the alkalis in cement to cause expansion and cracking of concrete. Suspect rocks are chert-bearing or silicified limestones and mudstones, and some volcanic rocks of acid or acid-intermediate composition. Many gravels will be unsuitable as concrete aggregate because they contain pebbles of deleterious rocks. The detection of reactive aggregates is lengthy, complex and uncertain, but the effect is reduced when low-alkali cements (those with less than 0·6 per cent by weight of total alkalis expressed as Na_2O) are used. British cements are generally of this type. The use of low alkali cements is the safest procedure if suspect and untried aggregates must be used, but this does not always give adequate protection. The natural and crushed flint gravel aggregates of south-east England are slightly reactive, as disclosed by chemical tests, but in practice they give little, if any, trouble. This might be because these aggregates consist wholly of flint whereas if they contained an admixture of inert material internal stresses leading to cracking might be developed. The problem of alkali-aggregate reaction is complicated (NBS).

15. Certain dolomitic limestones cause alkali-carbonate reaction, which may lead to cracking of concrete. It is usually associated with dolostones (or magnesian limestones) containing 40 to 60 per cent dolomite ($MgCa(CO_3)_2$) and 5 to 20 per cent of clay minerals and fine quartz particles. Prediction depends on laboratory observations on rock specimens immersed in alkaline solutions. The effect may be controllable by the use of low alkali cements.

16. *Dangerous salts.* Some soluble salts are dangerous, either when present in the aggregate or when introduced into the concrete after placing, by groundwater movement for example. Sulphates react with Portland cement, causing a volume change accompanied by cracking and disintegration. The commonest natural sulphate is probably gypsum (hydrous calcium sulphate), which often occurs as readily visible and distinctive crystals. More rapid attack however usually results from sulphates, such as those of magnesium and sodium, which have much greater solubilities and for this reason are rarely visible in the solid state, though they may be present in high concentrations in the groundwater. Chlorides (mainly as sodium chloride or common salt) cause efflorescence and may attack steel reinforcement, causing corrosion and consequential cracking of the concrete. In hot arid areas, sulphates and chlorides are often introduced into aggregate materials or concrete near the surface by the evaporation of rising groundwater.

The simultaneous presence of both contaminants increases the danger and the risk is greatest in the presence of a shallow saline water-table, as in coastal areas or inland desert basins. Aggregate stockpiles in these situations should be isolated from the ground by a suitable impermeable layer to prevent deleterious salts accumulating in them from the upward movement of capillary water. An excessive concentration of salts in the aggregate or mortar may cause cracking and spalling by the physical bursting action of crystal growth in the pores of the material.

17. *Suitable rock types.* The rock types suitable for crushed stone concrete aggregates are so diverse that no general principles for locating sources by their associated landforms can be given here. Reference should be made to Chapter 7. The best guide is a geological map of the area, if available. This can be misleading however if it does not show the superficial deposits, particularly those resulting from deep tropical weathering, which often make sound rock virtually inaccessible over considerable areas. In the absence of local experience, outcrops of igneous rock will normally be the first choice for consideration, because of their strength and their comparative uniformity in the usual absence of a layered structure.

Fig 106. **Surface** hardened limestone (calcrete), Bahrain

279

Fig 107. Surface layer of calcium carbonate (caliche) on sand, Kuwait

Metamorphic rocks are strong, but often closely layered and non-uniform. In an area of sedimentary rocks only, the older strata should be considered first, particularly the limestones since these tend to be stronger than other rocks of comparable age.

Limestones in hot arid regions, such as the southern Mediterranean and the Middle East, are often hardened for 1 metre or so below the surface, possibly due to the evaporation of capillary groundwater causing additional calcite to be deposited in the pores (Figure 106). Below this hardened surface or 'duricrust' they may remain soft and chalky. Their surface appearance therefore can be misleading if quarrying is proposed. On the other hand such a hardened crust is the only source of good quality aggregate in many areas. It is not confined to limestones and can form on the surface of any sufficiently calcareous rocks and soils (Figure 107). Where nothing else is available, as on most oceanic atolls, crushed coral can be used to make concrete (ME VOL XIV PT I 1974).

Crushed stone aggregates for bituminous surfacings and bases

18. Specifications for these aggregates are given in BS 594 1973: and BS 4987: 1973. Rock may be from any of the trade groups except flint and schist. Methods of testing are covered in BS 812: 1975. The necessary physical and mechanical

properties are essentially the same as those for concrete aggregates, with the additional requirement that in wearing surfaces certain values may be specified for the polished stone coefficient and the aggregate abrasion value, both as defined in BS 812. The UK Department of the Environment specify minimum polished stone coefficients of 0·59 or 0·62 (DOE 1969), depending on the conditions of use, a requirement which is more likely to be met in rocks of the gritstone and basalt trade groups. In Britain the flint (not accepted in BS 594), hornfels and limestone trade groups polish the most readily, with coefficients of 0·5 or less. In addition, aggregate abrasion values of not less than 10 or 12 per cent may be specified. Some gritstones tend to wear rapidly and may be incompatible with requirements in respect of both polishing and abrasion resistance. Aggregates with smooth surfaces such as flint, quartzite and coarsely crystalline rocks (especially those containing mica), may be more liable to stripping of the bitumen, but only experiment will determine this. A test for coating and stripping is specified in ASTM D 1664–69.

19. Bitumen is a relatively inert medium and aggregates for bituminous pavements are comparatively immune from the shrinkage, staining or reaction that might occur if the same materials were used in concrete. The aggregates should be free from soluble salts which are hygroscopic and may impair adhesion of the bitumen. This is only a problem in arid tropical areas.

20. Sources of suitable stone may be chosen on the same criteria as those suggested for concrete aggregates.

Natural coarse aggregates for concrete, bases and bituminous surfacings

21. A natural gravel should contain rock types conforming to the requirements already listed in paragraph 8. Most gravels will contain more than one type of rock and some variation in the composition is to be expected. The natural grading of a gravel seldom conforms to the requirements of BS 882 or BS 594 and some crushing and screening are therefore usually needed to achieve this. Some specifications for wearing courses (notably for airfields) require a minimum proportion of crushed faces in the aggregate.

22. The main sources of natural gravel can be listed as follows:

Beaches, including raised beaches

River channels, outwash fans and deltas

Submarine gravels

River terrace gravels

Lag and other residual gravels

Glacial gravels.

Since most of these deposits result from relatively recent erosional processes they have a close relationship with topography and can often be recognised by characteristic associated landforms (*see* Chapter 5).

23. Gravel beaches and raised beaches tend to contain one predominating size fraction, so that crushing and screening will be necessary to achieve an acceptable grading. The gravels are usually round and clean, with the weaker material largely eliminated by the abrasive effects of wave action. The main contaminants are shells and other marine organisms. A small amount of shell (perhaps up to 5 per cent by weight) is permissible in coarse aggregate for concrete, but any shell is undesirable in bituminous mixes, especially those for wearing courses. Salt in the concentrations normally expected of beach gravels can be tolerated in concrete, but some efflorescence must be expected.

24. River channels, outwash fans and deltas may be roughly stratified and a wider grading may be obtainable from a single source. Screening and recombination of the separate fractions may give an acceptable grading without crushing, but some wastage is likely. The maximum size of material will depend on the velocity of the river in flood and gravels are therefore more characteristic of the upper and middle reaches of a river's course. Further downstream, gravels are common below the finer surface materials of the modern flood plain, since they

Fig 108. Torrent bedded wadi gravels near Muscat, Oman

Shows wide grading of material and weak carbonate cementation, allowing face to stand vertically.

282

were very often deposited in early post-glacial times when the sea level was lower and the rivers more energetic than at present. These deposits are usually below the water-table but may be won by dragline or dredger. In arid areas where surface water flows intermittently, the gravels may be cemented by calcium carbonate ('calcrete' gravels). This effect is common in wadi bottoms and outwash fans at the bases of desert escarpments. The possibility of high sulphate and chloride concentrations under these conditions must be kept in mind. The smaller outwash fans tend to contain much material of local origin and the weaker rock fragments may not have been eliminated by abrasion (Figure 108).

Wadi gravels are often deposited under turbulent conditions with poor sorting of the particles. They may therefore be a useful source of all-in material, if uncontaminated by sulphates and chlorides, but there is often a high silt content to be separated and carried to waste.

25. Submarine gravels are often beach or river gravels submerged by a rising sea level and reworked by tidal currents. They are often free from fines, but may be heavily contaminated with shell, which is costly to remove.

26. River terrace gravels were laid down under the same general conditions as modern river gravels and have similar characteristics. They normally occur above the water-table, which simplifies working, but some terraces will have been exposed to weathering for a considerable period and the constituents could be partly decomposed. Silt and clay may also have been introduced by infiltration from above. Cementation, especially by calcium carbonate, is common, particularly in tropical environments.

27. Lag gravels are typical of flat desert terrain where wind has carried away the fines and concentrated the larger fractions at the surface. Over wide areas of the Arabian peninsula they form the only source of aggregate. The loose surface layer may be of high quality because only the most durable material has survived, but immediately beneath there may be heavy gypsum contamination which prevents mechanised working of the uncontaminated surface material (Figure 109). Crushing and screening is generally necessary to secure an acceptable grading. Other residual gravels are characteristic of wet or seasonally wet tropical climates where chemical weathering is rapid.

Some may consist of fragments of the parent rock which have survived the decomposition of the remainder in which case they are likely to be themselves decomposed and embedded in a clayey matrix. More useful are the concretionary gravels which have been formed in place by chemical weathering. A good example is the nodular ironstone which often covers the surface of lateritic soils and has been successfully used as a concrete aggregate. Exposed rock surfaces in tropical or arctic areas may be covered with a gravel formed in place by thermal or frost cracking. This will be angular, with an absence of fines.

Fig 109. Gypsum below surface armouring of lag gravel, Oman

28. Glacial gravels usually have a very wide grading and tend to be contaminated with silt. Washing is generally necessary to give good concrete or bituminous aggregates. A good supply of fine aggregate is usually also available simultaneously from glacial gravels. The deposits tend to be very variable and to contain lenses of silt and clay. They are not much worked if better sources are accessible.

Natural well-graded aggregates for pavement sub-bases, and low-grade bases

29. The stability which the sub-base provides is usually obtained by compacting a widely graded granular material to a maximum density at a controlled moisture content. It is not required to resist abrasion or polishing and the particles need not have the mechanical strength of those used in the base, so that materials containing relatively weak or weathered rock particles may be acceptable. Moisture content changes will impair stability unless the material passing the 425 μm sieve has a low plasticity. The UK Department of the Environment requires a plasticity index of less than 6 per cent for their granular sub-base Type 2 (DOE 1969). The gradings allowed in this specification are wide, within limits of 75 mm for the maximum size and a fraction of not more than 10 per cent

finer than 75 micron mesh. American specifications tend to allow an upper limit of 15 per cent on the passing-200 (75 μm) fraction for compacted natural material. Up to 25 per cent is allowed in the sand-clay bases as specified by the US Federal Aviation Administration (FAA 1968). The latter also gives useful design specifications using materials not common in Britain such as shell gravel and caliche. Caliche is the thin limestone deposit which is sometimes formed at or near the surface by weathering processes in hot arid climates. When the specified gradings cannot be met from natural sources a cement or bitumen stabilised base course may be necessary, especially for severe operating conditions.

30. Many of the sources already described for natural coarse aggregates will be suitable in the 'as dug' condition for sub-base or low-grade base materials; others may require some screening to eliminate unacceptable size fractions. Submarine or beach gravels are likely to be deficient in the smaller fractions, unless these are available from other local sources for blending. Many glacial gravels which may require too much screening and washing to be used for concrete or bituminous aggregates, make admirable natural sources of granular sub-base material. High level terrace and residual gravels, which may contain weathered material and be contaminated with silt and clay, are generally acceptable. They may be cemented with lime or gypsum in tropical latitudes, but often crush down on compaction to give a suitable grading. A high gypsum content however might make them unsuitable for use with concrete pavements. In arid climates watering may be necessary to control dust and achieve the required density.

31. Some materials other than natural sands and gravels may form a good sub-base. Examples are crushed coral, shell gravels, laterite and calcite-cemented beach and dune sands. Such materials are widespread in hot countries, throughout the Middle East for example and in the West Indies.

32. Most well graded granular soils are stable enough to form a temporary natural sub-base and even to serve as a pavement if properly maintained. With increasing clay content however, the wet weather stability becomes impaired. Many less well graded tropical soils are naturally stabilised by surface concentrations of mineral substances. Caliche and laterite are good examples. Laterite, in spite of its high clay content, often gives a surprisingly good wet weather performance. The salt marsh ('sabkha'), characteristic of many coastal areas of the Middle East, although it consists largely of silt and clay, often develops a good dry season wearing surface through the crystallisation of salts. Due to its hygroscopic properties, however, it tends to be slippery in the early morning or after light rain. In the wet season the naturally high water table rises further and dissolves the saline crust, which fails by pot-holing. This can be averted to some extent by building the sabkha into compacted embankments to keep the wearing surface above the water-table. The stable surface crust can be quickly remade by watering with brine.

Section 11.1

Natural fine aggregates for concrete

33. Aggregate passing the 5·0 mm British Standard sieve (*see* Chapter 5, Table 2) is described as fine aggregate. It should comply in respect of grading and other properties with the conditions laid down in BS 812. Methods of testing are defined in BS 882.

34. The specified gradings require a wide range of particle sizes down to material retained on the 75 μm British Standard sieve. Most sources of natural coarse aggregate will yield some parts of this range. Deficiencies in grading can be met with sands from other sources or with dust-free crusher fines. The latter are harsh and angular and tend to impair workability, but their use is unavoidable in many areas which have a shortage of good natural sands. The best gradings are likely to be given by river sands and river terrace deposits, or by outwash fans and deltas. Submarine and beach sands tend to have narrow gradings lying mainly in the finer part of the range. This is also generally true of British Tertiary and Mesozoic sands where these are sufficiently uncemented to be dug for concreting use. Good gradings can often be obtained from glacial sands and gravels after washing to remove the silt fraction.

35. Although the great majority of sands in the latitude of Britain consist predominantly of silica grains, beach and dredged marine sands may contain a proportion of shell. Many beach and marine sands in warmer latitudes, expecially off desert coasts, are carbonate and not silica sands and consist almost entirely of the worn debris of various shells and other marine organisms. For this reason they may have quite wide gradings.

Provided the shell fragments are reasonably rounded, a high shell content is of no great significance, though some reduction in workability may result. Shelly sands used in a dry condition will have a higher absorption than silica sands and the water-cement ratio may need adjustment to compensate for this. Since shell is a weaker material than silica it may be expected to cause some reduction in strength when used in concrete but this is unlikely to be important unless the highest quality structural concrete is required. However, the UK Department of the Environment (DOE 1969) precludes the use of fine aggregate containing over 25 per cent of calcium carbonate within the top 50 mm of pavement quality concrete, mainly on the ground of reduced frost resistance.

36. Due to evaporation, beach sands may contain high salt concentrations and together with all marine sands should be washed before use in reinforced concrete structures. For permanent reinforced concrete structures, fresh water should be used in the concrete to prevent corrosion of the reinforcement. For temporary structures and for mass concrete, sea water is acceptable, but some efflorescence should be expected.

37. Many sabkha areas consist of old marine sands and the loose dry surface layer may have to be dug for concreting sand in the absence of more suitable

286

material. It is usually heavily contaminated with gypsum and chlorides. The gypsum tends to occur in crystalline masses which can be partly removed by screening, after which washing, preferably with fresh water, must always be carried out to reduce the high chloride concentrations.

38. By an irony of nature, most desert areas tend to be deficient in concreting sands. Wind-blown desert sands are in general too fine and silty for use in concrete. The water-laid sands of wadi beds are usually better, but they may be contaminated with wind-blown dust which must first be removed. Gypsum is also a likely contaminant. Except in major wadis, these sands tend to be thin and in impersistent lenses not suited to large-scale working.

39. Mica is an undesirable constituent of concreting sands and it is often abundant in granitic and metamorphic terrain. Sands containing more than about 2 per cent of mica should be rejected.

Natural fillers for bituminous mixes

40. Requirements for fillers are specified in BS 594: 1973 and BS 4987: 1973. They should consist predominantly of silt-sized materials, with 85 per cent finer than 75 μm British Standard sieve. Only two types of filler are recognised by BS 594; both are processed and not naturally occurring materials. These are limestone dust (a by-product of the quarrying industry) and Portland cement. In practice however, most inert silt-grade materials can be used. If a source of crusher dust is not available, then a suitable local stone, preferably limestone, may be crushed in a ball or hammer mill and the passing 75 μm fraction separated by some form of pneumatic classifier. Alternatively a natural source of silt must be sought. The main problem, except in arid conditions, will be to find material with a low enough moisture content (say under 5 per cent) to be treatable in a classifier, since this is virtually the only method of processing the quantities required. Dried silts from river, lake or glacial sediments could be used, and desert sands normally contain a substantial silt fraction. Other possible sources might be loess and the dust created by traffic or earthmoving operations, especially on chalk or soft limestone surfaces.

Well-graded soils for embankment and sub-grade construction

41. *General requirements.* Since a wide variety of soils can be used to form compacted earth embankments and sub-grades, it is normal practice to make the best use of locally available material rather than to move soil from a distance to suit a predetermined design. The following soil properties must be specified for the rational design and construction of compacted earth fill: natural moisture content, particle size distribution, liquid and plastic limits, and the dry density/ moisture content relationships given by the British Standard compaction test (BS 1377: 1974). For measuring the strength of compacted sub-grade material the California Bearing Ratio (CBR) is generally used (also given in BS 1377: 1974).

In the design of an earth-fill dam more complex tests would be necessary which are beyond the scope of BS 1377. Details of such tests are given, for example, in BISHOP and HENKEL (1962).

42. The UK Department of the Environment has issued a useful specification for the method of placing embankment materials, which are classified in terms of their compaction requirements (DOE 1969). Three classes of material are defined on the basis of grading and moisture content and an appropriate compaction procedure is specified for each class, as defined below:

a. Clays and marls with up to 20 per cent of gravel and a bulk moisture content (i.e. for clay plus gravel) not lower than 4 per cent below the plastic limit. Well-known British examples include over-consolidated clays such as London, Gault and Kimmeridge Clays, and the more cohesive glacial tills. Chalk having a saturation moisture content of 20 per cent or more is also included in this category.

b. Well-graded granular soils and dry cohesive soils with over 20 per cent of gravel and/or a moisture content lower than the plastic limit minus 4 per cent. Materials such as sandy glacial tills, 'Keuper Marl' and the hard shaly clays of the Coal Measures would be typical. Also included in this group are well-graded sands and gravels (e.g. many non-cohesive glacial deposits) with a uniformity coefficient higher than 10, and chalk with a saturation moisture content in the range 15 to 20 per cent.

c. Uniformly graded sands and gravels with a uniformity coefficient of 10 or less, and all silts. This group includes most sands of marine origin, such as the Bagshot Sands or Folkestone Beds of southern England.

43. No upper limits of moisture content are specified, but it is suggested that for cohesive soils a working value of about 1·2 times the plastic limits could be used. For granular soils an upper limit from 0·5 to 1·5 per cent above optimum moisture content (as determined from the British Standard compaction test) may be used.

Materials unsuitable for fill

44. The following should be excluded: material from swamps, marshes or bog, peat, tree stumps, logs and other perishable material, clays with a liquid limit over 80 per cent and/or a plasticity index over 55 per cent, and soils in a frozen condition. Materials liable to oxidation, such as oil shale or coal, should either be avoided or well mixed with a large volume of inert soil. In addition, most normally consolidated alluvial clays would be unacceptable because their moisture contents would exceed the recommended upper limit. Clays placed too wet cannot be compacted and merely heave under the weight of the roller. Topsoil should not be used on load bearing surfaces and under pavements.

45. Whenever possible, the above materials should also be removed from an embankment area prior to construction if excessive settlements and possible shear failure are to be avoided.

Problems caused by special soils

46. Some materials behave unexpectedly during compaction. For example, volumetric changes can occur during the placing and compaction of many soils, so that a given volume of excavation does not necessarily yield exactly the same volume of compacted fill. Loose sands may show 5 per cent or more volume reduction on compaction. The opposite effect occurs with dense soils; a very stiff or shaly clay might show about 5 per cent volumetric increase and a dense well-graded sandy gravel perhaps 2 to 3 per cent increase.

47. The behaviour of chalk is described in Section 10.5.

48. Glacial tills usually make excellent compacted earth fill because of their wide gradings and low to moderate moisture contents and plasticities. However, contained boulders may be troublesome. It is generally permissible to include isolated boulders up to about $0 \cdot 1$ m^3 in earth fill for road works, but the UK Department of the Environment specify that no stone exceeding $0 \cdot 014$ m^3 shall be placed within $0 \cdot 6$ metres of formation level under carriageways and hard shoulders (DOE 1969).

49. Dry loose silts such as loess may undergo severe settlement on saturation and should either be removed from below an embankment or artificially consolidated by irrigation. Proper compaction of silts requires close control of water content because of their rather low permeability, which prevents the pore water pressures, built up under the weight of the compaction equipment, from being dissipated quickly enough. This tends to cause sponginess or heaving during the process.

50. Embankments and cuttings formed of or in expansive clay soils may be liable to excessive shrinkage or swelling with moisture content changes. These clays can generally be recognised from their high liquid limits and plasticity indices, but if these are below the limits specified above, under the heading 'unsuitable materials', serious trouble should be avoided. The 'black cotton soils' characteristic of poorly drained situations in many tropical areas in Central Africa, India, Australia and Central America, show this behaviour to a marked degree. These soils should not be used if alternative materials are available. They have been used successfully in construction by careful control of moisture content, and sometimes after stabilisation either with hydrated lime or by burning (CLARE 1957).

Embankment materials for special purposes

51. Selected granular materials may be needed for drainage or filter layers in earth dams or in road works. The requirements for earth dams are not very critical because deficiencies in grading can usually be compensated by increasing the thickness of the layer. The main requirement is freedom from silt and clay, combined with a wide grading through the sand and gravel range. Filter blankets between the embankment fill and the protective rip-rap can be formed from most

sandy gravels in the 'as dug' condition, provided the gravel fraction forms at least 50 per cent of the soil. More tolerance is generally allowable on the grading for the downstream filter layer, between embankment and foundation. Consequently most clean well graded sand-gravel mixtures can be used, likely sources being river and terrace deposits or outwash fans and deltas. Glacial sands and gravels could also be suitable but might need prior washing. The design of filter layers for use with rip-rap protection is considered in a research report (CIRIA RP 88). Filter materials used with concrete structures and in road drainage have to meet stricter grading requirements because they are used in thinner layers. Natural granular soils seldom meet the requirements and the filter materials have to be made by recombining screened natural sands and gravels or crushed rock.

52. Impervious fill for the cores of earth dams should be capable of compaction to a high density, to give strength combined with low settlement and permeability. It should be free from gross inhomogeneities such as boulders or abrupt changes in soil type which might initiate seepage paths. Materials of low plasticity are generally favoured, to minimise swelling and shrinkage. However, the cracking and leakage which has occurred in some clay cores after impounding is thought to have been associated with the use of cohesive core materials of low plasticity and therefore low swelling coefficient, in which any cracks formed by hydraulic fracturing, for example, might tend to remain open (VAUGHAN 1970). With this reservation, good material for impervious fill is a widely graded clay-sand-gravel mixture, such as some glacial tills; although many tills will contain oversize material which must first be removed. Alternatively, most over-consolidated clays make satisfactory core material from the aspect of permeability, but may need careful control of the placement moisture content to achieve optimum stability.

Many dams have to be designed to withstand shaking from earthquake or may even be built across potentially active faults; some faults may not be seen until the dam foundation is excavated. It is now established that the filling of large reservoirs can also trigger local earthquake.

The cores of embankment dams in such situations must be associated with wide transition zones built from materials without cohesion, which will automatically collapse to form a 'crack-stopper' should displacement occur. The core itself should preferably be made from cohesionless materials at least in the upper part of the dam, and fines of dispersive clay should be avoided (SHERARD, CLUFF and ALLEN 1974).

SECTION 11.2. **SELECTION AND OPERATION OF BORROW AREAS**

Selection

1. The borderline between borrowing and quarrying is not well defined, and quarries may also be opened for local usage; but the term 'Borrow pit' is normally restricted to local excavations where blasting is not required. Commercial workings for chalk, clay, brickearth and glass-sand are essentially similar, but are located where the deposit is found in economic quantities.

2. Borrow pits are local sources of supply for soils and naturally-occurring aggregates required for construction works. They may range in size from the small pits a few metres square, dug by hand alongside dirt roads to supply materials for repairing potholes, up to the very large pits worked by scrapers and draglines used to obtain the vast supply of materials needed for earth fill dams. For example at Kainji Dam, Nigeria, about 20 million tonnes of potential suitable borrow material was proved during the site investigations. The design of these large structures depends upon the adequacy of local supplies of suitable, easily won, borrow material. On the other hand, transport routes are usually designed to ensure a balance between cut and fill sections in order to minimise needs for handling imported fill and surplus excavation of suitable material. The maximum economic depth of excavation in borrow pits is related to the scale of the operation and to surface slopes. Generally the depth below access should not exceed about 6 metres.

3. Borrow pits should be selected on the basis of their content of suitable material, distance from the work site, drainage problems, conformity to planning regulations (where such exist) and interference with other usage. It may be practicable to keep a clay deposit adequately drained even if it is below the water-table. Gravel deposits are too permeable to be kept dry by pumping, and must be worked by dragline, suction pump, or similar underwater methods. Much of the fine material may be washed out in the process, but this may be an advantage for use as concrete aggregate. Clay pits above water-table must be provided with drainage or pumped to deal with rainfall. Reservoir schemes may use borrow pits in the reservoir area for environmental reasons or to increase water storage and the material may have to be excavated early on, before impounding commences, and stockpiled until needed. Suitable stacking areas with adequate bearing capacity and drainage must then be identified during the site investigation.

4. Suitable deposits are often covered with surface soil which must first be stripped. The site investigations should include an estimate of the stripping quantities and the availability of a stocking area for the overburden which may be used later to restore the ground level as far as practicable, to landscape the area, or to assist stability.

Suitable deposits

5. Terrace deposits (old beach or river gravels) may be a good source of coarse materials. Since they are often on hillsides, they tend to be above the water-table and well-drained, so winning will be relatively easy. Beach deposits are more likely to be uniform in character. All hillside deposits may be contaminated by fallen scree and soil creep on the slopes above, or may be covered with undergrowth which must first be cleared. Care should be exercised to ensure that any excavation does not lead to instability in the exposed slopes.

6. Flood plain deposits may be fine or coarse-grained according to the original river regime. They may be very variable within short distances, and it may be

necessary to zone borrow pits and mix the materials mechanically to obtain the final requirements. Such pits, although often water-logged and liable to flooding, are sources of washed gravels which may form satisfactory coarse aggregate for concrete. Since the deposits have been laid down by river action on an existing ground surface the underlying weathered rock materials may or may not be suitable as borrow material.

7. Glacial deposits are widespread in Northern Europe and North America where they may be the only available material. They tend to be very variable over short distances and are usually ill-sorted but are often mainly silty in spite of the name 'till (boulder clay)' so generally used. A discussion on moraine classification is given in McGOWN 1971. Eskers are good sources of sand and gravel whereas moraines, whether terminal or side, may be variable depending on the nature of the debris deposited on the glacier. Ground moraines are comparable with crushed rock of the same material as the bedrock, both they and outwash fans being mainly fine-grained. The mile-wide terminal moraine at Portage Mountain, British Columbia, was a source of anxiety until it was determined that it had once impounded a large glacial lake and so must be impermeable; this was demonstrated when the reservoir behind the 180 metre high W.A.C. Bennett Dam was impounded.

8. Residual and weathered *in situ* deposits are generally considered to be a poor source of borrow material, since the depth is uncertain and the ground variable, but may be suitable for small quantities. In tropical areas basic igneous rocks may be deeply weathered to form 'black cotton soil' which can be used for clay cores of dams and also as a component for clay grouts. In wet weather it becomes extremely difficult to work with mechanical equipment. True laterites in tropical areas may also be suitable for some purposes, especially road and airfield construction. The red lateritic clays of East Africa and similar areas, produced by weathering in regions of high temperature and rainfall but with good drainage, are of some engineering value. It has been pointed out that the standard classification tests give anomalous results for the clay content, since the clay particles tend to aggregate into silt-sized particles which are not dispersed during the test.

9. Similar aggregation of clay particles has been noticed in the Triassic 'Keuper Marls' of Britain, which have been much studied in the course of motorway construction.

Section 11.3. **SELECTION AND OPERATION OF QUARRIES**

Selection

1. The primary requirement for a quarry for constructional purposes is that it should provide a suitable source of the intended material, which may be rip-rap or armouring for dams, revetments or breakwaters, sub-base for roads, fill for

embankments, aggregate for concrete, ballast for railways, masonry for ashlar facing or bulk infill, or roadstone which may or may not be coated. Relevant properties of the material are usually:

a. Unconfined compressive strength.

b. Porosity, because the ability to absorb water may lead to loss of strength.

c. Isotropy for a better shape factor during crushing (freestone).

d. Joint spacing for ease in extraction and suitable block size.

e. Absence of colloidal silica and other undesirable constituents for aggregate.

f. Susceptibility to weathering.

g. Impurities which may cause staining in building stone.

h. Bitumen adhesion for roadstone.

i. Polished stone value for roadstone; some apparently suitable limestones have a poor resistance to polishing and can seldom be used for a road wearing course. Coarse sandstones may vary greatly in mechanical strength (FISH 1972).

2. Most quarries are opened to satisfy a particular requirement, but later they may be developed for general commercial use. Accessibility to transport will be a major factor in determining whether a quarry is likey to be developed in this way. A quarry specially opened for a new road or railway may become viable commercially because of the improved access which the project provides.

3. Geological factors affecting extraction once a suitable rock body has been located are joint spacing and direction, and also the dip of sedimentary strata. The quarry face should be orientated so that these preferential planes of splitting assist the excavation. If a bedded deposit can be worked in an up-dip direction much less explosive will be required. On the other hand, a scarp face is more suitable for early high output, which may mean down-dip extraction. Extensive shallow-dipping bedded deposits, or small igneous intrusions, unless they project above general ground level, may not provide natural steep outcrops into which a face may be driven, and the rock-mass may have to be followed downwards from a level surface. This is a more costly method of working, firstly because haulage of the broken rock to surface is involved and secondly because initial output is low, although reserves may be much greater. Sheet jointing may develop parallel to the sides and floor of the excavation due to the change in ground pressures during quarrying, and these may assist rock breakage. A quarry of the hillside type may sometimes take the form of a single long face parallel with the contours, advancing into the hill until limited by increasing overburden. When this becomes excessive, extraction of the deposit may be continued by mining methods.

293

4. The quality and extent of the rock must be verified before a working face is opened. In commercial quarries expensive crushing plant, and perhaps a tar-coating plant, may have to be installed. The economic justification for a 'project quarry' is somewhat different, since such a quarry is usually required to be suitable for rapid exploitation over a relatively short period.

5. A non-geological factor affecting the siting of quarries is haulage, particularly important for low-cost aggregates, roadstone and sub-base materials. Drainage may also be a problem, especially in open pits, which may be limited in economic depth by the costs of pumping. The acceptability of blasting will depend upon the locality; dangers from blasting may be reduced by careful siting of the face.

6. The amount of overburden and of weathered or otherwise unworkable rock which can be stripped is limited by economics. In their early stages commercial quarries are usually uneconomic, since an appreciable time elapses before the full designed output can be developed. Meanwhile costs are incurred, and the recovery of these over the life of the quarry as well as the actual costs of keeping overburden stripped in advance of the face usually for 30 metres or so—will determine the acceptable depth of waste material, which should be stockpiled so as not to interfere with later quarrying.

Operation

7. All quarries, both hillside and pit types, are usually worked by benches. A benched face has improved stability, since the stresses developed within the rock near the foot of a face depend on its height. Commercial quarry faces, using modern down-the-hole drills, can be 20–30 metres high, but faces of less than 10 metres may be acceptable and even desirable for project quarries. The upper face is worked first, and its height may be conditioned by the need to work weathered rock separately; the next face is kept at least 10 metres behind, to provide adequate working space on the bench.

8. Full advantage should be taken of preferred splitting planes to assist removal of the rock and the faces orientated accordingly. At the same time the stability of the face must always be studied (*see* Section 10.4), since those planes which most assist excavation are also those which tend to weaken the face.

9. The rock structure may vary on either side of a fault, and where such discontinuities occur the geology should be determined well in advance so that the whole excavation sequence can be planned. Wide fault zones, or dykes, may not be worth removing except at access points. Intervening beds of unusable material, such as shale in a sandstone quarry, may have to be removed separately, perhaps by a special bench, and stockpiled in the same way as overburden; if they are thick enough they may make the quarry uneconomic even though usable rock occurs below, but the possibility of such layers being lenticular should not be overlooked, nor of their use for other purposes.

10. Most quarried rock needs treatment before it is fit for ultimate use, and such treatment may form the major part of the quarrying operation. Unless the product is required for a low-cost operation such as embankment fill, it will at least require screening. Oversize fragments will also require crushing or even secondary blasting. Washing may be required to remove clay material encountered in faults or weathered zones.

11. The foregoing refers to small and medium-sized quarries as in use today. However it is possible that increasing demands for crushed aggregate, and limitations on account of environmental requirements, may necessitate future extraction from very large quarries where geological factors on a larger scale, such as regional folding, facies change, or buried erosional features, will have a significant effect. The problems inherent in such mammoth quarries have been discussed by Roberts and others (ROBERTS, HOEK and FISH 1972).

REFERENCE LIST—CHAPTER 11

ASTM C 88–73 —American Society for Testing and Materials. Standard method of test for soundness of aggregates by use of Sodium Sulfate or Magnesium Sulfate (1973).

ASTM D 1664–69 —American Society for Testing and Materials. Standard method of test for coating and stripping of bitumen aggregate mixtures (1969).

BISHOP A W and HENKEL D J, 1962 —*The measurement of soil properties in the triaxial test.* 2nd edn. E. Arnold, London p. 228.

BRE, 1972 —Building Research Establishment Special Report No. 18—The weathering of natural building stones (Reprinted 1972).

BS 594: 1973 —*British Standard 594, 1973.* Rolled asphalt (hot process) for roads and other paved areas.

BS 812: 1975 —*British Standard 812 Pt. I, 1975.* Methods of sampling and testing mineral aggregates, sands and fillers.

BS 882, 1201: 1973 —*British Standard 882, 1201 Part 2, 1973.* Aggregates from natural sources for concrete (including granolithic).

BS 1377: 1974 —*British Standard 1377, 1974.* Methods of testing soils for civil engineering purposes.

Chapter 11

BS 4987: 1973 —*British Standard 4987, 1973*. Coated macadam for roads and other paved areas.

CIRIA RP 88 —*CIRIA research report No. RP 88*. Rip-rap design for windwave attack. Construction Industry Research and Information Association, London.

CLARE K E, 1957 —Tropical black clays. In symposium on airfield construction on overseas soils. *Proc. Inst. Civ. Engrs*. 8 (November 1957) 223–231.

CP 110, 1972 —*Code of Practice 110*—The Structural Use of Concrete. British Standards Inst., London.

DOE, 1969 —UK Department of the Environment, Specification for road and bridge works. HMSO, London.

ENGINEER MANUAL, 1963 —*Engineer Manual* EM 1110–2–2904, Design of breakwaters and jetties. US Corps of Engineers.

FAA, 1968 —Standard specifications for construction of airports. US Federal Aviation Administration, Washington DC.

FISH B G, 1972 —Road materials and quarrying. *Q Jl. Engrg. Geol*. 5 195–204.

FRANKLIN J A and CHANDRA R, 1972 —The shake-durability test. Inst. J. Rock Mech. Min. Sci. 9, 1972. 325–341.

ME VOL XIV PT I, 1974 —*Military Engineering Volume XIV Part I*, Concrete Practice (Army Code No. 70955). Her Majesty's Stationery Office, London.

McGOWN A, 1971 —The classification for engineering purposes of tills from moraines and associated landforms. *Q. Jl. Engng. Geol*., 4 (1971), 115–130.

NBS —National Building Studies. Reactions between aggregates and cement. Research papers Nos. 14 (1952), 15 (1952), 17 (1952), 20 (1958), 25 (1958). HMSO (London).

ROBERTS D I, HOEK E and FISH B G, 1972 —The concept of the mammoth quarry. *Quarry Mgrs. Jl*. 1972, Vol. 56 No. 7, pp. 229–245.

SHERARD J L, CLUFF L S and ALLEN C R, 1974   —Potentially active faults in dam foundations. *Géotechnique Vol. 24. 3* (Sept.).

VAUGHAN P R, 1970   —Cracking of clay cores of dams. Surveyor, 30 January 1970. *See* also report of informal discussion in *Proc. Instn. Civil Engnrs.* 46 (May 1970), 115–117.

CHAPTER 12

GROUNDWATER AND ITS EXPLOITATION

Section 12.1. INTRODUCTION

1. Of all the water existing in vapour, solid and liquid form in association with the earth, the liquid fresh waters comprise less than 0·6 per cent and can be grouped in two broad categories—surface water and sub-surface water. These are distributed as follows:

$$\left.\begin{array}{l}\text{Lakes}\\\text{Rivers}\end{array}\right\}1\cdot7 \text{ per cent Surface waters}$$

$$\left.\begin{array}{l}\text{Vadose water}\quad 2\cdot1 \text{ per cent}\\\text{Groundwater } 96\cdot2 \text{ per cent}\end{array}\right\}\text{Sub-surface waters}$$

Although it may sometimes be possible to separate surface from sub-surface water, it is essential to appreciate that they are merely phases of the continuing system of water circulation that is known as the hydrological cycle (Figure 110).

2. On the large scale of both area and time there is a limited amount of water in circulation with no practical possibility of increasing the total quantity; and this has been so for much of geological time. Being a cycle, it has neither beginning nor end, though its description for the land portion of the earth may conveniently start with precipitation (snow, rain, hail, dew etc) and its three-way movement thereafter. Some infiltrates into the ground and some of this shallowly penetrating water is transpired by plants or moves laterally to a stream, while much of that penetrating deeper to become groundwater proper re-emerges in like manner, and the remainder passes directly to the sea or ocean. That part of the precipitation which does not infiltrate is either intercepted by plants or artificial works and is directly evaporated, or alternatively, flows over the ground surface towards a nearby stream.

3. While this system cannot be halted or the overall amount of water either increased or decreased, it is common for the intervention of man to modify the cycle locally so as to affect the individual components. Because groundwater exploitation can only rarely be carried out in isolation, in any area under investigation, some understanding of the relationship between groundwater and other phases of the cycle is essential, such as the effect of well abstraction on the groundwater component of streamflow.

4. The water in the hydrological cycle is called meteoric water, and a part of this is named connate water when circumstances in the geological past have

298

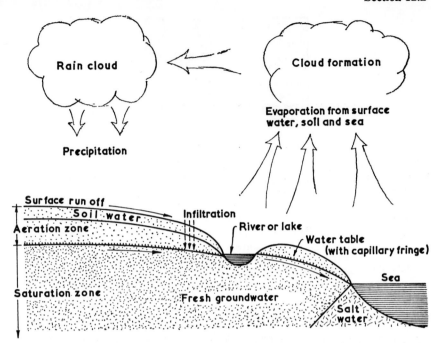

Fig 110.   Hydrological cycle

served to trap initially fresh or saline water so as to remove it temporarily from the circulation system until subsequently tapped by wells. Any water not having previously been part of the cycle is known as juvenile (or new) water, and is presumed to have had a magmatic or volcanic source. Such water is known to exist, but it is quantitatively insignificant in comparison with meteoric water. When encountered, it is likely that it will be excessively mineralised.

5. Groundwater is of engineering importance, in particular the amount available and its capacity for self-replenishment.

SECTION 12.2. GROUNDWATER OCCURRENCE

Porosity and permeability

1. The size of voids and their degree of interconnection determine which deposits function as reservoirs and conduits. In any materials the ratio of void space to total volume of the sample is known as the 'porosity', and the voids may be either primary or secondary in origin. Grain interstices (pores) were present at the original time of deposition of the rock and may therefore be regarded

299

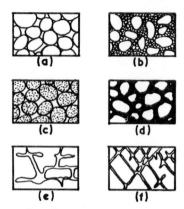

(a) Well-sorted sedimentary deposit having high porosity.
(b) Poorly-sorted sedimentary deposit having low porosity.
(c) Well-sorted sedimentary deposit consisting of pebbles that are themselves porous, so that the deposit as a whole has a very high porosity.
(d) Well-sorted sedimentary deposit whose porosity has been diminished by the deposition of mineral matter in the interstices,
(e) Rock rendered porous by solution.
(f) Rock rendered porous by fracturing.

Fig 111.   Types of voids

as primary, whereas secondary porosity is the product of subsequent fracture (faults and joints) or solution (fissures and caverns), which may be reduced by cementation (Figure 111).

2. The porosity of a deposit provides no indication of the amount of water that can be obtained therefrom, and so far as water supplies are concerned the capacity of material to yield water is of much more importance than its capacity to hold water. Even though water may fill completely the interconnected voids, not all of this water can be removed by drainage or pumping. Some water is held against gravity by molecular attraction, and the ratio of the volume of this water to the total water content is the specific retention. In complementary manner the ratio of drained water to total water is the specific yield (S_y), and the void space from which such water is removed represents the effective porosity of the material. The capacity of a deposit to transmit or to yield water is known as permeability and depends on the texture and lithology of the material being considered. In a clastic deposit, for example, permeability is a function of the size, sorting, shape and packing of the constituent geometry of the linked void spaces. In a consolidated, well-cemented rock, on the other hand, the size, shape

300

and degree of communication of the fissures or other open channels are of most significance. Thus, clay, gravel and chalk may each have a total porosity of about 40 per cent, but the permeability of the clay is low because its very small pore spaces offer great resistance to the flow of water. Conversely, the permeability of a well-sorted gravel is high because its pore spaces are relatively large and water can move more easily through them. However, the overall permeability of chalk is dependent on the degree of fissuring since unfissured chalk has a low permeability. Any classification of deposits as permeable or impermeable is only relative in a particular context. Truly impermeable strata are rarely encountered, and Figure 112 gives some indication of the range of permeabilities for the more common materials that comprise aquifers and aquicludes (*see* also Section 12.3, paragraph 4). Permeability may be anisotropic, for example in rock with predominant direction of jointing, or orthotropic in layered alluvium.

Intrinsic permeability (k) darcies

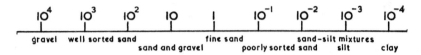

Hydraulic conductivity (K) metres per day (1md^{-1} = 1.16x10^{-5} m/s)

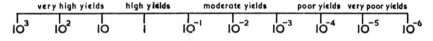

Fig 112.   Range of permeabilities

Folds, faults and joints

3. Any tectonic processes that result in a change of attitude in strata from their initial presumed horizontal position are of fundamental importance in hydrogeology. The processes range in scale from minor tilting to complex folding, with fracturing common to all but the least warped strata. These are two chief effects of folding:

a. The distribution of compression and tension so that the resultant fracture pattern provides local differences in the permeability of the strata.

b. The formation of types of structures that provide favourable hydrogeological conditions.

4. In general, anticlinal structures are regarded as likely to yield water more freely than synclines because of the development of tensional radial fractures in the upper parts of anticlines with complementary compression of material in the

301

synclines. At depth in large-scale folds the opposite will occur. Where the strata involved are non-homogeneous, such as in an alternating sequence of permeable and impermeable layers, then the form of the fold will control the direction of flow and subsequent availability of groundwater.

5. Joints or faults occur in all types of consolidated strata. Joints generally occur as intersecting sets or systems and invariably result in an increase in the permeability of the strata.

6. Faults differ greatly in extent and the amount of displacement. They can affect the water-bearing properties of an aquifer in a number of ways, thus influencing groundwater movement. They can also increase the permeability, of aquifer by the development of a zone of local fracture. Alternatively, injection of soft clayey material into the fault plane or associated shatter zone reduces the permeability and impedes the flow of groundwater. Faulting can lead to juxtaposition at the fault zone of permeable and impermeable strata. Fault planes may also serve as a line of weakness along which groundwater (sometimes mineralised or thermal) can rise (*see* Figure 128(g)).

Zones of aeration and saturation

7. In the normal sequence of the hydrological cycle, that part of precipitation which is not evaporated or does not flow away as surface run-off, penetrates into the ground to become sub-surface water. The process whereby water moves downward from the ground surface is known as infiltration and is restricted to the zone of aeration, which is that part of the ground in which the voids are usually not wholly filled with water. The thickness of this zone may range from less than a metre to, exceptionally, more than 100 metres. A large part of the water that passes below ground level is held on the surfaces of individual grains by molecular attraction in the form of hygroscopic and pellicular water which together with capillary water is called suspended water. The moisture content of this zone varies from a near saturated state close to the surface shortly after rainfall to a minimum after a lengthy period without replenishment. When saturated, the soil is said to be at field capacity and depletion of moisture as a result of withdrawal by plant roots will continue until the remaining water is no longer available, at which stage wilting of plants will occur.

8. The lower limit of the zone of aeration is the level below which all the void spaces are commonly fully occupied by water, called groundwater. The upper limit of this is called the water-table and below is the zone of saturation. Whereas infiltration in the zone of aeration is mainly downward in direction, the movement of water in the zone of saturation is governed by hydrostatic pressure and may be in any direction, with a lateral component often the most important. The term percolation describes the process involving groundwater movement and, in view of the previous definition, should therefore be restricted to the zone of saturation.

9. A diagrammatic representation of the divisions and sub-divisions of sub-surface water referred to in this section is given in Figure 113 together with an indication of the vertical distribution of types of water after a lengthy period without replenishment.

Zones and Sub zones		Water		Process	Division	Pressure
Aeration	Soil water	Vadose	Hygroscopic	Infiltration	Discontinuous capillary saturation	Gas phase = atmospheric
Aeration	Intermediate	Vadose	Pellicular	Infiltration	Semi-continuous capillary saturation	Liquid phase $<$ atmospheric
Aeration	Capillary	Vadose	Capillary	Infiltration	Continuous capillary saturation	$<$atmospheric
			Water	table		
Saturation	Phreatic zone		Ground water	Percolation	Unconfined ground water	$>$ atmospheric

Fig 113. Divisions and sub-divisions of sub-surface water

Unconfined and confined conditions

10. Any saturated material that yields water readily enough to be significant as a source of supply may be designated an aquifer, which may serve both as reservoir and conduit. Where it is unconfined, the water-table normally takes the form of a subdued replica of the topographic surface, rising beneath hills and falling towards valley bottoms where natural streamwater levels approximate to groundwater levels.

11. An aquifer is said to be 'artesian' or confined where it is overlain by an impervious horizon so that when pierced by a well the groundwater is under sufficient pressure to raise it above the base of the confining bed to a height dependent on the pressure head. The pressure surface in confined conditions is an imaginary surface which coincides with the static level of water in wells, and which commonly does not bear any direct relationship to the topographic surface. The term piezometric (phreatic) surface is applicable to both pressure surface and water-table. Over-flowing confined conditions are found where the pressure is above ground level (Figure 114). Where the static water level is below ground surface the well is a non-overflowing confined well.

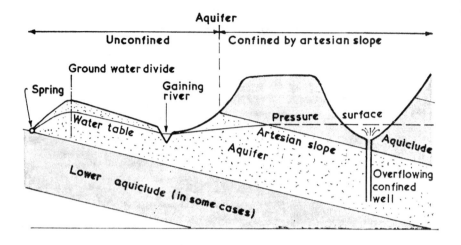

Fig 114.   Groundwater flow in association with artesian slope

12. A most common arrangement of strata is the simple 'artesian slope' illustrated in Figure 114. Such slopes provide natural reservoir capacity as well as affording protection from surface pollution. Examples of aquifers of this type that have been used extensively for groundwater supply in Britain are the Lower Triassic (Bunter) sandstones confined by Upper Triassic (Keuper) marl, the Lincolnshire Limestone overlain by Kimmeridge Clay, and the Lower Greensand beneath the Gault (SKEAT 1969).

13. Opposing artesian slopes form an artesian basin, the most usual kind being a synclinal structure with impermeable cover, such as is exemplified in Britain by the Chalk basins of London and Hampshire, and the Triassic basins of Cheshire and adjoining areas of Salop.

14. *Groundwater contours.* By using measurements of groundwater levels from wells and springs it is possible to construct contour maps showing the form and elevation of the piezometric surface. In an unconfined aquifer the difference between ground surface and the groundwater contours will indicate the depth of strata to be drilled before the saturated aquifer is reached. Of greater hydro-geological significance, however, is the fact that groundwater can be assumed to flow at right angles to the piezometric surface contours unless there are inter-mediate aquicludes. It is thus possible to determine regional groundwater flow patterns, to define the limits of groundwater catchment areas, and to express quantitatively the regional hydraulic gradient and associated flow lines (Figure 115).

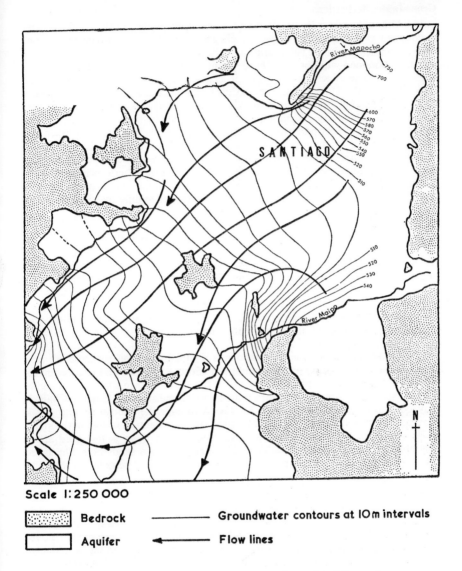

Scale 1:250 000

Bedrock   —————— Groundwater contours at 10m intervals

Aquifer   ◄—————— Flow lines

Fig 115.   Groundwater contour map and associated flow net

305

Section 12.2

Groundwater fluctuations

15. Although water-levels fluctuate commonly under both confined and unconfined conditions, the causes which give rise to such fluctuations are not always identical and may be natural or artificial. The most significant fluctuations in an unconfined aquifer are those resulting from seasonal infiltration (*see* Figure 116). The effects of pumping on changes in groundwater levels may be important locally. Under confined conditions the fluctuations caused by changes of hydrostatic pressure due to seasonal infiltration at outcrop are likely to be insignificant. There are additionally many minor fluctuations; for example, the water-table may reflect patterns of transpiration, and where adjacent to river and sea may respond to changes in river stage and to tidal level. Similarly, wells in confined conditions exhibit fluctuations that are related to changes in loading such as are brought about naturally by changes in atmospheric pressure, tides, earthquakes, and landslips, or even artificially by trains and explosions. The fluctuation of level in a well which penetrates a confined aquifer may be influenced more by atmospheric pressure than by seasonal infiltration.

16. From the point of view of water supply and resources the most significant cause of piezometric surface fluctuation is that part of precipitation which reaches the zone of saturation. Variations in the amount of effective rainfall, evaporation and run-off, provide irregular and intermittent seasonal changes to the zone of saturation and result in an undulating and fluctuating water-table. A rise in the water-table after a period of precipitation is delayed to the extent of the time taken for the water to travel from the surface of the ground to the zone of saturation, and is affected by the change in soil moisture content.

17. The magnitude of the fluctuations is a function of the permeability and specific yield of the aquifer and the volume of effective infiltration. For an equivalent amount of effective infiltration the rise of the water-table in a highly permeable Triassic sandstone may be no more than 1 metre, whilst in the English Chalk it may commonly be in excess of 8 metres. Although the sandstone and chalk have similar values of porosity, the amount of specific retention is high in the small voids of the chalk as compared with sandstones and there is less void space available per unit volume for additional water than in a sandstone. The more highly permeable strata allow relatively rapid infiltration and percolation resulting in shallow hydraulic gradients.

Relationship between surface water and groundwater

18. Between falling as precipitation and passing out of an area as stream-flow, water moves over or through the ground. Depending on the route of this movement, the discharge of a river can be divided into three components namely; surface run-off, interflow and groundwater flow. Surface run-off represents the water which travels over the ground without passing beneath the surface and also includes precipitation falling directly on the channel. Interflow is that part of the water which infiltrates the ground and is diverted by inhomogeneities in the upper layers to join the stream without having reached the water-table. The depth to

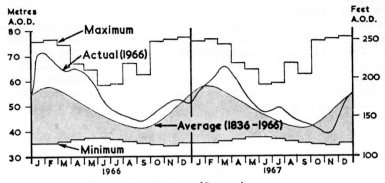

(a) Station: Chilgrove (Sussex)

Hydrometric area....Sussex rivers (41)   Based on monthly readings by Portsmouth Water Co.
Aquifer .... Chalk   Ground level ....77·2m A.O.D. 253·2 ft A.O.D.

Records from 1836-1966 used to produce the maximum,minimum,and average values

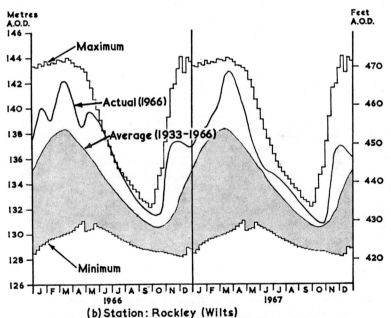

(b) Station: Rockley (Wilts)

Hydrometric area   Thames (39)   Based on weekly readings by Thames Conservancy
Aquifer............. .....   Chalk   Ground level.........146·8 m A.O.D. 481·0 ft A.O.D.

Records from 1933-1966 used to produce the maximum,minimum,and average values

Fig 116.   Groundwater level fluctuations

307

which it infiltrates is most variable and determines, in part, the time taken between infiltration and discharge to streams, thus contributing to total run-off. Ground-water flow is that part of stream-flow which is provided by issue of groundwater in the form of springs or seepages. Because the groundwater component of river flow has followed a more devious route than the other components, there is a greater time-lag between precipitation and discharge into the stream channel so that the water may be regarded as being in temporary storage. Thus, while groundwater discharge may make a negligible contribution to flood flows, it

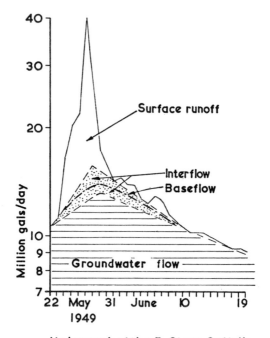

Hydrograph of the R. Stour, Suffolk

Hydrograph of the R.Stour, Suffolk, plotted on a logarithmic scale to show interflow and groundwater components of runoff. (From an original diagram by A.Bleasdale and others, in Conservation of Water Resources, Inst. Civil Engineers, pp. 121–136, London, 1963.)

Fig 117.   Surface run-off and base-flow components of stream discharge
(after BLEASDALE et al 1963)

can wholly maintain 'dry weather flows'. When the small part of interflow that penetrates more deeply towards the water-table is added to the groundwater flow the combined discharge to the river system is termed 'base flow' (Figure 117).

19. Quantitative determinations of the major components of river flow are best undertaken by analysis of the river discharge hydrograph, paying attention to the time-distribution of rainfall and to groundwater fluctuations in the aquifer. Hydrogeological conditions within a stream catchment will have an influence on the proportion of groundwater contributed to stream-flow and on the resultant form of the depletion curve (Figure 118).

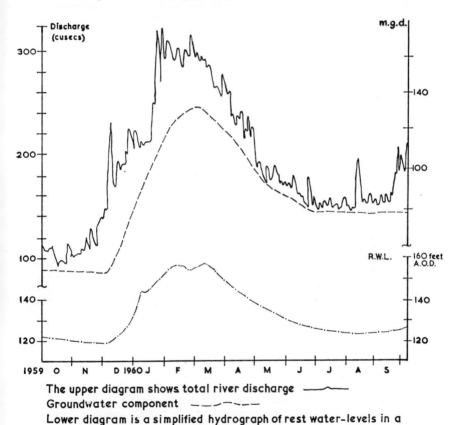

The upper diagram shows total river discharge ———
Groundwater component — — — — — — —
Lower diagram is a simplified hydrograph of rest water-levels in a chalk well at Hambledon, Hampshire

Fig 118.   Groundwater component of River Itchen, Hampshire, Oct 1959–Sept 1960 (after INESON and DOWNING 1963)

20. The nature of the surface deposits plays an important part in determining the flow regime of a stream and the proportion of groundwater discharge in total stream-flow. An important point to remember in connection with such 'gaining' streams, is that abstraction from wells must be at the expense of dry-weather stream-flow.

SECTION 12.3. GROUNDWATER MOVEMENT

Principles of groundwater flow

1. The modern concepts of groundwater movement were discovered in the mid-nineteenth century when it was demonstrated not only that the velocity of flow in small pipes was directly proportional to the hydraulic gradient, but also that discharge varied with changes in temperature. In 1856 Henri Darcy confirmed such observations by experimentally demonstrating their application to the vertical percolation of water discharging at atmospheric pressure through a horizontal bed of filter sand. Darcy's Law states that the rate of flow through porous media as exemplified by ideal aquifers is directly proportional to the head loss and inversely proportional to the length of the flow path. It is commonly expressed as:

$$V = \frac{Ch}{l} = Ci, \text{ where:} V = \text{velocity of flow } (LT^{-1})$$

$$C = \text{coefficient dependent on nature of aquifer } (LT^{-1})$$

$$h = \text{head loss } (L)$$

$$l = \text{length of flow path } (L)$$

$$i = \frac{h}{l} = \text{hydraulic gradient } (-)$$

2. The law is valid only for laminar flow, and the limits of its applicability depend on a number of factors, among which the ratio of inertial to viscous forces (Reynolds Number) is the most important. As a generalisation, pores and interstitial voids give rise to intergranular flow, while groundwater movement through joints, faults and solution voids may be described as fissure flow. Both such types of flow may lie in the laminar range, though an increase of velocity in excess of the critical limit can result in turbulent flow which does not obey Darcy's Law.

3. In strata where the flow is dominantly intergranular, groundwater velocities range widely from 2 metres/day to 2 metres/annum or less. Exceptionally, flow rates of the order of 2000 metres/day or greater may occur in highly fissured strata such as in some cavernous limestone regions. Strictly speaking, inasmuch as the water flows through the voids, the average void velocity $V = Q/A_V$, where $Q =$ discharge, and $A_V =$ cross-sectional area of voids in aquifer. Although actual flow velocities may be very important, for example in groundwater pollution studies, it is the total volume of flow that is usually required in most groundwater resource problems and this can be determined from a modification

of Darcy's Law in which the 'specific discharge' $V = Q/A$, where Q = discharge and A = cross-sectional area of aquifer. Since $V = Q/A$, then $Q/A = Ci$ and $Q = CiA$ where C and i are as previously defined.

Aquifer properties

4. A list of some relevant aquifer properties with an indication of their symbols and dimensions is given in Table 24, but the most important are those governing yield and storage.

TABLE 24. AQUIFER PROPERTIES

Term	Symbol	Unit
(a)	(b)	(c)
Hydraulic conductivity (Sec 12.3, para 4)	K	m³/day/m²
Intrinsic permeability (Sec 12.3, para 4)	K	darcies
Transmissivity (Sec 12.3, para 4)	T	m³/day/m
Storativity (Sec 12.3, para 4)	S	m³/m³
Hydraulic conductivity of aquiclude (Sec 12.9, para 13 and Fig 140)	K'	m³/day/m²
Saturated thickness of aquifer (Sec 12.3, para 4) ...	b	m
Saturated thickness of aquiclude (Sec 12.9, para 13) ...	b'	m
Leakance of aquiclude (Sec 12.9, para 13)	K'/b'	m³/day/m³
Leakage factor of aquiclude (Sec 12.9, para 13 and Fig 140)	L	m
Specific yield (Sec 12.2, para 2)	S_y	m³/m³
Drainage factor of aquifer (Sec 12.8, para 6)	D	m
Drainage delay index (Sec 12.8, para 6)	$1/\alpha$	days

The coefficient in the previous equations was described as one dependent on the 'nature' of the medium through which flow takes place and needs further consideration. Commonly known to engineers as 'permeability', it may be seen that the term has the dimensions of velocity (LT^{-1}). The permeability of a medium has previously been described as the ease with which a liquid may move through that material, and being proportional to the square of the void diameter it has dimensions of area. This is the property of the medium alone and is known as intrinsic permeability (k) expressed in darcies (1 darcy = 0.987×10^{-8} cm²). For groundwater purposes, hydraulic conductivity (K) is the more meaningful parameter since, being defined as $Cd^2 \dfrac{\gamma}{\mu}$, where γ = density of water and μ = viscosity, it takes account of the properties of the fluid being transmitted, and may be regarded as the volume of water that in an isotropic medium will move in unit time under a unit hydraulic gradient through a unit area measured at right angles to the direction of flow. This is shown in Figure 119 and is expressed as m³/day/m². The term hydraulic conductivity is equivalent to the geotechnical engineer's coefficient of permeability except that the latter is expressed in m/sec.

311

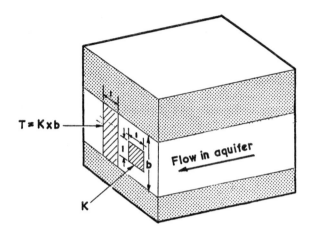

T = Kxb

Flow in aquifer

K

T and K relate to flows of unit hydraulic gradient

Fig 119.   Hydraulic conductivity (K) and transmissivity (T) of an aquifer

Another related term is transmissivity (T), this being the rate at which water is transmitted through a unit width of the aquifer under a unit hydraulic gradient. It may be easier to visualise this term as the product of the hydraulic conductivity (K) and the full saturated thickness of the aquifer (b). In all that follows permeability of an aquifer is assumed to be the same in all directions, i.e. isotropic, which is not always the case in nature.

5. The storage facility of an aquifer is described as the storativity (S) or coefficient of storage, which may be expressed as the volume of water removed from each vertical column of aquifer of height (b) and unit basal area when the head declines by one unit. As it is the ratio of a volume of water to a volume of aquifer, it is dimensionless and always less than unity.

SECTION 12.4. GROUNDWATER CHEMISTRY

1. Groundwater is never found in the chemically pure form of H_2O because prior to becoming groundwater it has been involved in chemical reactions with the materials comprising the atmosphere and zone of aeration. As a result of contact with a variety of gases in the air and a variety of minerals beneath the surface, most waters are complex dilute solutions by the time they reach the zone of saturation. Such waters are constantly in motion yet constantly attempt to achieve chemical equilibrium with their particular contact material. One of the dominant factors governing such change is the rate of groundwater movement.

The chemical processes responsible for actual changes in the composition of subsurface water are numerous: solution, precipitation, reduction, concentration, absorption and ion exchange may all be operating in different parts of the aquifer.

Water analysis

2. The composition of natural waters depends on the relative concentrations of a number of impurities which are best determined by undertaking quantitative analysis of representative water samples. A full analysis would take account of mineral matter, as well as quality and sanitary considerations, and an example is provided in Figure 120.

3. The major elements and radicals present in solution which have most significance in groundwater studies are calcium, magnesium, sodium, potassium, bicarbonate, sulphate, chloride, and nitrate. Fluoride and the metals iron, manganese, zinc, copper and lead may be troublesome in critical concentrations, while selenium, boron and cyanide cannot be tolerated even if present in trace concentration. The properties of water which give rise to its character include pH, hardness, alkalinity, total dissolved content (including salinity), turbidity, odour, taste and free carbon dioxide, which to a large extent are dependent on the ions in solution.

4. There are numerous methods of undertaking detailed chemical analysis and the reader is referred to such standard texts as RAINWATER and THATCHER 1960 and HOLDEN 1970. For the study of changes in composition, it is important that the total concentration and the relative proportions of all the major elements and ions mentioned previously should be determined.

TABLE 25. **REACTION COEFFICIENTS FOR SELECTED ELEMENTS**
(Reciprocals of equivalent weights)

CATIONS		ANIONS	
(a)	(b)	(c)	(d)
Ca	0·0499	CO_3	0·0333
Mg	0·0822	HCO_3	0·0164
Na	0·0455	SO_4	0·0208
K	0·0256	Cl	0·0282
		NO_3	0·0161
		F	0·0526

NOTE: To convert mg/l to mEq/l, multiply by reaction coefficient.

The most satisfactory method of reporting water analyses is to present a list of the chemical elements, both in units of 'milligrams per litre' (mg/l) and 'milligram equivalents per litre' (mEq/l), which provides an expression in chemically

WATER ANALYSIS

Sample taken on ....20.1.75........................ by .....P. E. GREEN...................................

Sample received in laboratory on .....21.1.75..............

Sample analysed on .....22.1.75................... by.....A. K. WHITE...............

RESULTS IN MILLIGRAMS PER LITRE (mg/l)

Appearance ......Bright with a few particles..

Colour......Nil... Turbidity .....Less than 3.....................

pH .....7·3.. Free CO_2 .....15.......................

Electrical conductivity .....475....μmhos/cm................ Odour ....nil..................

Chlorine present as chloride.....30..................... Residual chlorine .....—..............

Dissolved solids dried at 180°C .....305........................

Hardness: Total .....225.......... Carbonate .....185.......... non-carbonate .....40.....................

Alkalinity as Calcium Carbonate..........185..............................

Nitrate Nitrogen .....5·0.............................. Nitrite Nitrogen.....absent.................

Metals.....Iron Manganese Lead Zinc Copper all absent....................

Fluoride .....less than 0·1........................... Silica.....12......................

MINERAL ANALYSIS

	Cations			Anions	
	mg/l	mEq/l		mg/l	mEq/l
Ca...72		3·59	HCO_3...110		3·66
Mg...13		1·07	SO_4...18		0·37
Na...15		0·68	Cl...30		0·85
K...4		0·10	NO_3...27		0·45
Total Cat		5·44	Total An		5·33

$$\text{Percentage error} = \frac{\text{Tot Cat} - \text{Tot An}}{\text{Tot Cat} + \text{Tot An}} \times 100 = \frac{0·11}{10·77} \times 100 = 1·02\%$$

Fig 120. Chemical analysis sheet

equivalent terms i.e. cations should balance anions. This relationship among the ions in solution is obtained by multiplying the determined concentration of an ion by its reaction coefficient so that $mEq/l = mg/l \times \dfrac{valence}{atomic\ weight}$. A list of reaction coefficients for the more common ions is given in Table 25, and as an example one can refer to the mineral part of the analysis quoted.

5. It is usual to classify waters according to the predominant cation and anion in solution, and the calcium bicarbonate type of the previous analysis is common for groundwaters near outcrop conditions where they tend to show $Ca > Mg > Na$ and $HCO_3 > SO_4 > Cl$. Thereafter, as such waters move down gradient from areas of replenishment toward areas of discharge, there is a tendency for groundwater to change from bicarbonate water to sulphate water and then to chloride water where travel time is sufficiently long. The end point in the change will produce a water in which $Na > Mg > Ca$ and $Cl > SO_4 > HCO_3$.

Presentation of analytical data

6. There is a need to present analytical data in a form that can help the delineation of trends in groundwater chemistry within an aquifer, and this can be done visually by the use of pattern diagrams. Some of these diagrams are shown in Figure 121, where the units used may be either mg/l or mEq/l percentage

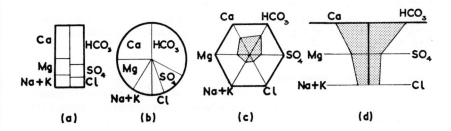

Fig 121.   Alternative types of pattern diagram

reacting values. An example of the lateral change in the proportion of constituents using one such diagram is shown in Figure 122 for part of the Santiago Basin in Chile, in which the trend of chemical changes in the direction of groundwater flow may be observed. An even more useful method of presentation is that of the trilinear diagram in which for an individual sample, percentage components expressed in terms of mEq/l of the cations and anions are plotted in separate triangular fields, and the respective points projected into a 'diamond' shape field to provide a single point position. This allows classification of the water, a visual indication of progressive changes in chemical character such as by

315

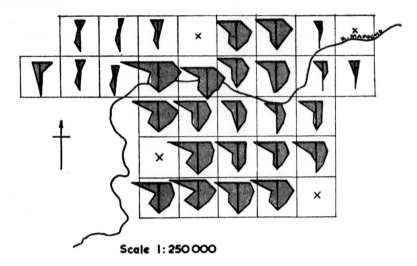

Scale 1: 250 000

Fig 122. Changes in groundwater quality in part of the Santiago Basin, Chile (based on pattern diagram shown at Figure 121(d)

NOTE: The area covered is the northern part of that shown in Figure 115.

solution, as well as conclusive demonstration of groundwater mixing. Figure 123 shows two chemically different waters, one a fresh groundwater and the second a brine; any mixtures of the two will lie on the straight line joining the two individual points.

Effects of chemical composition

7. The constituents of groundwater derived directly from solution of minerals in rocks are the cations and silica, while most of the anions are derived from other sources. The general relationship between the mineral composition of a water and that of the solid materials with which the water has been in contact may be simple, or so complicated as to be difficult to resolve fully.

8. The practical information required from the chemical analyses depends on the purpose for which the water is to be used, and will thus vary according to whether the water is for domestic supply, industrial boiler feed, cooling, laundry purposes or irrigation requirements.

9. The selection of lining materials, pipework or amount of open-area available for flow will be influenced by the corrosive or encrusting nature respectively of the groundwater. Similarly, prior knowledge of such parameters, as extreme hardness or high iron in water, will allow preparations to be made for softening or iron removal before the water is put into supply.

316

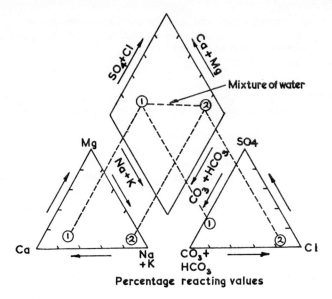

○1 Fresh groundwater

○2 Saline water

Fig 123.   Trilinear diagram

10. A pH value of 7·0 denotes a neutral reaction; in general for lesser values the reaction will be acid and the water will have a tendency to be corrosive, while values of pH in excess of 7·0 are described as alkaline in their reaction and will have a tendency to be encrusting. The word 'alkaline' should not be confused with 'alkalinity' which is a property determined by the amount of carbonate and bicarbonate present in the water.

Alkalinity is analogous to hardness, being similarly reported in equivalent parts per million of calcium carbonate that would have to be dissolved in distilled water to produce the same effect as that found in the natural water. Hardness is the property in water dominantly due to the presence of calcium and magnesium compounds, which as bicarbonate, sulphates and chlorides form insoluble curds as well as incrustations or 'scale' on metallic surfaces. Hardness determinations are commonly reported as total hardness, comprising carbonate (temporary) hardness and non-carbonate (permanent) hardness. Temporary hardness is removed by boiling.

11. The total amount of dissolved solids present in a water may be a function of a number of processes among which solubility, hydraulic conductivity and human environmental factors are the most important. The direct relationship

317

between ionic concentrations and specific electrical conductance means that simple measurement of conductance can be used to provide some indication of ion concentration or salinity of a water sample.

Salinity

12. Assessment of groundwater resources must consider the quality as well as the quantities required, though any definition of saline water will depend on the use to which the water is to be put. An arbitrary boundary between fresh and saline water could be drawn at a dissolved solids content of 1000 mg/l which is equivalent to a specific conductance of 1400 µmhos at 25°C. Although the World Health Organisation regards 1500 mg/l of total dissolved solids (TDS) as the 'maximum allowable' value; nevertheless, in the absence of anything better, water up to 3000 mg/l has to be used for drinking purposes in some places such as parts of the Middle East.

13. Groundwater can become saline in a number of different ways:

a. Evaporation from shallow water-tables in arid regions can produce an excessive build-up of salts.

b. Accumulation of salts in groundwaters associated with coastal areas may be accentuated by precipitation from on-shore winds.

c. Mixing with natural waters having a high salinity content such as connate waters or brines.

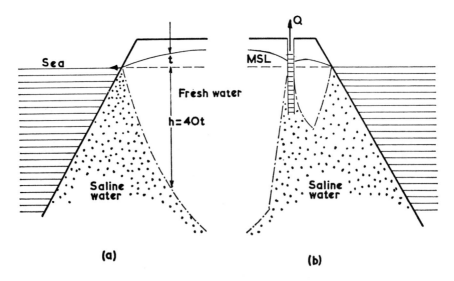

Fig 124.   Saline/fresh water relationships

d. Contamination directly or indirectly by activities such as faulty well construction, overpumping or inadequate control of the disposal of waste products.

14. One of the more common problems associated with saline water is that produced along coastlines or in small islands by the migration of sea water in response to abstraction from the aquifer. The basic principles governing this saline/fresh water relationship are well known and are illustrated in Figure 124(a) where t is the height of the water-table above mean sea level and h is the depth of fresh water below mean sea level. Under natural conditions there is a state of dynamic equilibrium and the 'interface' between the waters of differing density is in reality a transitory zone of mixed water. When pumping lowers the elevation of the water-table this is reflected in a rise in the position of the 'interface' as seen in Figure 124(b).

15. The principle determining the interface position is the amount of fresh water discharging from the aquifer into the sea as illustrated in Figure 125 and defined by the equation:

$$q = \frac{0 \cdot 013 \, (K \, h_o{}^2)}{L} \qquad \text{Eqn 12(1)}$$

where q = fresh water flow per unit width (L^2T^1)

K = hydraulic conductivity (LT^1)

h_o = piezometric head (L)

L = landward extent of saline water penetration (L).

16. Excessive abstraction from coastal aquifers produces the situation where the quantity 'q' is significantly depleted and the 'wedge' of saline water moves inland and renders unusable those existing production wells nearest to the coastline.

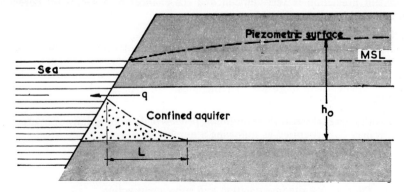

Fig 125.   Saline water wedge

Section 12.5

Safety of water for drinking

17. In conclusion it may be worth emphasising that a chemical analysis alone will give no indication of the potability or safety of that water for drinking purposes. A full quality analysis including bacteriological examination is desirable, and routine chlorination procedures should be mandatory.

SECTION 12.5. GROUNDWATER EXPLORATION

Purpose and approach

1. The purpose of exploration is to enable the hydrogeological conditions (whether they be for a region or an individual site) to be investigated in as detailed a fashion as is necessary to meet the requirement of the project. This proviso recognises the limitation of time and/or money that may restrict such work. Exploration begins with the framework of a broad idea which is confirmed with specific quantitative data, from which a detailed prototype or hydrogeological map, or a physical or mathematical model can be prepared. The exploration process is never-ending, all new information serving to refine the model.

2. Groundwater exploration implies an organised search using scientific methods, but it is necessary to be realistic and accept that the prevailing hydrogeological conditions determine the degree of speculation in the exercise. For example, in such a region as the Indo-Gangetic Plain every properly constructed borehole will be successful, with little or no risk involved, in providing a supply of 50 litres/sec when fitted with 50 metres of 150 millimetre diameter slotted screen. The other extreme is to be met in the hard rock areas of the same sub-continent where, as in so many countries of the world, the water is present in individual fractures, and then only to relatively shallow depths. Under such circumstances confident prediction is very difficult and the chances of a successful borehole may be no higher than 1 in 10.

3. It is claimed that certain individuals are able to locate water by dowsing, and although they may be sincere in their belief in the usefulness of such practices, the results are inconsistent and the technique should not be considered reliable when compared with more scientific methods.

4. Although exploration is a continuing process of data collection and interpretation, it may be recognised as following a number of stages:

   a. *Desk study*—collation and analyses of existing data.

   b. *Feasibility study*—collection and analyses of additional data.

   c. *Pilot study*—site investigation and subsequent action.

   d. *Development*.

5. In some countries such as Britain there already exists a fair amount of information on groundwater conditions, but in a developing country there may be little or no data to rely on at the start of a project. Even where some information is available it is frequently neither reliable nor relevant. The methods of exploration will depend on the particular problem and the financial and technical resources available for the investigation.

Field reconnaissance

6. *Topography.* A base map showing topographic features and elevation contours relative to some datum is a necessary prerequisite to field reconnaissance. Such a map may be largely dependent on aerial photography, but ground levelling to provide well top elevations is essential for regional studies.

7. *Geology.* Photogeological interpretation and field control allow preparation of a geological map that is representative of surface or near surface conditions. This assumes that sufficient strata outcrop at the surface so that their sub-surface extension can be estimated. The prime contribution of the hydrogeologist at this stage of the work is his knowledge of rocks and structural conditions, and his ability to recognise those materials that are likely to function as aquifers or aquicludes.

8. *Hydrogeology.* A 3-dimensional representation of the sub-surface hydro-geological regime would be greatly assisted by field determination and measurement of as many as possible of the following items:

   a. Precipitation and evaporation.

   b. Location, elevation and discharge of springs and seepages.

   c. Stream discharges.

   d. Evidence of saline and alkaline soils.

   e. Distribution of vegetation types.

   f. Location of wells and measurement of water levels and abstractions.

Surface geophysics

9. In groundwater work, geophysics is used to supplement surface data by providing sub-surface information that is sometimes of direct but mainly of an indirect nature. The geophysical techniques that have significance in groundwater exploration require a considerable amount of instrumentation and commonly need specialist staff both for operation of the equipment and interpretation of the results. As a consequence, geophysical methods of exploration may be expensive.

10. Although all the surface geophysical techniques, referred to in more detail in Chapter 6, have some potential application for groundwater work, the electrical and seismic refraction methods are more commonly used than the magnetic and gravity methods. Electrical methods were the first geophysical techniques to be adapted to groundwater exploration in the 1930's, in the measurement of the resistance to an artificially induced electrical potential within the ground, and are still the most widely used. It should be emphasised that resistivity values cannot be directly related to hydraulic characteristics such as permeability. Generally the resistivity decreases with increased total porosity of the strata, with saturation and with increased mineralisation of the contained water. In the saturated zone one can thus distinguish between highly resistive materials such as sands and gravels and poorly resistive materials such as clay and shale; and on this basis the more highly resistive zones would be regarded as having the better aquifer potential. The method is suitable for preliminary exploration of relatively small areas and the major limitations are its restriction to depths of about 150 metres and inability to elucidate geological conditions in the absence of supplementary data. Seismic methods are relatively accurate, particularly the refraction method, although this is generally restricted to the uppermost 100 metres. Some indication of yield may be gauged from measurements of seismic velocities under favourable conditions, but the method is most frequently used to determine depth to bedrock.

Exploratory drilling

11. The interpretation of the field surveys will remain largely theoretical until confirmed by data directly derived from sub-surface investigation. The purpose of exploration drilling is to obtain information about the physical nature, depth and thickness of the strata penetrated, to obtain samples of water and to derive indications of the hydraulic characteristics of potential aquifers. The tools used are the same as those described in Chapter 9. The choice of a particular method depends on the depth and diameter of the test hole and the nature of strata.

12. The resultant information on the geological succession is available in three forms of record or well log, two of which require samples while the third uses a detailed time record. The common drillers log is based on the instruction given to the driller about the sampling frequency and is simply his written description of the character of material sampled from each 2 metre sample interval or as specified. Additionally, it would record depth at which water was first struck and, where possible, rest water levels during the progress and at completion of drilling. When a geologist examines these samples, preferably during drilling, he compiles a lithological log wherein each distinctive stratum is identified and described in geological terms with thickness and depth below surface. An accompanying graphic representation of the strata is frequently prepared to provide visual appreciation, and proportion of sand may be similarly shown. A drilling-time log may be used in conjunction with rotary-drilling methods. It records the length of time taken to drill each metre of depth, and variations in geological strata are interpreted on the basis of changes in penetration rate.

Borehole logging

13. This term has a specialised meaning in well drilling and is generally used to describe sub-surface methods for determining the physical properties of the strata penetrated and the water intercepted by the borehole (*see* Chapter 6). The more commonly used logs in water wells come from one or other of three groups:

a. Electrical.

b. Radioactive.

c. Mechanical.

Electrical logging is one of the best techniques in borehole investigations, and usually consists of two basic types of log. The spontaneous potential or self potential (SP) log is that of the naturally occurring electrical potential difference,

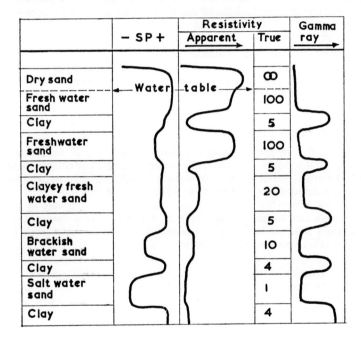

Artificial electric and gamma ray log approximating the appearance that an actual log would have in a sequence of clay beds and granular aquifers having good porosities. Aquifers are assumed to be non radio-active.

Fig 126. Borehole log—diagrammatic

323

which is mainly dependent upon the electro-chemical effect of adjacent strata. In particular the method measures electrical differences between the borehole fluid and that contained in each lithological unit penetrated during drilling. It is used to locate permeable horizons, the precise position of formation boundaries and indications of formation water quality. Apparent resistivity logs measure the effect of an artificially generated electric current, introduced into the geological formations by a combination of electrode arrangements so as to achieve varying distances of current penetration. Since the resistance of the fluid-saturated rock to the electric current is a function of the porosity of the rock, the salinity of the fluid and the temperature of the rock and fluid, resistivity logs may be used in conjunction with the SP log to distinguish lithological units and estimate the nature of the fluid. Figure 126 illustrates the appearance of some of these logs with some comments on the causes of particular effects. It is conventional in plotting electrical logs to place the SP curve on the left and one or more resistivity curves on the right side as demonstrated.

14. Unlike the electric logs, which are restricted to open holes, the gamma-ray sondes can provide information on cased holes; on the other hand the gamma-ray log is not affected by changes in water quality and on its own is therefore unable to distinguish between saline and fresh groundwater. Clays and shales contain higher concentrations of radioactive elements than sands, limestones or sandstones, and the essential purpose of gamma-ray logs is to distinguish argillaceous from other geological formations.

Mechanical aids to groundwater exploration are found in caliper logs, flow velocity and fluid conductivity, and temperature readings. The caliper log provides a record of bore diameter, which can assist with lithological interpretation since soft unconsolidated strata tend to cave, while consolidated rocks retain the size of the drill bit. When the hole dimensions are known it is possible by use of an impeller flow meter to identify the levels at which restricted fissure flow enters the bore, while fluid conductivity measurements enable the monitoring of variations in the salinity of the borehole water column.

15. In conclusion, unless hydrogeological conditions are simple, no single exploration method will provide the whole picture, but a combination of the techniques outlined above should yield information that will allow predictions to be made with some confidence.

SECTION 12.6. GEOLOGICAL FORMATIONS AS AQUIFERS

1. The occurrence and availability of groundwater are dependent on the interrelationship of a number of the geological factors that have been mentioned previously. These factors govern the volume of void space within any natural material and, more importantly, the ability of such voids to transmit water. Because of the dominant roles played by the processes of cementation, fracturing

and solution, one can identify at least three fundamental hydrological regimes on the basis of their flow patterns (Table 26) and demonstrate their relationship to various kinds of geological materials.

TABLE 26. GROUNDWATER FLOW PATTERNS

Serial No.	Dominant geological factor	Significant nature of material	Type of flow pattern
(a)	(b)	(c)	(d)
1	Absence of cement	Unconsolidated	Intergranular
2	Fracturing	Consolidated	Fracture
3	Solution	Carbonate	Fissure

Figures 127–130 show examples of aquifers in a variety of geological structures. The water-table in each case is represented by a broken line and small triangle.

2. The distribution of groundwater in the various geological formations is best considered on the basis of lithology as follows.

Sedimentary strata

3. *Arenaceous rocks.* The classification of arenaceous materials into consolidated and unconsolidated is both simple and practical in terms of water extraction techniques. The loose surface dune sands of deserts and sea shores do not usually provide significant quantities of fresh water, but can do so where favourable underlying geological conditions or localised underground drainage exists. Gravels and alluvium of glacial origin afford the predominant source of groundwater in unconsolidated sedimentary rocks on account of their high permeability and excellent storage potential. Abundant evidence from Arabian Gulf countries demonstrates that alluvial fans and gravels formed along the fringes of mountain ranges are normally good sources. So also, to a lesser extent because of the higher percentage of silt and clay, are the alluvial plains and wadis stretching farther away from the mountains. The nature of the bedrock affects the supply; a permeable bedrock will allow the water to penetrate to depth while an impermeable bedrock will retain the water in the gravels. In favourable circumstances it is remarkable that significant yields may be obtained from gravels in valleys which may only be flooded for a short time once every few years.

4. Alluvium associated with present-day rivers frequently has permeability as high as if not higher than any other geological material. It may reach a total thickness of hundreds of metres as in the extensive Indo-Gangetic plain where the huge groundwater reservoir provides prolific yields for only small drawdowns. On the other hand in Britain many tracts of alluvium are less than 3 metres thick

325

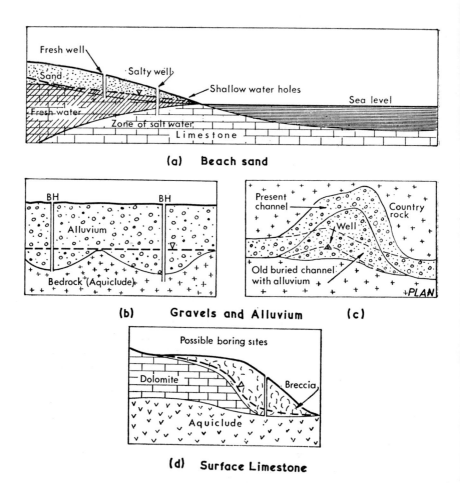

(a)   Beach sand

(b)   Gravels and Alluvium   (c)

(d)   Surface Limestone

Fig 127.   Aquifer examples I

326

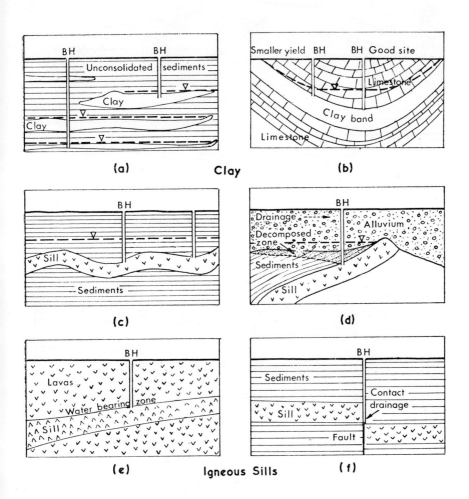

Fig 128. Aquifer examples II

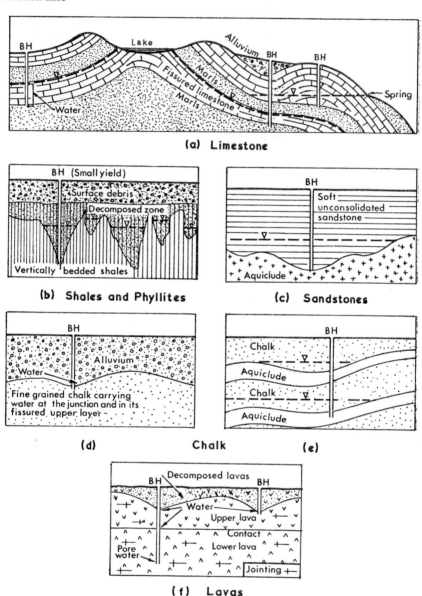

(a) Limestone

(b) Shales and Phyllites

(c) Sandstones

(d) Chalk

(e)

(f) Lavas

Fig 129. Aquifer examples III

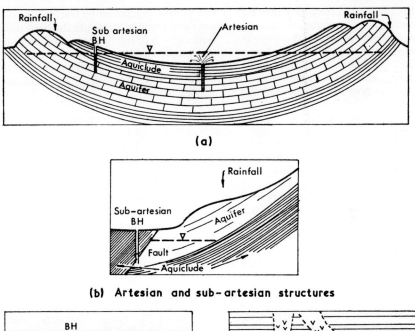

(a)

(b) **Artesian and sub-artesian structures**

(c)   **Dykes cutting drainages**   (d)

Fig 130.   Aquifer examples IV

and of such a limited extent that they may not be worth developing for reliable, large-scale resources.

5. The consolidated strata have suffered varying degrees of cementation and induration that reduce the volume of interstitial voids, but also make the material more amenable for the creation of additional void space through secondary factors. In many respects the consolidated sedimentary materials have more in common with igneous rocks than with their unconsolidated types. For example a

329

hard indurated sandstone may have a flow pattern dominated by fractures to the exclusion of intergranular flow. Sandstones and conglomerates are the hard rock equivalents of sand and gravel; mixed flow regimes of intergranular and fracture voids exist within the same formation and provide an improvement on the yield that could be expected from either regime alone. The Nubian Sandstone of North Africa and the Triassic Sandstones of Europe are examples of the excellent aquifers provided by consolidated arenaceous strata.

6. *Carbonate rocks.* This group of sedimentary rocks deserves special mention because its chemical composition makes it amenable to solution by groundwaters that are not fully saturated with calcium carbonate. The result is that preferred solution along fracture planes increases the size but decreases the number of fissures that transmit the greater part of the groundwater flow. The voids so produced may be exceedingly large as typified by the caves and channels of karstic regions, so that the flow regime rarely conforms to the normal pattern of of groundwater flow through a porous medium. Both the siting of wells and prediction of yields are fraught with difficulty, and must be regarded as highly speculative since success depends on the borehole intersecting a number of water-bearing fissures.

7. *Limestones.* Because of their fissured nature, massive crystalline limestones often contain considerable quantities of water, particularly where the limestones are underlain by an impervious layer. Such is the case in parts of the Middle East where extensive generalised water-tables extend for many kilometres, often cutting through complex geological structures which affect them only slightly or not at all.

8. Dolostones differ from limestones in that they contain decomposed portions and a general absence of extensive water-tables.

9. Chalk varies from the relatively unconsolidated variety of England to the consolidated type of Israel and Syria. The former provides excellent aquifers combining high transmissivities with good storage potential. Where the chalk has a higher clay content (e.g. Chalk Marl) or is more compact, then it is less perme-able, though it is often fissured and, as in Syria, may contain associated permeable bands of broken chert, conglomerates and sandstones.

10. *Argillaceous rocks.* The fine-grained clays, marls, shales and mudstones that comprise this group of strata are all relatively impermeable and therefore function as aquicludes. They may retain water in overlying aquifers or conversely restrict replenishment to underlying aquifers, but themselves provide no useful supplies.

Igneous rocks

11. Plutonic rocks, of which granites, granodiorites, diorites and gabbros are the most common, are in general not good aquifers. However, water can sometimes

be extracted from zones of jointing or decomposition, and the best yields are obtained where such zones are deepest and where drainage and catchment have been favourable to their supply.

12. Extrusive igneous rocks are commonly represented by those of volcanic origin and show considerable variation in aquifer potential. Lava flows tend to possess a blocky nature due to vertical cooling joints, and while the massive lava is usually impermeable, the layers of ashes, tuffs, and weathered surfaces in association with the vertical joints produce spring flows of considerable magnitude. Borehole supplies may be good if sufficient water-yielding horizons can be tapped.

13. Intrusive igneous rocks can for water supply purposes be restricted to sills and dykes. Sills are sheets of hard and usually impervious rock that have been injected along bedding planes. When they are injected into permeable formations they usually concentrate groundwater. Dykes are likewise generally impervious but, because of their near vertical orientation, they frequently act as barriers to groundwater movement, and can be exploited on this account. However, it occasionally happens that dykes weather more easily than the rocks into which they are intruded, so that the dyke itself becomes the groundwater reservoir, as with the pegmatite dykes which often occur in metamorphic terrains.

Metamorphic rocks

14. Gneisses like plutonic igneous rocks are not good aquifers. Areas of shattering and of decomposition are however relatively common; if, because of appropriate catchment they are aquifers, they may be located by means of resistivity measurements.

15. Schists and slates resemble shales in their water-bearing properties, and may be aquicludes when their planes of schistosity or of cleavage are horizontal. Sheared and crushed zones which are common in schist are frequently decomposed or weathered; these zones may alternate with unweathered zones of schist or gneiss. The weathered zones, particularly if the dip is high, may be tapped for water supplies down to the limit of decomposition.

SECTION 12.7. **GROUNDWATER EXTRACTION**

Well types

1. Water wells usually have the form of vertical bores or shafts, though they may have more of a horizontal nature where special circumstances require infiltration galleries or collector wells. Particularly well known from the Middle East in this connection are the gently inclined tunnels or 'qanats' that were excavated by hand in alluvial deposits to lead water below ground from the foothills of mountain regions across flat plains to points of supply. Many of these have been in existence for hundreds of years and in Iran and Afghanistan, for example, are still in use.

2. Wells are here defined as holes specifically constructed for the purpose of intercepting groundwater, and therefore include shafts and boreholes (tubewells), as well as combinations of both and various modifications. Throughout history

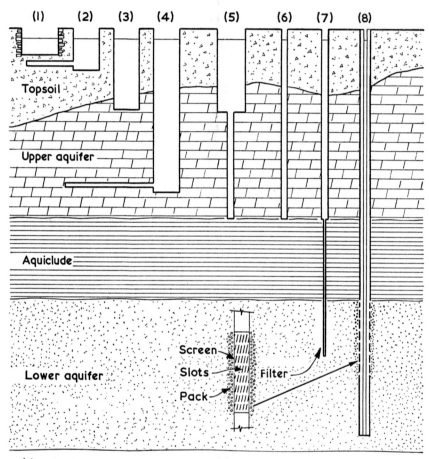

(1)	Tank	(2)	Radial collector well
(3)	Dug well	(4)	Shaft with adit
(5)	Shaft with bore	(6)	Bore open hole
(7)	Bore (reducing diameter)	(8)	Bore(with screen and pack)

Scale :   1cm = 10m−vertical
            1cm = 5m−horizontal

Fig 131.   Types of well

the availability of a means of lifting water to the surface has largely determined which type of well is most acceptable. Figure 131 illustrates the variety of well constructions that are currently used world wide to provide water and, with the exception of the radial collector well, they are arranged and numbered in chronological sequence.

The trend towards narrow deep holes at the expense of wide, relatively shallow excavations, is due partly to the replacement of manual by mechanical drilling techniques and partly also to the development of slim submersible pumps in place of the large reciprocating pumps, which themselves had replaced human and animal power. Although in developed countries boreholes are nowadays so common that shafts, either with or without adits, are only rarely excavated, nevertheless there are many developing countries where the absence or excessive cost of electricity necessitates the use of 'Persian wheels' or limited lift suction pumps in manually excavated shafts. The radial collector well, combining a shaft with lateral pipes, is a method of construction that is generally limited to unconsolidated strata (preferably sands and gravels) in hydraulic continuity with adjacent perennial rivers.

Well construction

3. Many methods of conventional well construction are available; shallow wells (<30 m) are commonly dug, augered or jetted, while deeper wells (> 30 m) are nearly always drilled. Alternatively drilling techniques are the percussion and rotary methods, each being particularly suited for drilling in certain geological and climatic conditions.

4. In principle, the percussion or cable-tool method uses a chisel-edged bit and a lift-and-drop motion to cut into the rock by impact. A string of tools provides weight at the end of the drilling cable. When drilling above the water-table, some water needs to be added to reduce friction on the bit and provide a slurry for the drill cuttings. These are removed from the hole by a bailer which is a length of pipe having a flap valve at the lower end.

5. Where the borehole sides have a tendency to collapse, some form of casing will be necessary, either as a temporary or permanent feature. In any event, some 15 metres of casing at the top of a well is desirable, and in some countries mandatory, combined with external cement grouting to prevent contamination from surface or near-surface water.

6. The rotary method of drilling is a faster procedure whereby a continuous rotary action is imparted to a hollow bit and connected drill rods. A continuous flow of drilling fluid is maintained to clean the face ahead of the bit and lift the rock cuttings to the surface. Casing is normally not needed during drilling because the drilling fluid forms a mud cake on the wall of the well thereby preventing collapse as well as limiting loss of fluid. The composition of the drilling mud will depend on a number of different requirements; in many circumstances use of a special bentonite clay, with or without additives, makes for easier drilling operations, though for water wells in Britain it is customary to

circulate plain water and allow the dispersion of natural clay material from the operation to form the drilling mud.

7. The normal circulation procedure is to pass mud of appropriate viscosity and density down the centre of the tools so that the cuttings are raised up the annulus between the tools and the mud sheath lining the walls. A variation of this method is that known as 'reverse circulation' whereby drilling water under a high head flows down the space outside the tools, with the cuttings moving up the hollow pipe assisted by air injection where necessary. Whichever method of circulation is adopted, the drilling fluid emerges at the surface and is discharged via a settling tank to the main pit or tank before recirculating down the borehole. The larger particles settle out and the finer particles remain in suspension.

8. Sampling of cuttings from the settling tank for preparation of a lithological log requires some allowances to be made both for the delay in travel time between the bit and emergence at the surface, as well as the loss of finer grained material which tends not to settle from the drilling fluid. Where the strata being drilled are reasonably hard and unbroken, it is possible by rotary core drilling to provide ideal material both for logging and laboratory testing.

Screens and packs

9. In consolidated strata having a stable nature, the bore may be finished in an open-hole condition (Figure 131, types (4) and (6)). However, in unconsolidated or weak strata, some dual-purpose lining is necessary to prevent collapse of the walls and movement of sand into the well, and yet provide easy access for water. Linings with some variety of slot (horizontal, vertical, bridge or louvre) or alternatively wire wrapped screens, fulfil such requirements (Figure 131, type (8)), and the size of the slot or opening should correspond to the mean size (D_{50} see Annex A) of the sand formation to pass through and create a relatively permeable zone outside the lining. The desirable slot size is best determined on the basis of mechanical grain-size analysis of the aquifer materials of all the different beds present.

10. Slots or screens alone may not keep sand from entering the bore where the aquifer is of fine grain. Under such circumstances it may be necessary to form a filter zone surrounding the screen by introducing an annular ring of pack material of a size appropriate to the particle size distribution of the formation. Although commonly described as 'gravel packs', the pack material is usually of a uniform sand size (i.e. <2 mm diameter), carefully chosen to relate the aquifer size distribution to the slot size, with a view to minimising sand movement and maximising well performance. Various criteria are used, and one much favoured in Britain for uniform artificial packs is that the D_{50} size value of the pack should be 5 times greater than the D_{50} size value of the aquifer material.

Well development

11. Too often development, the final phase of well completion, is ignored or inadequate; yet in as much as most drilling procedures necessarily impair the

walls, it should be an integral part of all water well construction. The purpose of well development is to produce as efficient a hydraulic structure as is feasible by the removal of fines from that part of the aquifer immediately surrounding the borehole, and the creation of a highly permeable zone. The most effective methods of development are (a) mechanical surging with a plunger and (b) injection of compressed air; in both cases the intention is to modify flow patterns so as to disturb any 'bridging' of grains and move the loosened fine material into the well, from where it may be conveniently extracted. Such completion procedures should ensure that the well will yield on a sustained basis the maximum quantity of sand-free water for the minimum drawdown of pumping water level. A 'performance test' as described later (*see* Scheme A of Figure 134) should allow quantification of any well losses caused by inefficient development.

Section 12.8. WELL HYDRAULICS

1. One of the most accurate methods of determining the aquifer properties is to monitor the response of water level in the surrounding aquifer to the abstraction of water from a pumping well. The radial flow of groundwater to the abstracting well may either achieve a steady state condition independent of time (equilibrium) or have a non-steady nature should the pumping duration be too short.

2. The steady state relationship illustrated in Figure 132 between the discharge (Q) and the change in head (h) in the area surrounding the discharge point, is given by the Dupuit-Thiem equation with h_2-h_1 represented by the more practical s_1-s_2 and having the form in consistent units of:

$$Q = \frac{2\,\pi\,K\,b\,(s_1-s_2)}{2 \cdot 303\,\log_{10}\left(\dfrac{r_2}{r_1}\right)}$$

Eqn 12(2)

where Q = discharge
K = hydraulic conductivity
(T = transmissivity)
b = saturated thickness
s_1, s_2 = drawdown of water level at specified distances
r_1, r_2 = radial distance of observation points from pumping well

It should be noted that the equation is applicable to both confined and water-table conditions, but in the latter case only where the drawdown is small in comparison with the saturated thickness of the aquifer.

3. Where steady state conditions are *not* attained, the resultant equation is much more complex since time becomes one of the dependent variables as indicated in Figure 133. The differential equation for radial flow in a confined aquifer is:

$$\frac{d^2s}{dr^2} + \frac{1}{r}\frac{ds}{dr} = \frac{S}{T}\frac{ds}{dt}$$

Eqn 12(3)

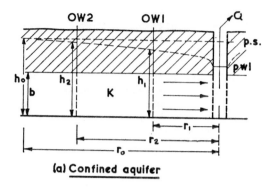

(a) Confined aquifer

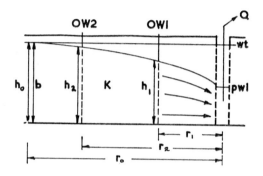

(b) Unconfined aquifer

OW = Observation well
Q = Discharge
wt = water-table
pwl = pumped water level
ps = piezometric surface (original)

Fig 132.   Steady state radial flow

The first solution to take account of the time factor and consider water derived from storage within the zone of influence of the well was the 'non-equilibrium formula' of THEIS (1935) in which:

$$s = \frac{Q}{4\pi T} \int_u^\infty \frac{e^{-u}}{u} \, du \qquad \text{Eqn 12(4)}$$

where s, Q and T are as defined above, and $u = r^2 S/4Tt$ with $S =$ storativity and $t =$ time.

336

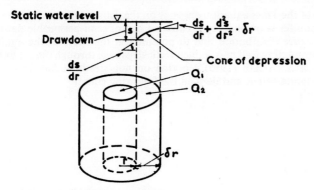

(a) The difference in the rates of flow Q_1 and Q_2 at the cylindrical faces is derived from storage

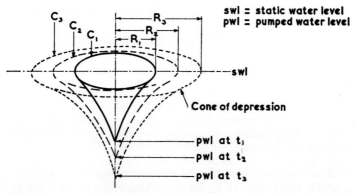

swl = static water level
pwl = pumped water level

(b) Extent and depth of cone of depression at three equal time intervals

Fig 133.   Non-steady state radial flow

The integral expression cannot be integrated directly, but its value is given by an infinite mathematical series as follows:

$$s = \frac{Q}{4\pi T}\left[\; -0.5772 - \log_e u + u - \frac{u^2}{2\cdot 2!} + \frac{u^3}{3\cdot 3!} - \frac{u^4}{4\cdot 4!} + \; \dots \right] \quad \text{Eqn 12(5)}$$

The value of the series included in square brackets is conveniently abbreviated and referred to as 'the well function of u' and expressed as $W(u)$. The determination of the relationship between $W(u)$ and u was tabulated by Theis and is given in most text books (TODD 1959; DAVIS and deWIEST 1966), and the equation can be simplified to:

$$s = \frac{Q}{4\pi T} W(u) \qquad\qquad\qquad \text{Eqn 12(6)}$$

337

4. Application of the Theis equation depends on the validity of certain assumptions. When the aquifer is overlain by a confined bed of semi-permeable nature, the change of pressure distribution due to pumping in that part of the aquifer surrounding the abstracting well may induce vertical leakage through the confining bed. HANTUSH and JACOB (1955) reported on non-steady radial flow in a leaky aquifer of infinite extent and derived the following equation:

$$s = \frac{Q}{4 \pi T} \int_u^\infty \frac{1}{y} \exp\left(-y - \frac{r^2}{4L^2 y}\right) dy \qquad \text{Eqn 12(7)}$$

and its simplification:

$$s = \frac{Q}{4 \pi T} W\left(u, \frac{r}{L}\right) \qquad \text{Eqn 12(8)}$$

Equation 12(8) may be seen to have the same form as Eqn 12(6) except that the integral function is known as the 'well function for leaky artesian aquifers' and has two parameters in the integral, u and $\frac{r}{L}$, where L is the leakage factor of the aquiclude. Values for the function $W\left(u, \frac{r}{L}\right)$ for given values of $\frac{r}{L}$ as u varies have been tabulated by HANTUSH (1956) and graphically presented by WALTON (1962).

5. The vertical flow component has even greater significance in water-table conditions where the dewatering by gravity drainage of part of the saturated thickness of the aquifer produces the complications both of a reduction in aquifer thickness and variations in the rate of drawdown. Three distinct segments may be seen in the drawdown behaviour. Initially the aquifer response is only of elastic storage, gradually being replaced by the dewatering of the strata in the cone of depression until such time as gravity drainage is completed when the specific yield becomes almost constant and the well responds comparably to pumping from confined conditions. BOULTON (1963) proposed a complicated solution to take account of such changes and expressed it as follows:

$$s = \frac{Q}{4 \pi T} \int_0^\infty \frac{2}{x} \left[1 - e^{-\mu_1}\left(\cosh \mu_2 + \frac{\alpha t r (1-x^2)}{2\mu_2} \sinh \mu_2\right)\right] J_0\left(\frac{r}{vD}\right) dx$$

$$\text{Eqn 12(9)}$$

6. By analogy with the Theis solution the function may be called the 'gravity well-function' and the equation may more simply be expressed as:

$$s = \frac{Q}{4 \pi T} W\left(u_{ay}, \frac{r}{D}\right) \qquad \text{Eqn 12(10)}$$

where $u_a = \dfrac{r^2 S_a}{4 T t_a}$ and $u_y = \dfrac{r^2 S_y}{4T t_y}$

with S_a, S_y = storativity for early and late time respectively

t_a, t_y = specified instants of time from early and late data.

$$D = \sqrt{\frac{T}{\alpha Sy}} \text{ where } \frac{1}{\alpha} \text{ is a delay index used to determine cessation of}$$

drainage effects and D is the drainage factor of the aquifer.

7. Despite the apparently restrictive assumptions upon which the equations in this section have been based, it is surprising how often field data are amenable to analysis. The limited scope of the section has not allowed for discussion of the differing effects of recharge and barrier boundaries either alone or in combination, nor of partial penetration effects on pumping or observation wells, and the well-storage effect produced by large-diameter wells in aquifers of low permeability. Similarly it may be necessary to correct the field data; loading effects such as those due to atmospheric pressure and tidal changes can occur under confined conditions, while seasonal trends, the pumping of nearby wells and hydraulically inefficient wells may each necessitate adjustment of field data derived from both confined and unconfined conditions. For assessment of the problems and means of solution the reader is referred to such texts as HANTUSH (1964) and KRUSEMAN and DE RIDDER (1970).

Section 12.9. PUMPING TEST ANALYSIS

1. *Purpose*. While it is true that quantitative work in groundwater hydrology depends upon knowledge of the relevant aquifer properties, it should be emphasised that the purpose of test pumping is not simply to enable such hydraulic parameters to be calculated. A well-planned pumping test is a multipurpose exercise designed for the determination of some or all of the following points:

a. The hydrological parameters and hydraulic properties of contributory aquifers.

b. The effect of present and future abstraction from the well on groundwater conditions in the aquifer.

c. The yield characteristics and potential of a well, and prediction of operating conditions.

d. The efficiency of the pumping well performance as an indication of its hydraulic condition.

2. Items a and b are best derived from a constant-rate test of reasonable duration and preferably utilising observation wells, while items c and d require a different test procedure in which the abstraction rate is varied in increasing

fashion. In order to provide as much information as possible, two combined procedures are therefore necessary; a constant-rate aquifer test and a variable-rate performance test. Since most formulae used in aquifer tests for pumping wells assume the well to be a 100 per cent hydraulically efficient structure, it is desirable to conduct the performance test prior to the aquifer test so that some quantitative determination can be made of the condition of the well; alternative schemes are given in Figure 134.

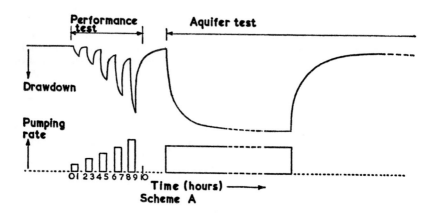

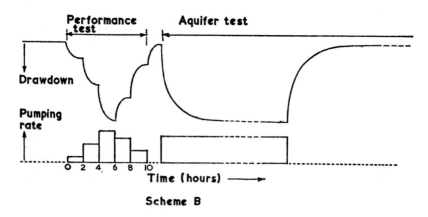

Fig 134.   Pumping test procedure

3. *Procedure.* In principle the pumping test is simple; it involves abstracting water from a well at known discharge rates and observing the resulting behaviour of the water levels in the well and in the aquifer. Observation points are highly desirable for proper evaluation of an aquifer and should preferably be more than two in number, radially arranged at varying distances from the pumping well. The distances will depend upon the duration and the abstraction rate of the test, as well as the nature of the aquifer and its expected hydraulic properties, but in general are likely to fall in the range of 15 to 1500 metres.

4. *Water level measurement.* When autographic water level recorders are not in use, it will be necessary to prepare a time schedule for manual water level measurements on the basis of the logarithmic scale that will subsequently be used for data plotting, for example:

Every one minute up to 10 minutes.
Every five minutes from 10 to 50 minutes.
Every ten minutes from 50 to 100 minutes.
Every thirty minutes from 100 to 310 minutes.
Every sixty minutes from 310 minutes to end of test.

Basic equipment for measuring water levels should consist of two electric well dippers, a pocket steel tape, stop watch and proforma sheets to include data indicated in Figure 135.

5. *Discharge measurement.* The rate of discharge should be measured accurately and recorded periodically throughout the test. Precise and continuous measurements are achieved by use of 90 degree V-notch with autographic recorder or manual reading, using hook gauge in conjunction with a water meter built into the discharge pipe along with a constriction to ensure that the pipe is kept full. However, either will suffice alone, and other alternatives include a circular orifice plate with associated manometer and even a simple container of known volume such as an oil drum. The subject of flow measurement is treated in some detail in ME VOL VI, 1975 to which reference should be made for more detailed information.

Performance test

6. After sufficient development of the newly bored well has been completed, a step drawdown test should be performed so as to evaluate well characteristics. As illustrated previously this consists of pumping the well at various discharges, each for a fixed time period of about 100 minutes (Figure 134). An example of the required information is given in Figure 137.

7. Two methods of analysis are available; in the first, values of drawdown are plotted against the different rates of abstraction on arithmetic scale and a line of best fit interpolated so as to produce the yield-drawdown graph illustrated in Figure 136. The straighter and more upright this line, the more 'efficient' is the well, and one can confidently make predictions regarding drawdown at any

341

PUMPING TEST DATA SHEET
ABSTRACTION*/OBSERVATION* WELL

Project..

Name of abstraction well .... Location............................

Name of observation well .... Location............................

Distance of Observation well from abstraction well..

Depth .... Diameter..

Lining details: casing ....

perforations*/screen*..

Date of test: Start.................................... Finish................................

Pump: Type .... Depth of pump inlet ....

Initial water level:.................................... above*/below* datum

Datum point:.................................... above*/below* ground level

Ground level (O.D.) .... Datum point (O.D.)............................

Units of measurement: t = ... Q = ... s = ... r = ...

Observers: ....

Remarks: ....

....

....

Actual time / Date	Elapsed time (t)	Water level below datum	Change in water level(s)	Corrections			Corrected water level	Remarks / Q
				Trend	Barometer* dewatering	Other		

*Delete alternative

Fig 135.   Pumping test pro-forma

abstraction rate that lies within the range of values used to construct the curve. Not infrequently however, the line is curved and a greater degree of curvature may be taken as an indication of the relative inefficiency of the pumping well as a hydraulic structure. For the Chalk of England a more accurate approach is given in an article on yield-drawdown curves by INESON (1959).

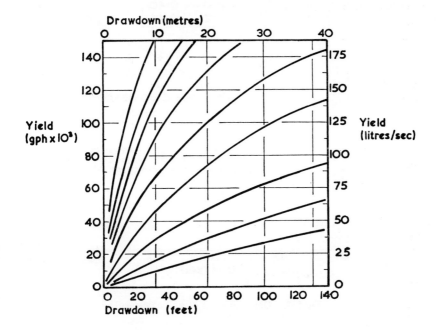

Fig 136.   Yield-drawdown curves

8. A simplified step-drawdown solution attributed to Jacob makes use of the equation $s = BQ + CQ^2$, and this divided through by the discharge Q gives $\dfrac{s}{Q} = B + CQ$, which can be solved graphically by plotting $\dfrac{s}{Q}$ against Q. The points C and B are then determined respectively as the slope of the line of best fit through the plotted points and the intersection of that line with the $\dfrac{s}{Q}$ axis.

Figure 137 numerically demonstrates this method which is applicable to confined and unconfined conditions. The efficiency of a well is not determined merely by the value of drawdown, but by the rate of discharge per unit height of drawdown. The ratio of s to Q is also known as specific drawdown of the well and is fairly constant for given hydrological and topographical conditions.

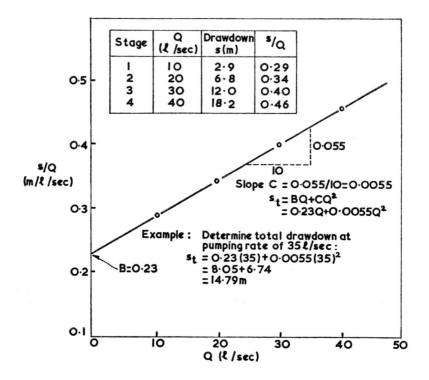

Stage	Q (ℓ/sec)	Drawdown s(m)	s/Q
1	10	2·9	0·29
2	20	6·8	0·34
3	30	12·0	0·40
4	40	18·2	0·46

Slope C = 0·055/10 = 0·0055

$s_t = BQ + CQ^2$
$\quad = 0·23Q + 0·0055Q^2$

Example: Determine total drawdown at pumping rate of 35 ℓ/sec:

$s_t = 0·23(35) + 0·0055(35)^2$
$\quad = 8·05 + 6·74$
$\quad = 14·79 m$

B = 0·23

Fig 137. Determination of well efficiency

Aquifer test analysis

9. Methods available for analysis of aquifer test data have been comprehensively described in the literature (HANTUSH 1964, KRUSEMAN and DE RIDDER 1970), and it is intended in this section to restrict the treatment to the more commonly used type curve solution of the equations derived in Section 12.8.

Non-leaky confined conditions

10. In the Theis non-equilibrium equation previously described, although the integral expression cannot be integrated directly, its value is given by the convergent series of Equation 12(5) which may be conveniently abbreviated to $W(u)$ and expressed in the form:

$$s = \frac{Q}{4\pi T} W(u)$$

344

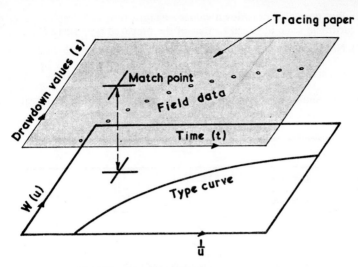

Fig 138.   Matching field data to type curve

11. Tabulated functions of $W(u)$ and u are available, and a graphical solution based on a 'type curve' was devised by THEIS (1935) whereby a plot of field data s/Q versus r^2/t was compared with a plot of $W(u)$ versus u, both plots being on logarithmic paper. In practice, the computation is greatly simplified for a constant abstraction rate and fixed observation point if the field data logarithmic plot of s versus t is matched to the logarithmic plot of $W(u)$ versus $\dfrac{1}{u}$ (*see* Figure 138). A tabulation of values of $W(u)$ for values of $\dfrac{1}{u}$ from 10^{-1} to 10^{10} is provided in most text books and can be used to construct the bounding or limiting type curve of Figure 143. Rearrangement of the equations in consistent units as follows enables solutions to be derived for both transmissivity and storativity for any individual observation well:

$$\text{Transmissivity } T = \frac{Q}{4\pi s} \, W(u)$$

$$\text{Storativity } \quad S = \frac{4Tut}{r^2}$$

Only the former can be derived from pumping well data. The necessary procedure is itemised in the following sub-paragraphs as well as an example to illustrate the method.

Procedure

(1) Adjust field data for any corrections that may be necessary (*see* Section 12.8, paragraph 7).

345

(2) Plot corrected drawdown values s against time t on tracing paper with a logarithmic scale identical to that of the prepared type curve (Figure 143).

(3) Note constants Q and r, as well as aquifer thickness b where known.

(4) Superimpose field-data plot on type curve of $W(u)$ versus $\frac{1}{u}$, and, keeping co-ordinate axes parallel, match as many data points as possible to the curve.

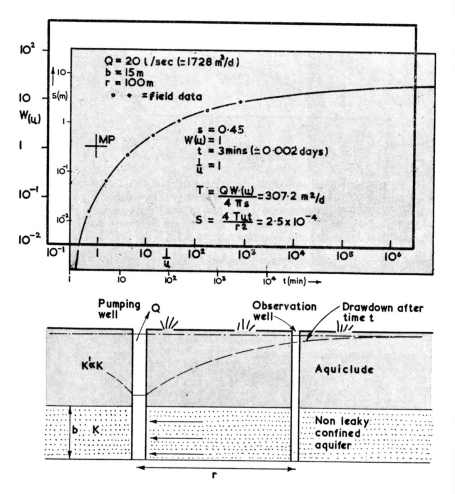

Fig 139.   Non-leaky confined test example

346

(5) Select convenient 'match point' (MP) at any intersection of major axes and note the values of s, $W(u)$, t and $\frac{1}{u}$.

(6) Substitute appropriate values into rearranged equations, taking care to ensure consistent units, so as to solve for T and S; K can be determined when b is known.

Example

12. The time-drawdown field data from an observation well situated 100 metres from a pumping well abstracting at a rate of 20 l/sec from a confined aquifer 15 metres thick, is plotted and superimposed on to the type curve as in Figure 139 so as to produce the following match point values, $s = 0.45$ metres, $W(u) = 1.0$, $t = 3$ minutes (equal to 0.002 days), $\frac{1}{u} = 1.0$. Substitution of these values into the relevant equations along with $Q = 20$ l/sec $= 1728$ m³/day provides aquifer property values as indicated in the figures.

Leaky confined conditions

13. Hantush and Jacob's equation for these conditions (Eqn 12(8)) has the same form as the Theis non-leaky equation except that there are two parameters in the integral $\left(u, \frac{r}{L} \right)$ where L is known as the leakage factor and solution follows much the same pattern. HANTUSH (1956) compiled a tabulation of values for $W\left(u, \frac{r}{L} \right)$ as u varies for given values of $\frac{r}{L}$, but WALTON (1962) developed a family of type curves using modified tabulation of $W\left(u, \frac{r}{L} \right)$ as $\frac{1}{u}$ varies for given values of $\frac{r}{L}$. The relevant rearranged equations are as follows:

$$T = \frac{Q}{4\pi s} W\left(u, \frac{r}{L} \right) \qquad S = \frac{4\,Tut}{r^2}$$

$$K' = \frac{T b' \left(\frac{r}{L} \right)^2}{r^2} \qquad L = \sqrt{\frac{Tb'}{K'}}$$

Procedure

(1) Adjust field data for any corrections that may be necessary (*see* Section 12.8, paragraph 7).

(2) Plot corrected drawdown values (s) against time (t) on tracing paper with logarithmic scale identical to that of Figure 143.

(3) Note the constants Q, r, b'.

(4) Superimpose field data plot on type curves, and keeping co-ordinate axes parallel, match as many data points as possible to an individual curve, and note $\dfrac{r}{L}$ value of the curve.

(5) Select convenient match point at any intersection of major axes and note values of s, $W\left(u, \dfrac{r}{L}\right)$, $\dfrac{1}{u}$ and t.

(6) Substitute the above mentioned values and those of Q, r, b' and $\dfrac{r}{L}$ into the rearranged equations so as to solve for T, S, K' and L.

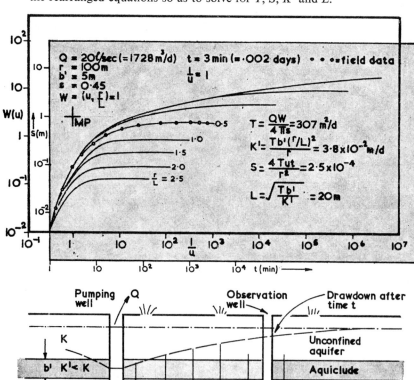

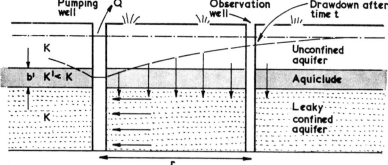

Fig 140. Leaky confined test example

348

Example

14. The time-drawdown field data from an observation well situated 100 metres from a pumping well abstracting at a rate of 20 l/sec from a leaky confined aquifer 15 metres thick overlain by an aquiclude 5 metres thick, is plotted and superimposed onto the family of type curves as in Figure 140 so as to produce the following match point values: $\dfrac{r}{L} = 0.5$; $s = 0.45$ metres; $W\left(u, \dfrac{r}{L}\right) = 1.0$;

$t = 3$ mins $= 0.002$ days; $\dfrac{1}{u} = 1.0$. Substitution of these values along with Q into the relevant equations yields aquifer property values as indicated in the figure.

Water-table conditions

15. One of the complications in unconfined aquifers is the delay in gravity drainage of the volume of aquifer being dewatered in the cone of depression. This is the phenomenon of 'delayed yield' for which BOULTON (1963) made some allowance in his solution for water-table aquifers. Again the form of the equation and the method of solution resemble those previously considered, and the necessary functions of $W\left(u, \dfrac{r}{D}\right)$ where D is drainage factor, for various values of $\dfrac{1}{u}$ have been tabulated by Boulton and are graphically represented in the family of type curves in Figure 143. To appreciate the significance of these type curves it is necessary to understand that the behaviour of water levels in response to pumping from a water-table aquifer initially conforms to that of the confined aquifer, then shows some of the characteristics of a leaky aquifer in the retardation of drawdown, to be followed by gradual increase in the rate of fall of the water level and once gravity drainage has been completed an ultimate conformation to the behaviour of a confined aquifer once again. This explains why the bounding or limiting curves of Figure 143 are simply the Theis type curve for confined conditions displaced with respect to one another. Furthermore, it demonstrates that there are two such families of curves which lie respectively to the left and right of the values $\dfrac{r}{D}$. The left-hand set of curves which has as co-ordinate axes the left ordinate $W\left(u_a, \dfrac{r}{D}\right)$ and the upper abscissa $\dfrac{1}{u_a}$, is identical to the set of type curves produced by Walton for leaky confined conditions. The right-hand set of type curves utilises the right ordinate $W\left(u_y, \dfrac{r}{D}\right)$

349

Section 12.9

and the lower abscissa $\dfrac{1}{u_y}$. The relevant equations rearranged from Equation 12(10) are:

$$T = \frac{Q}{4\pi s}\, W\left(u_{ay},\, \frac{r}{D}\right) \qquad\qquad S_a = \frac{4Tu_a t}{r^2}$$

$$S_y = \frac{4Tu_y t}{r^2} \qquad\qquad \frac{r}{D} = \sqrt{\frac{T}{aS_y}}$$

$$\alpha = \frac{\left(\dfrac{r}{D}\right)^2 \left(\dfrac{1}{u_y}\right)}{4t} \qquad\qquad \eta = \frac{S + S_y}{S}$$

Procedure

(1) Adjust field data for any corrections that may be necessary (*see* Section 12.8, paragraph 7).

(2) Plot corrected drawdown values (s) against time (t) on tracing paper with logarithmic scale identical to that of family of type curves.

(3) Note constants Q and r, also b where available.

(4) Superimpose field-data plot on type curves keeping axes parallel, whence it should be seen that the data fit partly to the 'a' set of curves and partly to the 'y' set of curves.

(5) Initially match as much of the early field data as possible to one of the 'a' set of curves, and note the $\dfrac{r}{D}$ value of that curve.

(6) Select convenient match point at intersection of major axes and note co-ordinate values of s, $W\left(u_a,\, \dfrac{r}{D}\right)$, t and $\dfrac{1}{u_a}$.

(7) Substitute above mentioned values into rearranged equation so as to solve for T and S_a.

(8) Move the field-data plot horizontally, and match as much of the late data as possible to one of the 'y' set of curves, which theoretically should have the same value of $\dfrac{r}{D}$ as that selected in step (4). Vertical shifting is sometimes required.

(9) Select a second match point, again at intersection of major axes, and note co-ordination values of s, $W\left(u_y,\, \dfrac{r}{D}\right)$, t and $\dfrac{1}{u_y}$.

(10) Substitute the previously-mentioned values into rearranged equations so as to solve for T and S_y.

(11) Solve for η by substituting determined values for S and S_y into the relevant equation. When η = 100, the line joining the early and late data plots is essentially horizontal; when η is between 10 and 100 this line is not horizontal, and early T differs from late T, though the method still provides a good approximation.

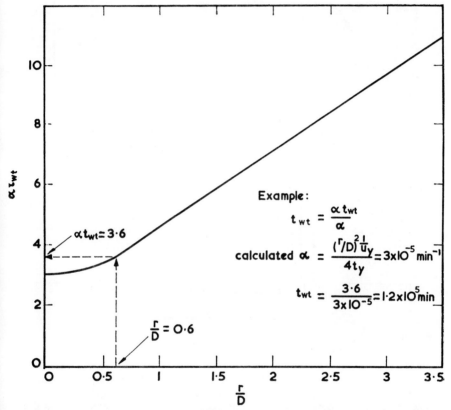

Fig 141.   Curve for estimation of time when delayed yield ceases to influence drawdown (after BOULTON 1963)

(12) For estimation of the time t_{wt}, when gravity drainage has ceased to be effective, some value of α is necessary. This requires the use of the delay index curve of Figure 141. Knowing $\frac{r}{D}$, find equivalent αt_{wt} from curve, which value divided by α, calculated from the late data, will provide t_{wt}.

351

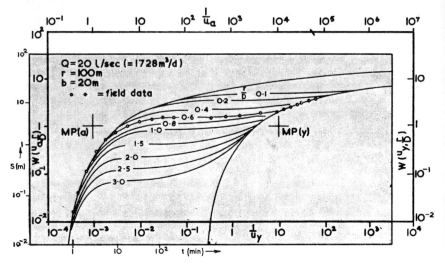

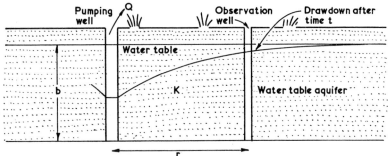

Fig 142.   Water-table test example

Example

16. The time-drawdown field data from an observation well situated 100 metres from a pumping well abstracting at a rate of 20 l/sec from a water-table aquifer 20 metres in saturated thickness, is plotted and superimposed onto the type curve as in Figure 142 so as to produce the indicated match point values. The match point MP(a) for early time-drawdown data has values of $s = 0.45$ metres, $W\left(u_a, \dfrac{r}{D}\right) = 1.0$, $t = 3$ minutes $= 0.002$ days, and $\dfrac{1}{u_a} = 1.0$, giving $T = 307$ m²/day and $S_a = 2.5 \times 10^{-4}$. The late data match point MP(y) has identical horizontal axis values but $t_y = 30\,000$ minutes $= 20.8$ days and $\dfrac{1}{u_y} = 10$, so that the transmissivity of the aquifer is confirmed, $S_y = 0.26$, and

352

$\eta = 1040$. The calculation of α and t_{wt} relevant to the example is given in Figure 141.

17. In concluding this section it is emphasised that the determination of aquifer parameters as described is not an end in itself. Knowledge of transmissivity, hydraulic conductivity and storativity should be further compared and verified against available geological, topographical and other water balance parameters to clarify and adjust for outstanding anomalies and inconsistency. After this the aquifer behaviour and potential, as well as the problems of the well itself, can be more accurately assessed.

SECTION 12.10. GROUNDWATER RESOURCES

Assessment of resources

1. Any quantitative determination of groundwater resources depends on the delineation of the spatial characteristics of the aquifer under study and the acceptance of a long-term scale. Sub-surface information is essential for definition of the lateral and vertical dimensions of the aquifer, which can be provided by the borehole data mentioned previously in this chapter and also by other logging techniques. In appraising the resources potential of the groundwater it is necessary to distinguish between the portion replenished on a seasonal basis (often defined as a perennial yield) and the more or less constant volume which remains below the level of average minimum conditions, the exploitation of which would mean over-pumping the aquifer. This latter volume can be approximated by the value of the aquifer's effective porosity, i.e. the total volume of voids containing the amount of water permissible for exploitation. In most resources studies the specific yield or effective porosity is the relevant property.

2. The aquifer's replenishment can be assessed by at least four methods but in practice it is as well to remember that the nature of the available data and surveying techniques make the assessment more of an estimate than a measurement. The assessment techniques briefly surveyed here are those determined **a.** from infiltration, **b.** using the modification of Darcy's equation (*see* Section 12.3, paragraph 4), **c.** from groundwater component of stream flow and **d.** from differences in the hydrological balance:

a. The nature of the ground surface and underlying strata will obviously determine how much of the precipitation that reaches the land surface actually infiltrates the ground. Sufficient knowledge of the infiltration capacities of the various hydrogeological units within the area is assumed to be available.

b. One modification of Darcy's equation shows that $Q = Tiw$ and it is possible by use of this equation to estimate the steady-state discharge (Q) across a groundwater contour for widths of flow section (w), having a known or assigned transmissivity (T) and a measurable hydraulic gradient (i) between adjacent contours.

353

c. In any natural groundwater catchment where a significant proportion of the streamflow is derived from groundwater, a quantitative estimate of the groundwater component of annual discharge will represent the depletion of groundwater resources over that period of time (adjusted for the time lag in accordance with the intensity of infiltration), and can be taken as equivalent to average annual effective infiltration.

d. The hydrological budget represents the natural and artificial inputs to and outputs from a catchment over a specified period of time. In theory the natural inputs and outputs should equate in the form Input = Output + Change in storage, and when in a state of balance, may be expressed in the simple equation $P = I + E + R_1$ where P = precipitation, I = infiltration, E = evaporation and R_1 = surface runoff, or in more detail:

$$P = E_d + t_v + t_g + R_1 + R_2 + R_3 \pm S_v \pm S_g \pm U$$

where the notation and relationship is as given in Figure 144.

Reliance should not be placed upon one method of budgeting alone.

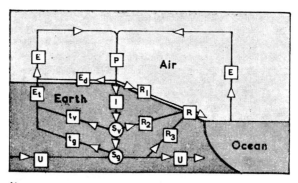

Key

P	=	Precipitation
I	=	Infiltration
E	=	Evaporation
E_d	=	Direct evaporation
E_t	=	Transpiration
t_v	=	Vadose water transpiration
t_g	=	Groundwater transpiration
R	= $R_1 + R_2 + R_3$ =	River discharge
R_1	=	Surface flow
R_2	=	Interflow
R_3	=	Groundwater flow
U	=	Underflow
S_v	=	Vadose water storage
S_g	=	Groundwater storage

Fig 144.   Hydrological budget

354

Resources management

3. The traditional viewpoint would limit the resources to potable waters, but groundwaters by definition include saline, brackish and polluted waters, so that non-potable waters are part of the hydrological cycle and cannot be ignored in any management practice. It follows that management needs to pay as much attention to the artificial as to the natural aspects of the hydrological cycle; it needs, for example, to be able to apply the same purpose and action to maintaining control of the quantity and quality of sewage effluent as to the provision of the potable water that initiated the human-use cycle. By virtue of the seasonal distribution of the replenishment factor and the differing physical nature of the ground, it is possible to modify the natural circulation of water significantly by a series of delays and short cuts. Basically, management techniques need to prolong the time taken for run-off to reach the sea, and to reduce time taken for groundwater to do the same.

4. The water management of an area should take into account all the possible demands for water and be prepared to use a variety of methods or devices to meet them such as the following:

Demands	*Devices*
a. Domestic water supply	1. Constraint
b. Industrial water supply	2. Re-cycling
c. Agricultural water supply	3. Impounding
d. Flood protection	4. Abstraction
e. Hydro-electricity	5. Transfer
f. Waste control	6. Use of waste
g. Recreation	7. Artificial recharge
h. Navigation	8. Vegetation control
i. Amenity	9. Desalination

(The letters and numbers refer to Table 27)

Depending on hydrogeology and climate, these devices may rely solely on one or other of the two major source groups (i.e. surface or sub-surface), or by integration of methods can achieve the optimum use of both.

5. Examples of management schemes, showing the demands made and the devices used, are given in Table 27.

TABLE 27. **EXAMPLES OF MANAGEMENT SCHEMES**
(Key to letters and numbers is in paragraph 4 above)

Scheme	Demands	Devices A = Surface B = Sub-surface
(a)	(b)	(c)
Kufra Oasis, Libya	a–c———	4B
Jordan Project, Israel	abc—f——	3A, 4A, 4B, 5A, 5B, 6A, 6B
Lambourn Valley, UK	abcd–fghi	4A, 4B, 5A, 7A
Kuwait	abc—f—i	2B, 4B, 5B, 9A, 9B
Santiago Basin, Chile	abcdef——	4A, 4B, 5A, 6A

Section 12.10

Management techniques

6. Management involves continuous decision making based upon the advice provided by the hydrogeologist. Modern hydrogeology is essentially quantitative, and seeks to supply answers to questions that have a predictive nature, such as involving yields or the drawdown of water level in the vicinity of pumping wells. We have seen previously how such predictions can be made for simple hydrogeological conditions involving a few wells within a small area. However, where the conditions are complicated by non-isotropic materials, variable infiltration, dissimilar boundary limits, and all manner of natural and artificial effects ranging from influent or effluent streams to well abstraction and irrigation seepage, then the conventional method of solution is so difficult and time consuming as to be impractical. Under such circumstances the simulation of the aquifer by a 'model' affords an indirect method of varying the inputs so that by a measurement of the subsequent response of the interrelated system one can provide quantitative estimates of the effects of a wide variety of conditions.

7. In groundwater studies 'models' are simulations of field prototype conditions, and those of greatest use as predictive tools are the dynamic models that may be grouped into three classes consisting of mathematical models, and two varieties of analogue models. The physical principles relating models in general are given in Table 28.

TABLE 28. PHYSICAL PRINCIPLES RELATING TO MODELS

Water	Electricity	Viscous flow
(a)	(b)	(c)
Flow of viscous liquid through porous medium	Flow of electricity through electrical conductor	Flow of viscous liquid through narrow channel
Darcy's Law $V = -K\dfrac{dh}{ds}$	Ohm's Law $V = -k_c\dfrac{dE}{ds}$	Poiseuille's Law $V = -k_p\dfrac{dh}{ds}$
V = velocity	V = velocity	V = velocity
K = hydraulic conductivity	k_c = electrical conductivity	k_p = conductivity of slot
h = head	E = voltage potential	h = head
s = distance	s = distance	s = distance

8. The two kinds of analogue model which have been most often used in groundwater studies are the electrical model and the viscous flow model. The

electrical model incorporates a network of resistors and capacitors to which is applied current 'pulses', and the resultant voltage response observed by oscilloscope or traced by plotter. Once the model is verified, it is possible to simulate variation in the duration and rate of a large number of inputs and outputs, so as to produce changes in voltage (= water-level) with time at any point within the network mesh. A graphical representation of the sequence and form of the results is demonstrated in Figure 145.

9. The viscous-flow (or parallel-plate) model represents a slice of the aquifer at right-angles to the groundwater contours, by a slot of width appropriate to the permeability of the aquifer and the viscosity of the oil generally used in the model. The particular advantage of this type of model is that oils can be used to represent the flow of waters of differing density. This makes the viscous-flow model ideal for simulation of conditions involving fresh and saline waters and prediction of the interface movement under changing conditions of abstraction and replenishment.

10. The groundwater flow systems that result from the interaction of a variety of hydrogeological conditions are too complicated to be amenable to conventional analysis, but their solution, although mathematically daunting, has been made practicably feasible by the development of the computer. The digital model is a useful method of simulating aquifer behaviour, thus providing the means of calculating the effects of recharge and withdrawals and the general water balance of the region. The digital model may be particularly useful in conjunction with an analogue model, when it may be employed in the preliminary study to test the known behaviour of the aquifer and so to confirm the design of the analogue model before construction. In the two-dimensional case, the aquifer is divided into rectangular or polygonal elements based on nodal observation wells. The two physical properties of each element which have to be specified are transmissivity and storage. At each element the replenishment and withdrawals must be estimated, along with the flows into and out of the boundary elements. The equations of flow across all contiguous faces of the elements are then computed to achieve balance. This is checked against sets of known conditions of the aquifer, and the computer programme can then be used to calculate the effect of any change of data.

Conclusions

11. *The work of the hydrogeologist can be seen in two phases.* Firstly, the exploratory studies which assemble data on the groundwater conditions and result in statements of the presence and availability of groundwater, and proposals for the use of the ground for water supply, water storage and drainage. Secondly, the development of techniques for groundwater management, especially the use of analogue and digital models. There will be environmental aspects of these studies in some cases. It is worth emphasising that the results from model studies cannot be more accurate than the initial data which the hydrogeologist assembles.

357

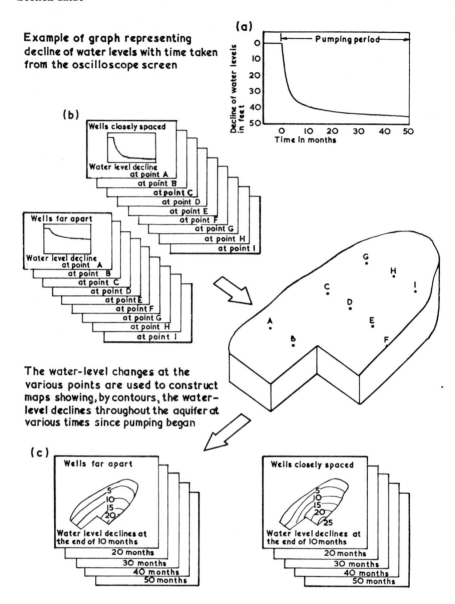

Fig 145.   Results of analogue model analyses (after ROBINOVE 1962)

BLEASDALE A & A L, 1963   —Conservation of water resources. *Inst. Civ. Engrs*, pp. 121–136, London.

BOULTON N S, 1963   —Analysis of data from non-equilibrium pumping tests allowing for delayed yield from storage. *Proc. Inst. Civ. Engrs.* 26 pp. 469–482 (Nov).

DAVIS S N and DEWIEST R J M, 1966   —*Hydrogeology*. John Wiley, New York. 463 pages.

HANTUSH M S, 1956   —Analysis of data from pumping tests in leaky aquifers. *Am. Geophys. Union Trans.* 37(6) 702–714.

HANTUSH M S, 1964   —*Hydraulics of wells in advances in hydroscience.* Chow V T (Ed), Academic Press. 281–442.

HANTUSH M S and JACOB E C, 1955   —Non-steady radial flow in an infinite leaky aquifer. *Am. Geophys. Union, Trans.* 36(1) 95–100.

HOLDEN W S (Ed), 1970   —*Water treatment and examination.* J & A Churchill. 513 pages.

INESON J, 1959   —Yield-depression curves of discharging wells. *Inst. Wat. Engrs, Jl.* 13, 119.

INESON J and DOWNING R A, 1964   —The groundwater component of river discharge and its relationship to hydrogeology. *Jl. Inst. Water Engrs.* Vol. 18, pp. 519–541.

KRUSEMAN G P and DE RIDDER N A, 1970   —Analysis and evaluation of pumping test data. *Intern. Inst. Land Reclamation*, Wageningen. Bull. 11, 200 pages.

ME VOL VI, 1975   —*Military Engineering Volume VI, Water Supply* (Army Code No. 71045). Her Majesty's Stationery Office, London.

RAINWATER F H and THATCHER L L, 1960   —Methods for Collection and Analysis of Water Samples. *U.S. Geol. Surv.* Wat. Sup. Paper 1454. 301 pages.

ROBINOVE C J, 1962   —Groundwater studies and analog models. USGS Circular 468, 12 pages.

SKEAT W O (Ed), 1969   —*Manual of British Water Engineering Practice*, Volumes I–III. 4th Edition. Chap 2. Inst. Wat. Engrs. London.

THEIS C V, 1935   —The relation between the lowering of the piezometric surface and the rate and duration of discharge of a well using groundwater storage. *Am. Geophys. Union Trans.* 16, 519–524.

TODD D K, 1959   —*Ground Water Hydrogeology.* J Wiley & Sons, New York.

WALTON W C, 1962   —Selected analytical methods for well and aquifer evaluation. *Illinois State Water Survey Bull.* 49, 81 pages.

ANNEX A

UNIFIED CLASSIFICATION SYSTEM FOR SOILS

(Reference Section 5.2, paragraph 7 and Section 5.3, paragraph 1)

1. The Unified Classification System (UCS) divides soils into two main categories, coarse grained and fine grained soils. These are defined as follows:

  a. Coarse grained soils are soils in which more than 50 per cent by weight of a representative sample of the soil is retained on the 75 μm (BS No. 200) sieve.

  b. Fine grained soils and soils in which 50 per cent or more by weight of a representative sample of the soil passes the 75 μm (BS No. 200) sieve.

2. The results of a sieve analysis are needed to place the soil in one or other of these categories. The procedure for classification is then carried out under the appropriate category as described below.

Classification of coarse grained soils

3. The procedure for classifying coarse grained soils is outlined in diagrammatic form in Figure 146.

4. By definition the coarse grained soil particles stopped by the 5 mm ($\frac{3}{16}$ in) sieve are gravels, and those passed are sand. The results obtained from a sieve analysis show whether the sample contains more gravel or sand, and this determines the first letter G (for gravel) or S (for sand) for the sample.

5. The sample is then placed in one of three cases depending on the percentage of the sample passing the 75 μm (BS No. 200) sieve as follows:

  a. *Case 1.* Less than 5 per cent of the sample passes the sieve.

  b. *Case 2.* Between 5 and 12 per cent passes the sieve.

  c. *Case 3.* Between 13 and 49 per cent passes the sieve.

6. The results of the sieve analysis enable the grading coefficients for the soil sample to be calculated. These are the coefficient of uniformity (Cu) and the coefficient of gradation (Cg) and are obtained from the formulae:

$$Cu = \frac{D_{60}}{D_{10}}$$

$$\text{and } Cg = \frac{D^2_{30}}{D_{60}D_{10}}$$

where D_{60}, D_{30} and D_{10} are the particle size diameters at which 60, 30 and 10 per cent of the soil passes.

7. The second letter of the classification is obtained as follows:

  a. *Case 1.* Determine the values of Cu and Cg for the sample as described in paragraph 6 and using Figure 146 determine the second letter.

  b. *Case 2.* Soils in this category are given dual classification by first determining the classification as for Case 1, and then as for Case 3. The soil sample is then classified using both pairs of letters, e.g. GW/GM.

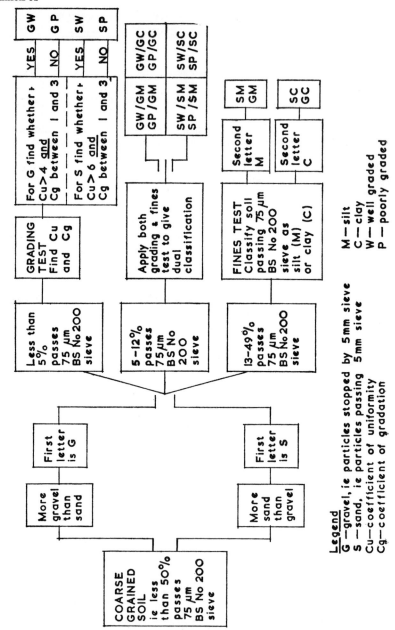

Annex A

Fig 146. Unified classification system—coarse grained soils

362

c. *Case 3.* Classify the soil passing the 75 μm (BS No. 200) sieve by means of a fines test which will determine whether the fines are clay or silt and this will decide whether M (silt) or C (clay) is to be used.

8. Soils can also be placed into high (H) or low (L) liquid limit groups depending on whether they have a liquid limit of more or less than 50, and this can be shown by the addition of H or L to the classification group (*see* Figure 147).

9. The UCS allows the addition of a third letter to soils of types GM and SM to indicate the construction quality of the soil, as desirable or undesirable. If the liquid limit is 28 or less and the plastic limit 6 or less the letter *d* is added, i.e. *GMd*; whereas if the liquid limit exceeds 28 or the plastic limit exceeds 6, the sample is undesirable and the letter *u* is added, i.e. *GMu*.

Classification of fine grained soils (fines test)

10. The classification of fine grained soils is determined by comparison of the plasticity index and the liquid limit of the sample. The liquid limit, the plastic limit, and the plasticity index are determined by tests. The plasticity index is plotted against the liquid limit on the plasticity chart (Figure 147) and the classification of the fine grained soil read off.

11. If the fine grained soil contains appreciable organic material, the soil is given the first letter *O*. The percentage of organic material should be quoted and the other constituents identified.

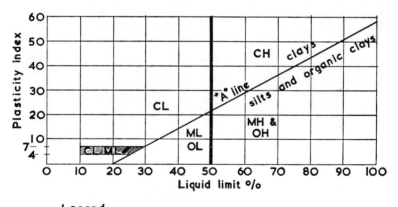

Fig 147. Plasticity chart (UCS)

INDEX

NOTE: *Roman numerals refer to Glossary*

Index

Index

Index

Printed in Great Britain by William Clowes & Sons Limited, London, Beccles and Colchester